					He^2 $1s^2$
B^5 $2s^2 2p$	C^6 $2s^2 2p^2$	N^7 $2s^2 2p^3$	O^8 $2s^2 2p^4$ 24	F^9 $2s^2 2p^5$	Ne^{10} $2s^2 2p^6$
Al^{13} $3s^2 3p$ 1.18	Si^{14} $3s^2 3p^2$ (7.1)	P^{15} $3s^2 3p^3$ (5.8)	S^{16} $3s^2 3p^4$	Cl^{17} $3s^2 3p^5$	Ar^{18} $3s^2 3p^6$

Ti^{22}	V^{23}	Cr^{24}	Mn^{25}	Fe^{26}	Co^{27}	Ni^{28}	Cu^{29}	Zn^{30}	Ga^{31}	Ge^{32}	As^{33}	Se^{34}	Br^{35}	Kr^{36}
$3d^2 4s^2$	$3d^3 4s^2$	$3d^5 4s$	$3d^5 4s^2$	$3d^6 4s^2$	$3d^7 4s^2$	$3d^8 4s^2$	$3d^{10} 4s$	$3d^{10} 4s^2$	$4s^2 4p$	$4s^2 4p^2$	$4s^2 4p^3$	$4s^2 4p^4$	$4s^2 4p^5$	$4s^2 4p^6$
0.4	5.4	312	96	1040	1390	628		0.85	1.08 (7.5)	(5.3)	(0.5)	(6.9)		

Zr^{40}	Nb^{41}	Mo^{42}	Tc^{43}	Ru^{44}	Rh^{45}	Pd^{46}	Ag^{47}	Cd^{48}	In^{49}	Sn^{50}	Sb^{51}	Te^{52}	I^{53}	Xe^{54}
$4d^2 5s^2$	$4d^4 5s^2$	$4d^5 5s$	$4d^6 5s$	$4d^7 5s$	$4d^8 5s$	$4d^{10}$	$4d^{10} 5s$	$4d^{10} 5s^2$	$5s^2 5p$	$5s^2 5p^2$	$5s^2 5p^3$	$5s^2 5p^4$	$5s^2 5p^5$	$5s^2 5p^6$
0.61	9.25	0.92	7.8	0.49			0.52	3.4	3.72	(3.5)	(4.3)			

Lu^{71}	Hf^{72}	Ta^{73}	W^{74}	Re^{75}	Os^{76}	Ir^{77}	Pt^{78}	Au^{79}	Hg^{80}	Tl^{81}	Pb^{82}	Bi^{83}	Po^{84}	At^{85}	Rn^{86}
$4f^{14} 6s^2$	$5d^2 6s^2$	$5d^3 6s^2$	$5d^4 6s^2$	$5d^5 6s^2$	$5d^6 6s^2$	$5d^9$	$5d^9 6s$	$5d^{10} 6s$	$5d^{10} 6s^2$	$6s^2 6p$	$6s^2 6p^2$	$6s^2 6p^3$	$6s^2 6p^4$	$6s^2 6p^5$	$6s^2 6p^6$
0.1	0.13	4.47	0.015	1.70	0.66	0.11		4.15	2.38	7.19	(8.5)				

IMPERIAL CHEMICAL INDUSTRIES LIMITED,
CORP. LAB. P.O. BOX 11
THE HEATH, RUNCORN, CHESHIRE.

LONG RANGE ORDER IN SOLIDS

ROBERT M. WHITE

Xerox Corporation
Palo Alto Research Center
Palo Alto, California

THEODORE H. GEBALLE

Department of Applied Physics
Stanford University
Stanford, California
and
Bell Laboratories
Murray Hill, New Jersey

ACADEMIC PRESS · New York San Francisco London
1979

A Subsidiary of Harcourt Brace Jovanovich, Publishers

ACADEMIC PRESS, INC.
111 Fifth Avenue, New York, New York 10003

United Kingdom Edition published by
ACADEMIC PRESS, INC. (LONDON) LTD.
24/28 Oval Road, London NW1 7DX

LIBRARY OF CONGRESS CATALOG CARD NUMBER: 55–12299

ISBN 0–12–607775–4

PRINTED IN THE UNITED STATES OF AMERICA
79 80 81 82   9 8 7 6 5 4 3 2 1

To Sara and Francis
who have always known that short range fluctuations
lead to long range order.

Contents

vii

VII. Impurities and Long Range Order

VIII. Domain Structures

IX. Order Amidst Disorder

Preface

This book represents notes used by us in a seminar course at Stanford University which focused on long range order in solids. This subject is typified by magnetism and superconductivity, but by no means limited to these phenomena, as we try to show. The level of this material is intermediate between a graduate course and a research seminar. Thus, it is intended for students who have already had formal courses dealing with the general principles of solid state physics but who have had little exposure to real materials.

We believe a great deal can be learned by looking carefully at experimental data, and that the experience of doing so will develop judgment and intuition. Such analysis may even lead to discovery, for clues to entirely new phenomena are often buried in existing data. Thus, one of our purposes is to review a wide selection of data in a systematic manner. In fact, the first "draft" of this book consisted entirely of figures of experimental data which served as the basis for classroom discussion.

Another purpose of this course was to give students a feeling for the generality and applicability of the concepts underlying long range order. To this end the "theoretical" material has been organized by concepts rather than subjects. Chapter I, for example, deals with the usefulness of the concept of an order parameter; Chapter II with the role of symmetry in describing phase transitions; and Chapter VIII with inhomogeneous long range order. It is our hope that this organization will enable the reader to make associations which will lead to new ideas.

In keeping with the flavor of these pedagogical purposes, we have included problems in the text to encourage certain directions of thought or to have the reader supply intermediate steps in our analyses.

The material also contains the elements of a research review. There have been a number of new theoretical developments in long range order, such as renormalization group techniques for describing critical phenomena and the density functional formulation for calculating exchange and correlation energies. We have tried to sketch the basic ideas underlying these developments. The reader should not, however, expect to learn how to employ these techniques himself from our discussions—these constitute courses in themselves. Our purpose is merely to indicate where these new developments fit into the conceptual framework we have constructed. As with any broad-brush picture, if one looks too closely one looses the overall impression.

We have chosen the data mainly from materials under current study, such as the A-15 compounds, layered compounds, and granular and amorphous materials. In this way, we hope to provide a bridge to the current literature. Although the data frequently involve magnetism and superconductivity, other examples of long range order, such as charge density waves, are also used.

We would like to thank a number of our colleagues for their comments on various parts of this manuscript: Jim Allen, Phil Allen, John Bardeen, Mac Beasley, Jim Boyce, Marvin Cohen, Frank DiSalvo, Peter Fulde, Rick Greene, Lars Hedin, Myron Salamon, Doug Scalapino, Bob Schrieffer, and Mike Thorpe. We would also like to thank the Xerox Corporation for supporting this project, particularly in the preparation of the figures and the typing of the manuscript. We are especially grateful to Bonnie Cook, Gail Pilkington, Rosemary Burnett, Jane Nielson, and Debbie Bernsen for their cheerful cooperation in what must have appeared as a never-ending typing task; and to Joe Leitner for his superb work on the figures. We also thank Giuliana Lavendel and her library staff for their invaluable help. One of us (T.H.G.) is also indebted to the John Simon Guggenheim Memorial Foundation and the Air Force Office of Scientific Research for support.

ROBERT M. WHITE
THEODORE H. GEBALLE

LONG RANGE ORDER IN SOLIDS

I. Phase Transitions and Order Parameters

Long range order refers to the situation in which the value of some *property* at a given point in a system is correlated with its value at a point infinitely far away. Nature displays a variety of long range order. Perhaps the most familiar is that of a crystalline solid. In this case that *property* is the mass density. In a liquid the motion of the atoms produce fluctuations in the density that destroy long range correlations. In a crystal the atoms are constrained to sites that are periodically related. In this case the density–density correlations do not vanish even for large separations. The nature of this atomic long range order in the crystalline state has been thoroughly explored by X-ray and electron diffraction and neutron studies. The transition from the solid state to the liquid state, as well as the liquid state itself, are not nearly as well understood. In this monograph, however, we shall restrict ourselves to long range order within the solid state. For the most part we shall consider other long range order superimposed on crystalline solids, although in Chapter IX we shall also consider solids in which the atoms are in a nonequilibrium amorphous state. In some cases we shall find that the presence of long range order in another property of the system will also modify the crystalline order.

A. DISORDER TO ORDER

At sufficiently high temperatures the forces of interaction between the existing molecular species are overwhelmed by thermal energy. This situation can simply be described macroscopically by the Gas laws, and on a microscopic scale, by classical kinetic theory. As the temperature is reduced towards absolute zero, a system approaches its ground state. If this state is nondegenerate the entropy approaches zero (third law of thermodynamics) and we speak of the system as being ordered. In practice this can happen in many different ways, some of which will be of great interest to us. It is worth noting, and we shall return to this point later, that although a system may be ordered, the ordering may not be long range.

1. Ordering with No Transition

In the absence of interactions there will be no discontinuity in the free energy function or any of its temperature derivatives as the system is cooled from its high-temperature state to its low-temperature state of zero entropy. This situation is approached in simple metals where the current carriers behave almost like a free-electron gas (Fermi liquid). The behavior of the heat capacity of such a system is illustrated in Fig. I.1a. T^* is the temperature above which the electrons behave classically. This is $k_B T^* \sim E_F$ which is of the order of 1 eV or 10^4 K for ordinary metals, which, of course, means that most metals would vaporize before their electron gas departs from the linear, low-temperature region of Fig. I.1a. A realizable departure is that of a doped semiconductor such as phosphorus in Ge. For 10^{19} donors/cm^3 the curve in Fig. I.1a applies, with the ordinate scaled down by a factor of perhaps 10^3.

In the low-temperature region, interactions may cause a transition into a new state which generally produces a change in the electronic heat capacity. Kohn and Luttinger,[1] in fact, claim that a weakly interacting system of fermions *cannot* remain normal down to absolute zero, no matter what the form of the interaction. This conclusion is due solely to the sharpness of the Fermi surface.

Another way in which one can achieve order without a phase transition is characterized by a Schottky anomaly. Consider, for example, a two-level atomic or molecular species with an energy splitting Δ. At high temperatures, where $k_B T \gg \Delta$, both levels are equally occupied. At low temperature, $k_B T \ll \Delta$, only the lowest level of each particle is occupied. For such a simple system the heat capacity is

$$C_{\text{Schottky}} = R \left(\frac{\Delta}{k_B T} \right)^2 \frac{\exp(\Delta/k_B T)}{[1 + \exp(\Delta/k_B T)]^2}. \tag{1.1}$$

This is illustrated in Fig. I.1b. It follows from the functional dependence of Δ/T that the same magnitudes are found for systems with electronic energy splittings (10's of millielectron volts) in the decades of temperature around room temperature as for systems with nuclear Zeeman splittings in the millidegree range. In practice residual interactions tend to sharpen the Schottky peak and broaden the base. Transitions which occur at low temperatures are more easily studied because other degrees of freedom do not compete.

One might expect that if a substance remained liquid it might not order at all. The conditions for a substance to remain liquid are that the zero

[1] W. Kohn and J. Luttinger, *Phys. Rev. Lett.* **15**, 524 (1965).

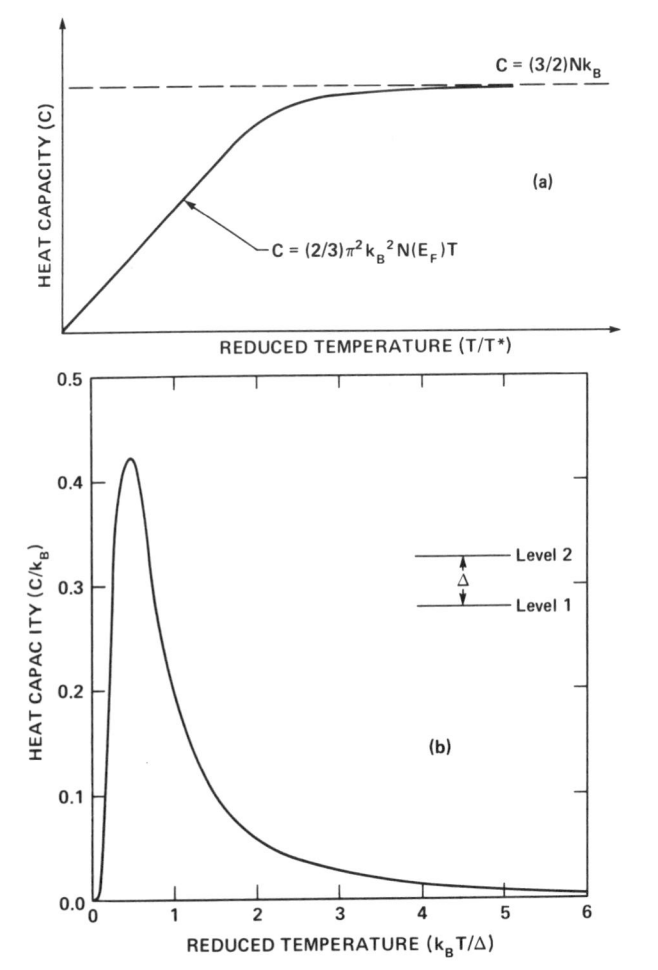

FIG. I.1. Heat capacities for (a) free electron gas and (b) a two-level system. The fact that the heat capacity of the two-level system goes to zero as T increases is due to the finite number of energy levels.

point energy, E_0, be greater than the energy gained by crystallization. E_0 can be estimated from the Uncertainty Principle, $E_0 \sim h^2 n^{2/3}/8m$ per particle, where n is the number density and m is the mass. Only the two isotopes of helium, ^{3}He and ^{4}He, have a small enough mass and weak enough interactions to remain liquid as $T \to 0$. In both cases, however, a superfluid transition representing an ordering in momentum space takes place at 2.17 K for ^{4}He and $\sim$2 mK for ^{3}He.

2. Ordering via a First-Order Transition

A first-order transition is characterized by a latent heat, a finite change in volume, and hysteresis. As most gases are cooled, they condense in response to the intermolecular forces via a first-order transition. While nothing as simple and universal as the Schottky relation can describe the liquid vapor transition, there is Trouton's rule, a convenient order of magnitude empirical relation, which states that the change in entropy of such a transition is approximately 21 cal/mol deg. This is very convenient, for example, in estimating the heats of vaporization of various liquids. The extent to which this "rule" is obeyed is illustrated in Table I.1.

Problem I.1

Trouton's rule is approximately followed by normal substances that obey a reduced equation of state in which the boiling point appears in units of the critical point. It is useful for making order-of-magnitude estimates without resource to tables. An example is the thermal capacity of a liquid H_2 reservoir. Breakdowns of the rule signal nonnormal behavior that can be understood microscopically. Entropies of vaporization can be too small either because the liquid has more entropy than a reduced equation of state would indicate or because the vapor has less. Comment in these terms upon why He and P have small entropies of vaporization (see Table I.1). Suggest why H_2O has such a large entropy of vaporization.

Notice that in gas–liquid transitions the *symmetry* of the system does not change—in both states the point symmetry is isotropic. Consequently, such first-order transitions occur along a line in the temperature–pressure plane that terminates at a *critical point*. It also follows that there are no critical points in liquid–solid and vapor–solid transitions where the symmetry does change.

TABLE I.1[a]

	T_c	ΔH (kcal)	ΔS (cal/mol deg)
CCl_4	350	7.17	20.5
CO	81.6	1.44	17.68
Cl_2	239	4.878	20.4
Pb	2023	43.0	21.3
N_2	77.34	1.33	17.2
H_2O	373.16	9.717	26.04
Au	2933	74.21	25.3
He	4.21	0.02	4.7
P	553	2.97	5.4

[a] Trouton's Rule states that the entropy change of vaporization is 21 cal/mol deg. In the last column we show the change in entropy associated with the vaporization of various substances obtained from the observed heat of vaporization, ΔH, according to $\Delta S = \Delta H / T$.

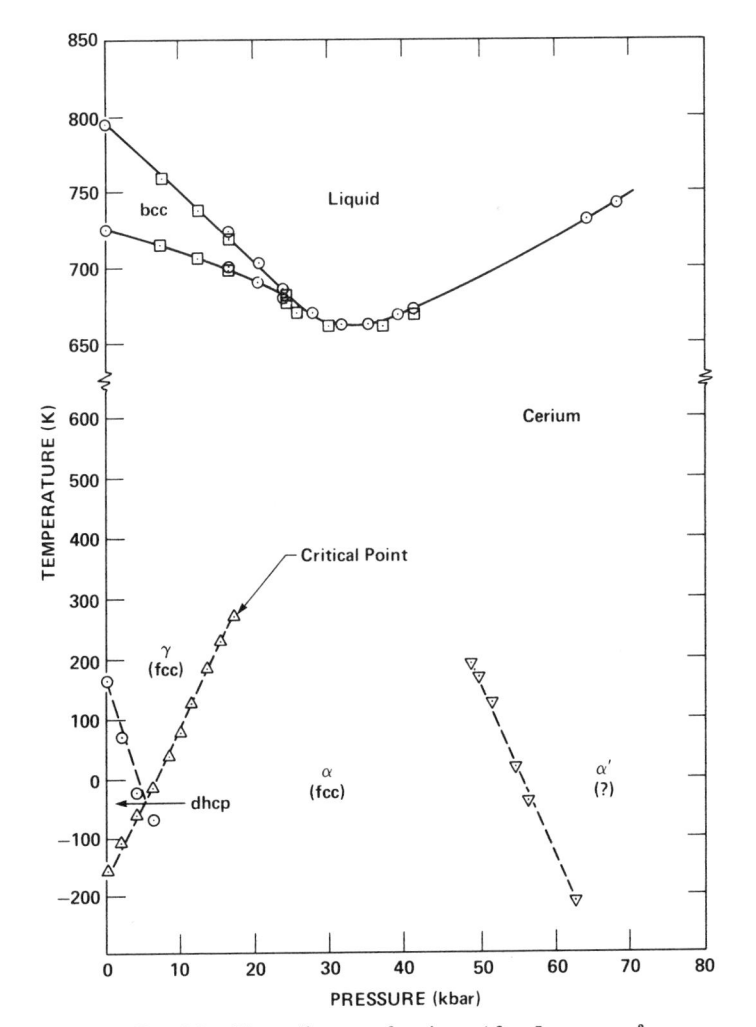

FIG. I.2. Phase diagram of cerium. After Jayaraman.[2]

An interesting example of a first-order transition that terminates in a critical point is seen in the phase diagram of Ce, (Fig. I.2).[2] Above the critical temperature it is possible to proceed continuously from the α to the γ phase since both phases are fcc.

This phase diagram has a number of other features that should be noted. The highest temperature stable phase at atmospheric pressure is the body centered cubic (bcc) phase. It is a common occurrence that the

[2] A. Jayaraman, *Phys. Rev.* **137**, A179 (1965).

close-packed phases of metals that are stable at $T = 0$ undergo first-order transitions to the bcc phase at higher temperatures. As pointed out by Zener this presumably is due to the increased lattice entropy of the bcc phase.

The face centered cubic (fcc) γ-phase with one 4f electron per Ce trans-forms to the collapsed α-phase (also fcc) in which the 4f electron has gone into the conduction band state. The pressure dependence of the melting transition in Ce is negative and extends to a pressure which is approxi-mately that which would be reached by the extrapolated $\gamma - \alpha$ phase boundary. From the Clausius–Clapeyron equation, $dP/dT = \Delta S/\Delta V$, Jay-araman[2] has interpreted the necessarily negative ΔV (i.e., $V_{liq} - V_{sol} < 0$) as being due to the fact that more of the 4f electrons are localized in the solid than in the liquid for $P < 30$ kbar.

A negative dP/dT also occurs in Ge and other elements of groups IV and V of the periodic table. Consider the phase diagram of Ge shown in Fig. I.3.[3] Here the liquid is metallic-like and more closely packed than in

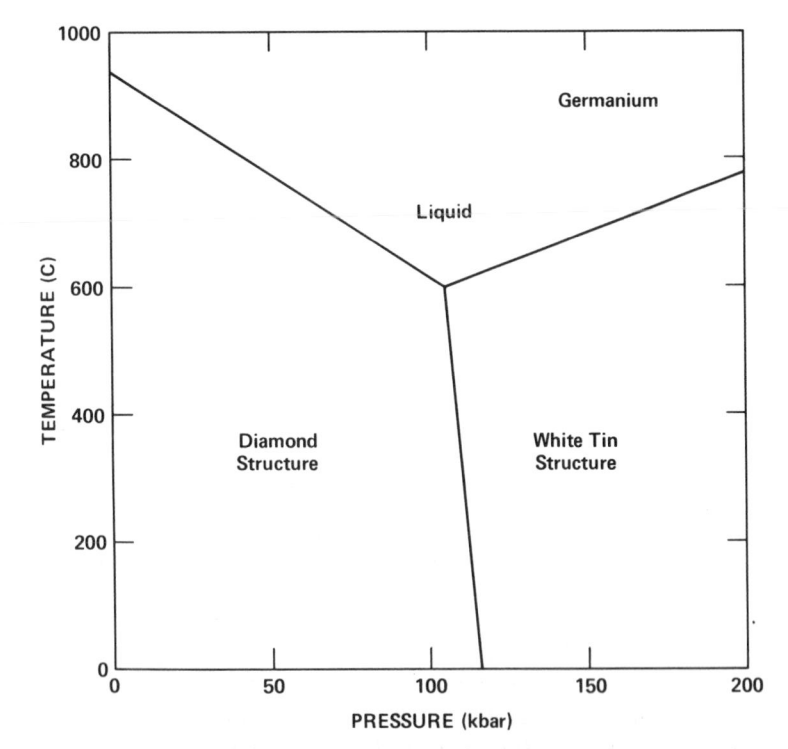

FIG. I.3.   Phase diagram of germanium. After Klement and Jayaraman.[3]

[3] W. Klement and A. Jayaraman, *Prog. Solid State Chem.* **3**, 289 (1966).

the covalently bonded solid. Hence, $\Delta V = V_{\text{liq}} - V_{\text{sol}} < 0$ is a result of the packing of the atoms (rather than in the change of size of the atoms as in Ce). Improved packing in the liquid state also occurs in the water–ice transition.

A more unusual negative dP/dT occurs in ^{3}He below 0.4 K where ΔV is >0, but $\Delta S < 0$ (Fig. I.4). The ^{3}He nucleus has a spin $S = \frac{1}{2}$. The (nuclear) spin–spin exchange interaction is greater in the liquid than in the solid because of the greater overlap of the nuclear wave functions resulting from the increased particle motion. Hence, the nuclear spin system is more ordered in the liquid than in the solid, sufficiently so to overcome the normal entropy of melting. This increase of entropy upon solidifying with compression is the basis of the useful Pomeranchuk cooling procedure. Temperatures less than 2 mK are achieved by following the liquid–solid equilibrium along an isentropic compression.

At a first-order transition there is frequently a discontinuity in the specific heat, in addition to a latent heat. This, of course, occurs because new modes of excitation become accessible. Typically the liquid has a

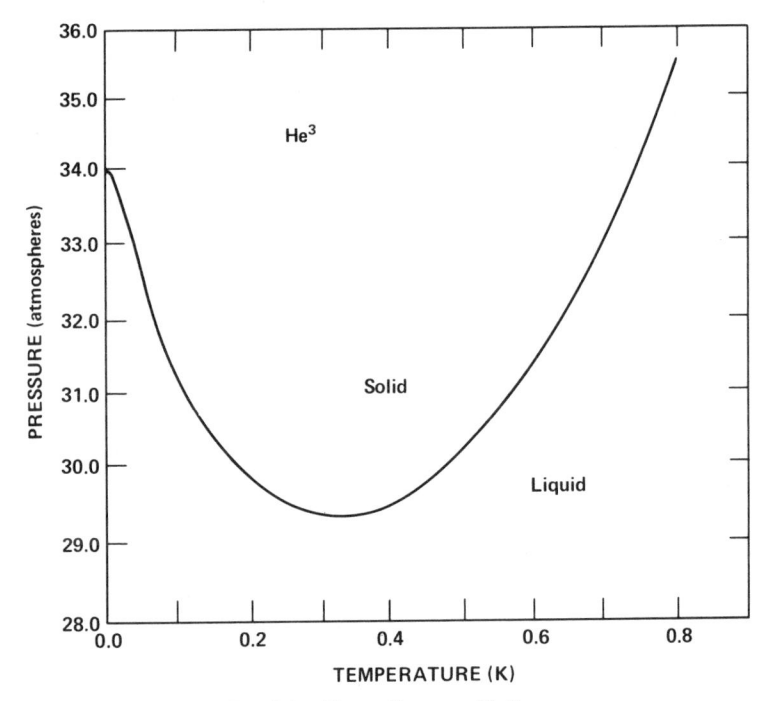

FIG. I.4. Phase diagram of helium.

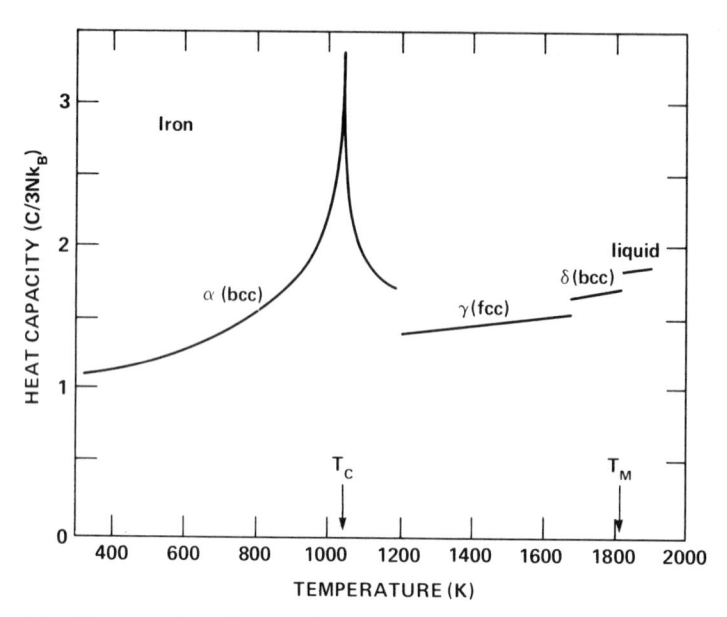

FIG. I.5.   Heat capacity of iron as a function of temperature. T_c is the Curie point. After Orr and Chipman.[4]

higher specific heat than the solid as illustrated by Fe in Fig. I.5.[4] In solid–solid transitions the discontinuities can be of either sign. The heat capacity for the fcc (γ) form of Fe is smaller than the adjacent magnetic phases because of the disappearance of the magnetic degrees of freedom.

3. Ordering via a Second-Order Transition

Second-order transitions have no latent heat nor hysteresis. They are characterized by discontinuities or singularities in the heat capacity, which occur in superconducting, superfluid, magnetic, ferroelectric, order–disorder, and special kinds of structural transitions.

The heat capacity of a superconductor in zero field has finite slopes as the transition is approached from above and below. This is illustrated by gallium as shown in Fig. I.6.[5]

4. The Ehrenfest Classification

Throughout our discussion we have been distinguishing between first- and second-order transitions by whether or not they involve a latent heat

---

[4] R. L. Orr and J. Chipman, *Trans. Metall. Soc. AIME* **239,** 630 (1967).
[5] N. E. Phillips, *Phys. Rev.* **134,** 385 (1964).

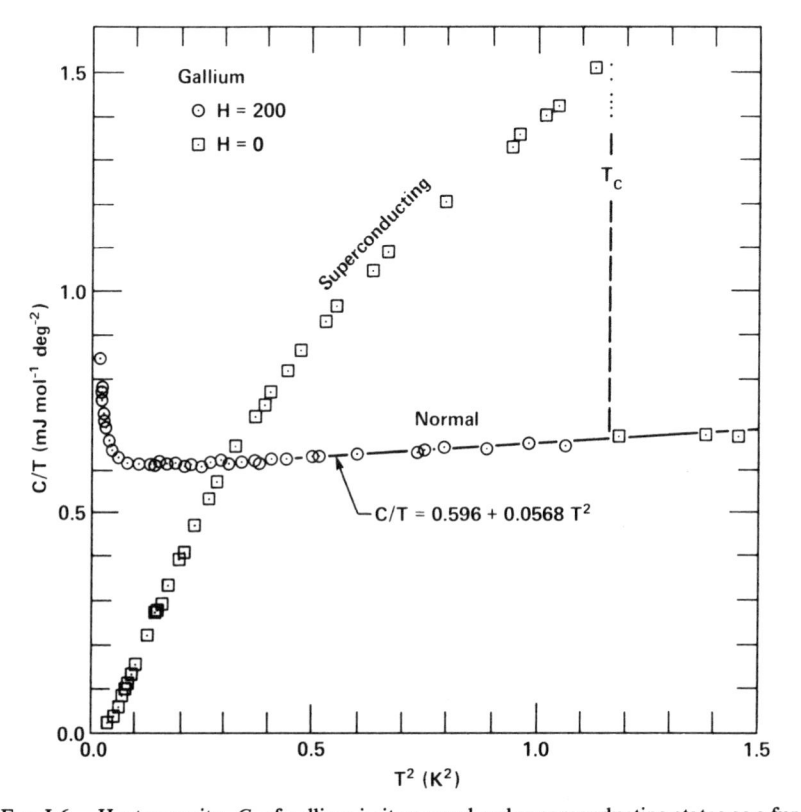

FIG. I.6.   Heat capacity, C, of gallium in its normal and superconducting states as a function of temperature. After Phillips.[5]

and hysteresis. In 1933 Ehrenfest suggested a more rigorous classification based on thermodynamics. According to the First Law of Thermodynamics, the change in internal energy of a system, ΔE, is equal to the work done on the system, ΔW, plus the heat received by the system, ΔQ,

$$\Delta E = \Delta Q + \Delta W. \qquad (1.2)$$

For a reversible process the heat is related to the change in entropy at temperature T by $\Delta Q = T\Delta S$. The work done on a system by a pressure P (an intensive variable) that changes the volume V (an extensive variable) by ΔV is $\Delta W = -P\Delta V$. For a magnetic system the work done depends upon how the system is defined.[6] Let us define our system as the magnetic

---

[6] For a discussion of magnetic work, see F. Reif, "Statistical and Thermal Physics," pp. 439–444. McGraw-Hill, New York, 1965; also L. Landau and E. Lifshitz, "Electrodynamics of Continuous Media," Sects. 30 and 31. Pergamon, Oxford, 1960.

sample enclosed by the surface S in Fig. I.7. Then Poynting's theorem tells us that the energy flowing across this surface in the time Δt is

$$\Delta W = -(c/4\pi) \int dS \cdot (\mathbf{E} \times \mathbf{H})\Delta t$$

$$= -(c/4\pi) \int dV \nabla \cdot (\mathbf{E} \times \mathbf{H})\Delta t \qquad (1.3)$$

$$= (c/4\pi) \int dV[\mathbf{E} \cdot (\nabla \times \mathbf{H}) - \mathbf{H} \cdot (\nabla \times \mathbf{E})] \, \Delta t.$$

Since we have defined our system to exclude the free currents producing the field, and if we assume that the displacement $\mathbf{D}$ within the volume bounded by S does not change,

$$\nabla \times \mathbf{H} = (4/c)\mathbf{j}_{\text{free}} + (1/c)(\partial \mathbf{D}/\partial t) = 0. \qquad (1.4)$$

Using the Maxwell equation, $\nabla \times \mathbf{E} = -(1/c)(\partial \mathbf{B}/\partial t)$, the work becomes

$$\Delta W = (V/4\pi)H\Delta B. \qquad (1.5)$$

The First law, Eq. (1.2), then becomes

$$\Delta E = T\Delta S - P\Delta V + (V/4\pi)H\Delta B. \qquad (1.6)$$

Notice that if we define the quantity $F = E - TS$, called the Helmholtz free energy, its differential is

$$\Delta F = -S\Delta T - P\Delta V + (V/4\pi)H\Delta B. \qquad (1.7)$$

Thus F has the property that $\Delta F = \Delta W$ in an isothermal reversible process. Equation (1.7) shows that F is a thermodynamic function of the variables T, V, and B. In most experimental situations one controls the pressure and the field H rather than the volume and the induction B.

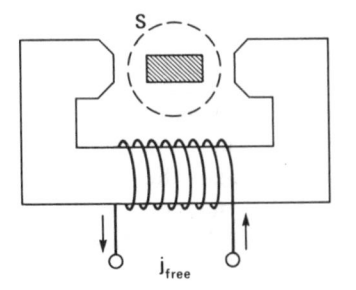

FIG. I.7.   Illustration of the surface S which defines the magnetic system for which the Gibb's free energy, Eq. (1.8), applies.

Therefore, it is convenient to define a new (Gibbs) free energy according to

$$G = F + PV - (V/4\pi)HB. \tag{1.8}$$

Then

$$\Delta G = -S\Delta T + V\Delta P - (V/4\pi)B\Delta H. \tag{1.9}$$

This expression shows that if the temperature, pressure, and field are maintained constant, the dependent variables adjust themselves so that the Gibbs free energy is an extremum. Furthermore,

$$\left(\frac{\partial G}{\partial T}\right)_{P,H} = -S, \quad \left(\frac{\partial G}{\partial P}\right)_{T,H} = V, \quad \left(\frac{\partial G}{\partial H}\right)_{T,P} = -\frac{V}{4\pi}B. \tag{1.10}$$

Differentiating again we obtain:

the heat capacity at constant pressure $= C_p = -T\left(\frac{\partial^2 G}{\partial T^2}\right)_{P,H}$

the isothermal compressibility $= \kappa = -\frac{1}{V}\left(\frac{\partial^2 G}{\partial P^2}\right)_{T,H}$ (1.11)

the expansion coefficient $= \beta = \frac{1}{V}\left(\frac{\partial^2 G}{\partial T\partial P}\right)_{H}$

the isothermal permeability $= 1 + 4\pi\chi = -\frac{1}{V}\left(\frac{\partial^2 G}{\partial H^2}\right)_{T,P}$.

These relations led Ehrenfest to suggest that phase transitions could be characterized by discontinuities in the derivatives of the Gibbs free energy. Thus a *first-order* transition would be one in which a first derivative of the Gibbs free energy is discontinuous. This is illustrated in Fig. I.8.

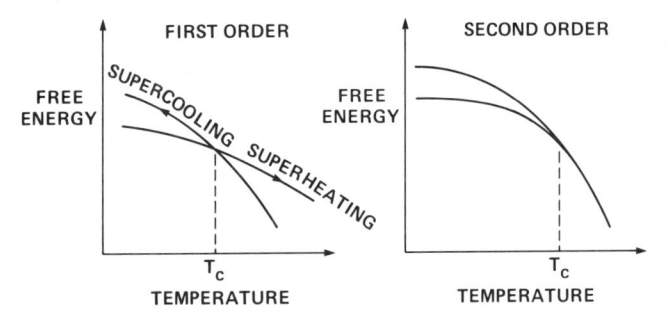

FIG. 1.8. General behavior of the free energies as functions of temperature for first- and second-order transitions.

Notice that a discontinuity in the entropy implies a latent heat that depends upon the change in entropy. In the case of very small latent heats it is difficult to distinguish experimentally between a first-order transition and one in which the heat capacity on both sides of the transition becomes very large. In such cases the second characteristic feature of a first-order transition becomes useful in its identification: hysteresis. That is, the transition occurs at a different temperature depending upon whether the temperature is increasing or decreasing. In very pure materials the thermal hysteresis becomes an intrinsic property understandable in terms of nucleation theory. For example, pure water can be supercooled to $-40°$ C before it crystallizes to ice; pure superconducting Al, driven normal by an applied field, can be "supercooled" in this normal state by reducing the applied magnetic field to 5% of its critical value before the superconducting state appears.

A second-order transition would be characterized by a discontinuity in a second derivative of G. This is also illustrated in Fig. I.8. In particular, the heat capacity would be discontinuous. The superconductivity transition in gallium (Fig. I.6) was such an example.

However, not all transitions satisfy the Ehrenfest criteria. Figure I.9[7] shows the heat capacity of the antiferromagnet $CoCl_2 \cdot 6H_2O$. The insert indicates that the heat capacity increases logarithmically close to the Néel temperature. The eventual rounding off of the transition is thought to be associated with sample inhomogeneities. This logarithmic behavior means that all Ehrenfest's derivatives *diverge*. Such "λ-type" anomalies appear in other systems as listed in Table I.2. In fact, the transition to superfluidity in ^{4}He at 2.17 K is referred to as the λ-point because of the shape of the logarithmic singularity in its heat capacity. Such divergences also appear in theoretical analyses. Figure I.10 shows Onsager's exact solution, obtained in 1944, for the heat capacity of the two-dimensional Ising model. This model is characterized by the Hamiltonian

$$\mathcal{H}_{\text{Ising}} = - 2J \sum_i \sum_{j>i} S_i^z S_j^z \qquad (1.12)$$

where S_i^z is the z-component of the spin at the ith site which takes on the values $\pm\frac{1}{2}$, and J is the exchange interaction between neighboring spins. Onsager's calculation shows that the free energy contains a term proportional to $(T - T_c)^2 \ln (T - T_c)$ which leads to a logarithmic divergence in the specific heat.

---

[7] J. Skalyo and S. A. Friedberg, *Phys. Rev. Lett.* **13**, 133 (1964).

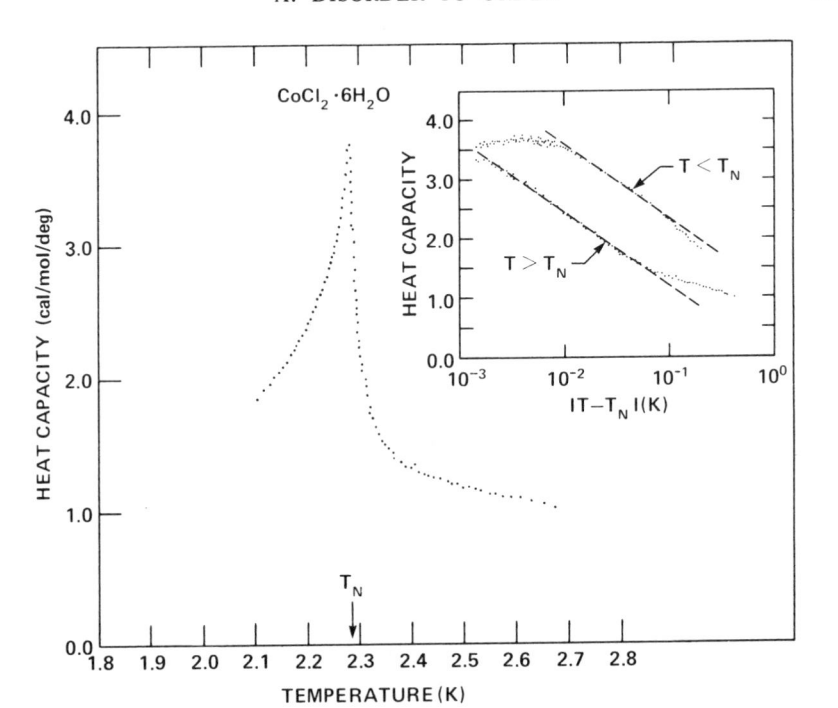

FIG. I.9. Heat capacity of the antiferromagnetic $CoCl_2 \cdot 6H_2O$ as a function of temperature. After Skalyo and Friedberg.[7]

Thus it appears that we need a description of phase transitions that is capable of dealing with divergences, not merely discontinuities. Before introducing such a description let us first indicate how these divergences are characterized.

TABLE I.2

EXAMPLES OF SYSTEMS EXHIBITING PHASE TRANSITIONS
WITH λ-LIKE SPECIFIC HEATS

System	Example	T_c
Liquid–gas	Ar	87 K
Ferromagnets	Fe (Fig. I-5)	1033 K
Antiferromagnets	$CoCl_2 \cdot 6H_2O$ (Fig. I-9)	2.28 K
Order–disorder	CuZn (β-brass)	742 K
Order–disorder ferroelectrics	KH_2PO_4	120 K
Superfluid	He^4	2.17 K
Superionic conductors	$RbAg_4I_5$ (Fig. I-12)	209 K

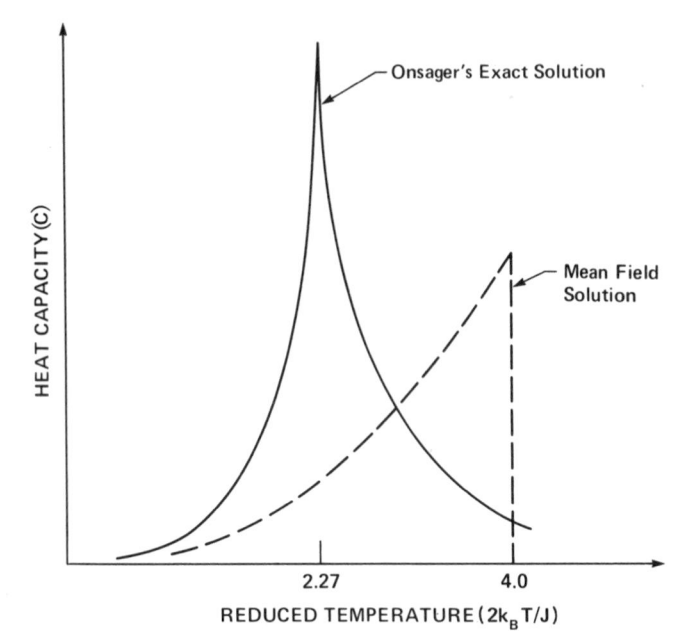

FIG. I.10.   Heat capacity as a function of temperature for the two-dimensional Ising model.

B. CRITICAL EXPONENTS

As we approach a second-order transition, either from above or below the critical temperature T_c, we expect the physical properties of the system to behave in a smooth fashion. We therefore assume that near T_c the temperature dependence of these physical properties may be described by a power series expansion in the parameter $\epsilon \equiv (T - T_c)/T_c$. As we get closer and closer to the transition temperature the behavior of the physical properties are dominated by the leading term in this expansion. Thus the exponent of ϵ for this leading term acquires a particular significance and is referred to as the *critical exponent*. For example,

$$
C \propto \begin{cases} \epsilon^{-\alpha} & T \to T_c^+ \\ (-\epsilon)^{-\alpha'} & T \to T_c^- \end{cases}
$$

$$
M \propto (-\epsilon)^\beta \qquad T \to T_c^- \qquad\qquad (1.13)
$$

$$
\chi \propto \begin{cases} \epsilon^{-\gamma} & T \to T_c^+ \\ (-\epsilon)^{-\gamma'} & T \to T_c^- \end{cases}
$$

defines the critical exponents α, β, and γ. These quantities happen to be *static* properties. Critical exponents have also been defined for *dynamic* properties such as transport coefficients.

Since the various physical properties are all related to the free energy, there exist relations between the critical exponents. The fact that the specific heat must be positive, for example, leads to the so-called Rushbrook inequality,

$$\alpha' + 2\beta + \gamma' \geq 2. \tag{1.14}$$

Problem I.2

It is sometimes convenient to define yet another chemical potential, let us call it A, in which the magnetization M is the variable,

$$A \equiv G + MHV + (1/8\pi)H^2V$$

$$\Delta A = -S\Delta T + V\Delta P + VH\Delta M.$$

The heat capacity at constant M is then

$$C_M \equiv T\left(\frac{\partial S}{\partial T}\right)_M.$$

a. Using the properties of functions of several variables prove the relation

$$\chi_T(C_H - C_M) = T\left(\frac{\partial M}{\partial T}\right)_H^2.$$

b. Heat capacities must be positive if a system is to be stable against thermal fluctuations. Use this fact to derive the Rushbrook inequality.

If the exponents associated with a model satisfy these thermodynamic inequalities as equalities, the model is said to exhibit "scaling." The two-dimensional Ising model is such an example.

Interest in critical exponents is due to the fact that, although the corresponding exponents may differ slightly from material to material (β ranges from 0.33 to 0.37), they seem to depend primarily on fundamental parameters such as the dimensionality of the system. This independence of the critical exponents on the nature of the system is referred to as "universality."

C. ORDER PARAMETERS

Let us now turn to a more general description of phase transitions than that provided by Ehrenfest. This description is based on the fact that *most* phase transitions are characterized by the appearance of some nonzero quantity in the ordered state. In a ferromagnet this quantity is the spontaneous magnetization. Such a quantity is called the *order parameter*. The concept of an order parameter is extremely convenient although its identity is not always apparent. In the remainder of this chapter we shall give examples of order parameters for various systems.

1. Magnetic Order

In order to identify the order parameter in a magnetic system it is convenient to consider the generalized susceptibility $\chi(\mathbf{q})$ that determines the response of the magnetization with the wave vector $\mathbf{q}$ to an applied magnetic field having the same wave vector. In Chapter IV it will be shown that this susceptibility takes the form

$$\chi(\mathbf{q}) = \frac{C/T_c}{(T - T_c)/T_c + (1 - J(\mathbf{q})/J(\mathbf{Q}))} \tag{1.15}$$

where C is the Curie constant, $C = (N/V)g^2\mu_B^2 S(S + 1)/3k_B$, $T_c = CVJ(\mathbf{Q})/g^2\mu_B^2$, and $\mathbf{Q}$ is the wave vector for which the Fourier transform of the exchange interaction (which is discussed in Chapter IV) has a maximum. For the nearest neighbor exchange coupling $J > 0$ and $\mathbf{Q} = 0$. In this case $\chi(\mathbf{q} = 0)$ diverges at $T = T_c$. This divergence means that the assumption of an isotropic ground state, which went into the derivation of $\chi(\mathbf{q})$, is invalid. If an instability develops on the high-temperature side of the transition with some wave vector $\mathbf{Q}$, this generally implies that the new ground state below the transition will also be characterized by this wave vector. Thus a divergence in the uniform susceptibility signals a transition to a ferromagnetic state. When $J < 0$, $\mathbf{Q}$ becomes $\pi(1, 1, 1)/a$ for a simple cubic system with lattice parameter a. In this case the direction of the magnetization in the ordered phase alternates from site to site along the body diagonal. Thus, in such antiferromagnetic ($J < 0$) systems the order parameter is said to be the "staggered" magnetization, $\mathbf{N} = \Sigma_i \eta_i \mathbf{S}_i$ where $|\eta_i| = 1$ and negative for nearest neighbors along the direction of $\mathbf{Q}$. In complex lattices where there is more than one exchange interaction, the nature of the order parameter must be determined from microscopic experimental probes such as neutron scattering. For magnetic interactions that confine the magnetization to a plane (i.e., a planar ferromagnet), the order parameter may then be written in the complex form, $M_x + iM_y = M e^{i\phi}$ where the phase ϕ is arbitrary.

There are various features of the order parameter that, as we shall see, have important implications for the collective excitations of the system. Consider, for example, the commutator of the y component of the ith spin with the z component of the total spin, $\mathbf{S} = \Sigma_i \mathbf{S}_i$,

$$[S_i^y, S^z] = i\hbar S_i^x. \tag{1.16}$$

If $\delta\theta$ corresponds to an infinitesimal rotation of the spin coordinates about the z-axis, then $\delta S_i^y = S_i^x \delta\theta$. Therefore

$$\delta S_i{}^{\nu} = -i[S_i{}^{\nu}, S^z/\hbar]\delta\theta. \tag{1.17}$$

This defines $S^z/\hbar$ as the *generator* of spin rotations about the z-axis. This means that the unitary transformation that rotates all spins through an angle θ about the z-axis has the form

$$U(\theta) = e^{iS^z\theta/\hbar}. \tag{1.18}$$

If our magnetic system is characterized by the isotropic Heisenberg Hamiltonian,

$$\mathcal{H} = -2J \sum_i \sum_{j>i} \mathbf{S}_i \cdot \mathbf{S}_j, \tag{1.19}$$

then using Eq. (1.18) gives

$$U(\theta)\mathcal{H}\,U^{-1}(\theta) = \mathcal{H}, \tag{1.20}$$

which means that $\mathcal{H}$ is rotationally invariant in "spin space." It is also interesting to note that

$$[\mathbf{S}, \mathcal{H}] = 0. \tag{1.21}$$

Thus we have the results that for a ferromagnet; (a) the order parameter is also the generator of a continuous group of transformations [Eq. (1.17)] and (b) it is a conserved quantity [Eq. (1.21)]. Notice that the order parameter for an antiferromagnet, the staggered magnetization $\mathbf{N}$, is *not* a conserved quantity. In fact, the ferromagnet is one of the few systems in which the order parameter is conserved.

Another feature of the order parameter is its *number of components, n.* In the case of a cubic Heisenberg ferromagnet the magnetization can lie along any of the three directions in space. Therefore $n = 3$. For an Ising ferromagnet $n = 1$. If one keeps only the "transverse" part of Eq. (1.19), this is called the $X - Y$ model and its ferromagnetic order parameter has two components. If the long range order is characterized by a finite wave vector $\mathbf{k}$, as in an antiferromagnet, then the number of components is further increased by the number of equivalent directions in $\mathbf{k}$-space. For example, consider the cubic antiferromagnet MnO whose ordered state consists of ferromagnetic (111) planes coupled antiferromagnetically. There are four vectors which constitute what is called the "star" of $\mathbf{k}$: $\mathbf{k}_1 = [\frac{1}{2}\frac{1}{2}\frac{1}{2}]$, $\mathbf{k}_2 = [-\frac{1}{2}\frac{1}{2}\frac{1}{2}]$, $\mathbf{k}_3 = [\frac{1}{2}\frac{1}{2} - \frac{1}{2}]$ and $\mathbf{k}_4 = [\frac{1}{2} - \frac{1}{2}\frac{1}{2}]$. The sublattice magnetization may lie in either of the two transverse directions to $\mathbf{k}_i$. Therefore the order parameter has eight components! In Chapter II we shall see that this number of components is intimately connected with the fact that the magnetic transition in MnO is first order.

One final aspect to the order parameter is the fact that it may give rise to a "broken symmetry." In a ferromagnet, for example, if $\langle S_z \rangle \neq 0$, then rotations about the x- and y-axes are no longer symmetry operations. In an antiferromagnet the same two symmetries are broken. In some cases, of course, the "order parameter" may not lead to a broken symmetry. In the liquid–gas transition, for example, the "order parameter" is simply the density difference $\rho_{liq} - \rho_{gas}$. In such cases we expect to have a critical point.

A more bizarre type of magnetic order is found in the so-called "spin glass." This describes the state of certain dilute magnetic alloys such as CuMn or AuFe in the concentration range 0.1–10 at. %. In this state the local moments are frozen into particular, but random, directions but do not exhibit long range order. The origin of the spin glass lies in the fact that the exchange interaction between moments in a metal has an oscillatory spatial dependence. Thus moments that are randomly distributed throughout a metallic host will experience a distribution of exchange interactions. Edwards and Anderson[8] have taken the order parameter to be

$$q \equiv \overline{|\langle \mathbf{S}_i \rangle|^2} \tag{1.22}$$

where $\langle \cdot \cdot \cdot \rangle$ represents a thermal average and the bar a spatial average. Note that the magnetization, $M \equiv \overline{\langle S_i \rangle}$, may be zero even if q is not. When q is nonzero *and*

$$\langle \mathbf{S}_i \rangle \cdot \langle \mathbf{S}_j \rangle \to 0 \qquad \text{as} \qquad (\mathbf{R}_i - \mathbf{R}_j) \to \infty \tag{1.23}$$

we have a spin glass. Sherrington and Kirkpatrick[9] have proposed an idealized model of a spin glass that allows an exact solution. This model consists of an Ising model in which the spins are coupled by infinite-range random interactions independently distributed with a Gaussian probability density. The paramagnetic-to-spin glass phase boundary is characterized by a cusp in the low field susceptibility as experimentally observed in Fig. I.11.[10] It is interesting to note that the resistivity does *not* show an anomaly at the spin-glass "freezing" temperature, T_0. The impurity resistivity appears to be increasing linearly with T in the vicinity of T_0. The low-temperature thermodynamics of this model, at least in the mean-field approximation, are peculiar in the sense that fluctuations of q persist for all $T < T_0$. This difficulty has not yet been resolved. It may have to do with the fact that q may not be the appropriate order parameter, or that spin-glass behavior is purely a non-equilibrium effect.

[8] S. G. Edwards and P. W. Anderson, *J. Phys. F* **5**, 965 (1975).

[9] D. Sherrington and S. Kirkpatrick, *Phys. Rev. Lett.* **35**, 1792 (1975).

[10] V. Cannella, *in* "Amorphous Magnetism" (H. O. Hooper and A. M. de Graff, eds.), p. 195. Plenum, New York, 1973.

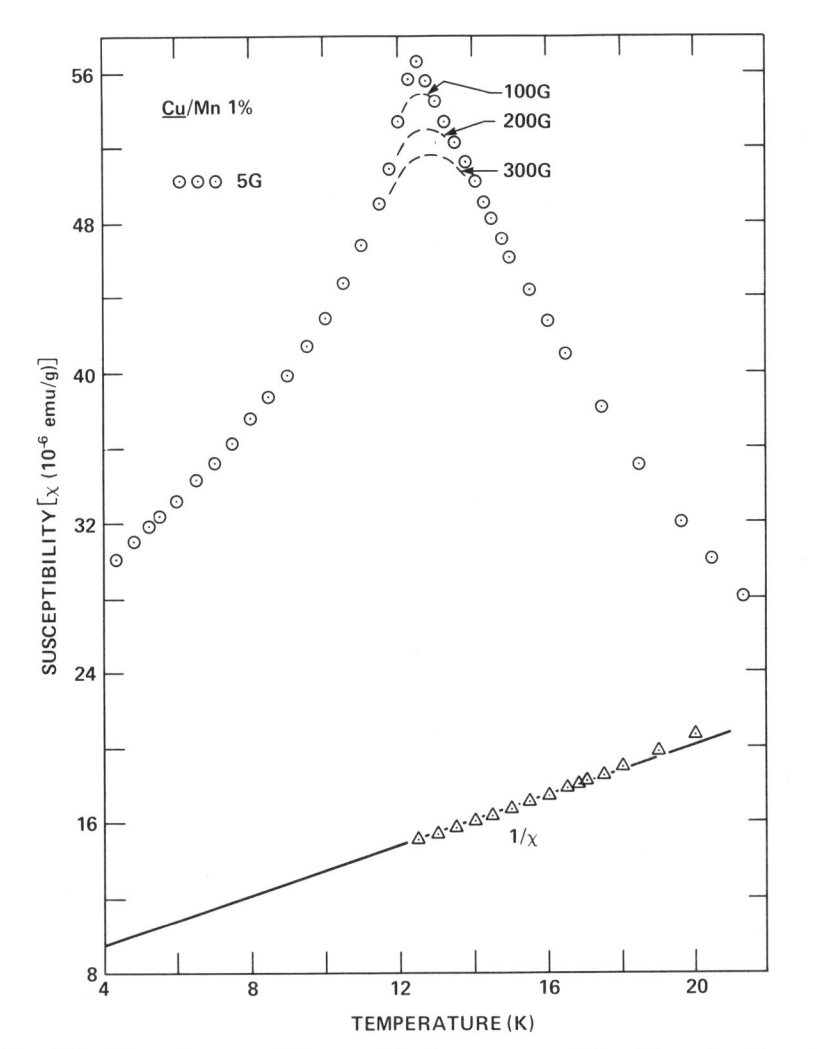

FIG. I.11. Magnetic susceptibility χ and the inverse susceptibility, $1/\chi$, as functions of temperature for copper containing 1% manganese in a field of 5 G. Data for χ at local fields of 100, 200, and 300 G are also indicated. $1/\chi$ is in arbitrary units. (After Cannella.[10])

2. Crystallization

As we have already mentioned, crystal lattices represent the most ubiquitous example of long range order. But what is the order parameter? A crystal is characterized by the appearance of a diffraction pattern. The existence of such a pattern depends upon the presence of Fourier components in the electron density. In particular, if we expand the charge density as

$$\rho(\mathbf{r}) = \sum_G \rho_G \, e^{i\mathbf{G}\cdot\mathbf{r}} \qquad (1.24)$$

where $\mathbf{G}$ is a reciprocal lattice vector then the scattering amplitude will be proportional to ρ_G. This suggests that we take the set of ρ_G's as our order parameter. Thus our order parameter has infinitely many components. The electronic energy associated with the presence of the periodic potential, i.e., the deviation from that of a free electron gas, is given by[11]

$$\sum_G V(\mathbf{G})\chi(\mathbf{G})|\rho_G|^2$$

where $V(\mathbf{G})$ is the Fourier transform of an effective atomic potential (the pseudopotential) and $\chi(\mathbf{G})$ arises from screening considerations. $V(\mathbf{G})$ decreases rapidly with $\mathbf{G}$. Therefore, only a few of the components of the order parameter enter the energy which means that it is not very sensitive to long range order.

Another point to note, however, is that the energy does include higher order terms, in particular, third order terms. In Chapter II we shall see that the presence of such third-order terms leads to first-order transitions. This is why crystallization is generally first order.

In certain compounds, such as silver salts, we can have the situation in which one of the sublattices "melts" within the remaining lattice. In such a case, the ionic conductivity has an anomalously high value, comparable to that of an ionic melt. An example of such a "superionic" conductor is $RbAg_4I_5$. As shown in Fig. I.12,[12,13] there is a first-order transition at 122 K associated with a structural charge which both increases the number of mobile Ag ions as well as reduces their activation energy for hopping. At 209 K there is another structural transition which further reduces the Ag activation energy. This latter transition appears to be weakly first order. However, analysis of the specific heat suggest it is Ising-like.[14]

3. Superconductivity

The case of superconductivity is more obscure. In 1934 Gorter and Casimir proposed a "two-fluid" model for superconductivity in which they

[11] See, for example, V. Heine and D. Weaire, *Solid State Phys.* **24**, 249 (1970).

[12] W. V. Johnson, H. Wiedersich, and G. W. Lindberg, *J. Chem. Phys.* **51**, 3739 (1969).

[13] B. B. Owens and C. R. Arque, *Science* **157**, 308 (1967).

[14] F. Lederman, M. Salamon, and H. Peisl, *Solid State Commun.* **19**, 147 (1976).

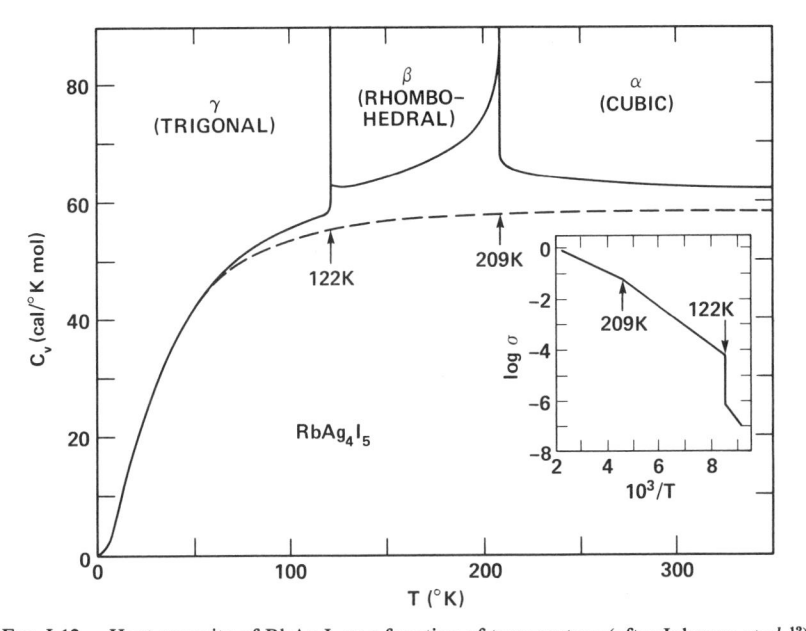

FIG. I.12. Heat capacity of $RbAg_4I_5$ as a function of temperature (after Johnson *et al.*[12]). Insert shows the electrical conductivity (after Owens and Arque[13]).

assumed that a fraction n_s/n of the electrons condensed below T_c into a superconducting state with a fraction $1 - n_s/n$ remaining "normal." Thus n_s/n becomes the order parameter, and their theory predicts a temperature dependence

$$\frac{n_s}{n} = \left(\frac{T_c - T}{T_c}\right)^4 = |\epsilon|^4. \tag{1.25}$$

The richness of superconducting phenomena stems from the true order parameter being not the simple scalar quantity of the two-fluid model, but rather a *complex* one just as in the case of the planar ferromagnet. (In fact, Anderson[15] has shown that the Hamiltonian for superconductivity resembles that of a planar ferromagnet.) It was only with the appearance of the Bardeen, Cooper, and Schrieffer (BCS) microscopic theory of superconductivity in 1957 that the superconducting order parameter was identified as the pair wave function. We shall assume the reader has encountered the BCS theory and merely review here those aspects relevant to the identification of the order parameter.

[15] P. W. Anderson, *Phys. Rev.* **112**, 1900 (1958).

As in most problems dealing with metals, it is convenient to work in the formalism of second quantization. In this formalism one introduces a so-called field operator

$$\psi(\mathbf{r}) = 1/\sqrt{V} \sum_k e^{i\mathbf{k}\cdot\mathbf{r}} c_{k\sigma}$$

where $c_{k\sigma}$ now carries the operator information. In a real metal, in addition to the kinetic energy, there are Coulomb interactions between the electrons themselves as well as between the electrons and the ions. The BCS theory singles out the interaction between electrons with equal but opposite momenta and opposite spins. When this Hamiltonian is expressed in second-quantized form it becomes the BCS "reduced" or "pairing" Hamiltonian,[16]

$$\mathcal{H}_p = \sum_{k,\sigma} \epsilon_k c_{k\sigma}^{\dagger} c_{k\sigma} + \sum_{k,k'} V_{kk'} c_{k\uparrow}^{\dagger} c_{-k\downarrow}^{\dagger} c_{-k'\downarrow} c_{k'\uparrow}, \tag{1.26}$$

where ϵ_k is the eigenvalue of a Bloch electron. This explicitly neglects exchange and Hartree self-energy corrections. Since we restrict ourselves to interactions between such pairs, we expect the ground state to be a coherent superposition of states in which the Bloch states are occupied in pairs. This assumption has the consequence that the quantity $\langle c_{-k\downarrow} c_{k\uparrow} \rangle$ is nonzero and approximately a constant, b_k. This result suggests replacing $c_{-k\downarrow} c_{k\uparrow}$ by its average value b_k in $\mathcal{H}_p$. If we also introduce the quantity

$$\Delta_k = -\sum_{k'} V_{kk'} b_{k'}, \tag{1.27}$$

then the "mean field" BCS Hamiltonian becomes

$$\mathcal{H}_p^{MF} = \sum_{k\sigma} \epsilon_k c_{k\sigma}^{\dagger} c_{k\sigma} - \sum_k (\Delta_k c_{k\uparrow}^{\dagger} c_{-k\downarrow}^{\dagger} + \Delta_k^* c_{-k\downarrow} c_{k\uparrow} - \Delta_k b_k^*). \tag{1.28}$$

The Hamiltonian may be diagonalized by introducing new fermion operators γ_{k0} and $\gamma_{k1}^{\dagger}$ according to

$$c_{k\uparrow} = u_k^* \gamma_{k0} + v_k \gamma_{k1}^{\dagger}$$
$$c_{-k\downarrow}^{\dagger} = -v_k^* \gamma_{k0} + u_k \gamma_{k1}^{\dagger}. \tag{1.29}$$

The operator γ_{k0} corresponds to a quasiparticle composed of an electron (**k** ↑) with amplitude u_k and a hole ($-\mathbf{k}$ ↓) with amplitude v_k.

---

[16] For a derivation of this Hamiltonian and its diagonalization, see G. Rickayzen, "Theory of Superconductivity." Wiley, New York, 1964.

Problem I.3

a. Substitute the expansions (1.29) into the Hamiltonian (1.28) and show that the coefficients of $\gamma_{k1}\gamma_{k0}$ and $\gamma_{k0}^{\dagger}\gamma_{k1}^{\dagger}$ vanish if

$$\left|\frac{v_{\mathbf{k}}}{u_{\mathbf{k}}}\right| = \frac{E_{\mathbf{k}} - \epsilon_{\mathbf{k}}}{|\Delta_{\mathbf{k}}|}$$

where

$$E_{\mathbf{k}} = (\epsilon_{\mathbf{k}}^2 + \Delta_{\mathbf{k}}^2)^{1/2}.$$

b. Using the normalization requirement, $|u_{\mathbf{k}}|^2 + |v_{\mathbf{k}}|^2 = 1$, solve for the coefficient $|v_{\mathbf{k}}|$. Plot $|v_{\mathbf{k}}|$ as a function of electron energy (recall that $\epsilon_{\mathbf{k}} = 0$ corresponds to the Fermi energy).

The form of $E_{\mathbf{k}}$ shows that a gap has opened up in the excitation spectrum. The states formerly in this gap are "pushed aside." Since the number of states remains the same, the superconducting density of states can be found by equating $N_s(E)dE = N_n(\epsilon)d\epsilon \approx N(0)d\epsilon$. Thus,

$$\frac{N_s(E)}{N(0)} = \begin{cases} \dfrac{E}{(E^2 - \Delta^2)^{1/2}} & E > 0. \\ 0 & E < 0. \end{cases} \tag{1.30}$$

In this case we see that $\Delta_{\mathbf{k}}$ can be physically identified with an energy gap. In the presence of magnetic impurities, however, as we shall see later, the energy gap is not simply the quantity $\Delta_{\mathbf{k}}$ defined above. This distinction can lead to a situation referred to as "gapless" superconductivity.

The quantity $\Delta_{\mathbf{k}}$ must be determined self-consistently. Using Eq. (1.29)

$$\langle c_{-\mathbf{k}\downarrow}c_{\mathbf{k}\uparrow}\rangle = u_{\mathbf{k}}\omega_{\mathbf{k}}\langle 1 - \gamma_{k0}^{\dagger}\gamma_{k0} - \gamma_{k1}^{\dagger}\gamma_{k1}\rangle. \tag{1.31}$$

The quasiparticles are also fermions. Therefore $\langle\gamma_{k0}^{\dagger}\gamma_{k0}\rangle = f(E_k)$ where $f(E)$ is the Fermi–Dirac distribution function and the equation for Δ_k becomes

$$\Delta_{\mathbf{k}} = -\sum_{\mathbf{k}'} V_{\mathbf{k}\mathbf{k}'}\Delta_{\mathbf{k}'}[1 - 2f(E_{\mathbf{k}'})]/2E_{\mathbf{k}'}. \tag{1.32}$$

When $\Delta_{\mathbf{k}} = 0$ we recover the normal Fermi liquid. When $\Delta_{\mathbf{k}} \neq 0$ it can be shown that the system exhibits superconducting properties. Therefore, we identify $\Delta_{\mathbf{k}}$ as the order parameter.

The superconducting ground-state wave function $|\phi\rangle$ is determined by the equations

$$\begin{aligned} \gamma_{k0}|\phi\rangle &= 0 \\ \gamma_{k1}|\phi\rangle &= 0. \end{aligned} \tag{1.33}$$

It is easy to verify that these conditions are satisfied by the BCS pairing state

$$|\phi\rangle = \prod_{\mathbf{k}} (u_{\mathbf{k}} + v_{\mathbf{k}} c^{\dagger}_{\mathbf{k}\uparrow} c^{\dagger}_{-\mathbf{k}\downarrow})|0\rangle \qquad (1.34)$$

where $|0\rangle$ is the vacuum state. Notice that $|\phi\rangle$ consists of a superposition of states containing different numbers of particles, i.e., 0, 2, 4, etc. This introduces some uncertainty in the number of pairs in the "condensate." This, however, is small, for[17]

$$n \equiv \left\langle \phi \left| \sum_{\mathbf{k}} (c^{\dagger}_{\mathbf{k}\uparrow} c_{\mathbf{k}\uparrow} + c^{\dagger}_{\mathbf{k}\downarrow} c_{\mathbf{k}\downarrow}) \right| \phi \right\rangle = \sum_{\mathbf{k}} 2|v_{\mathbf{k}}|^2$$

and

$$\delta n \equiv \sqrt{(n - \bar{n})^2} = 2 \sqrt{\sum_{\mathbf{k}} u_{\mathbf{k}}^2 v_{\mathbf{k}}^2}.$$

The sums over $\mathbf{k}$ are proportional to the number of electrons, N. Therefore, $\delta n/\bar{n} \sim 1/n^{1/2} \sim 10^{-10}$ corresponding to a distribution sharply peaked around $\bar{n} = \bar{n}$. The BCS state $|\phi\rangle$ is therefore analogous to the so-called "coherent state" in the quantum theory of light in which the number of photons is uncertain, but the expectation value of the quantized electromagnetic fields have their classical values.[18]

The diagonalization procedure described above only determines the magnitude of the complex coefficients $u_{\mathbf{k}}$ and $v_{\mathbf{k}}$. Since the wave function may always be multiplied by an overall phase factor (i.e., is gauge invariant), we may choose one of these, say $u_{\mathbf{k}}$, to be real. But $v_{\mathbf{k}}$ may differ by a relative phase, i.e., $v_{\mathbf{k}} = |v_{\mathbf{k}}|e^{i\phi_{\mathbf{k}}}$. The essential feature of the BCS state is that the pairs all have the same phase, i.e., $\phi_{\mathbf{k}} = \phi$. Then the order parameter, which is proportional to $u_{\mathbf{k}} v_{\mathbf{k}}$, will be characterized by this phase ϕ.

It is interesting to project out of $|\phi\rangle$ that component involving N particles. This is accomplished by the integral

$$|N\rangle = \int_0^{2\pi} d\phi \, e^{-iN\phi/2}|\phi\rangle. \qquad (1.35a)$$

This may be inverted to give

$$|\phi\rangle = \sum_{N} e^{iN\phi/2}|N\rangle. \qquad (1.35b)$$

[17] See, for example, M. Tinkham, "Introduction to Superconductivity," p. 23. McGraw-Hill, New York, 1975.

[18] See, R. Loudon, "The Quantum Theory of Light," Chapter 7. Oxford Univ. Press, London and New York, 1973.

It can be shown [19] that the number operator N can be regarded as the generator for *gauge transformations*. Thus the choice of the phase of v_k is equivalent to a choice of gauge. In the case where we are dealing with a single, homogeneous superconductor this phase has no physical consequences. However, if we have two superconductors weakly coupled together their relative phase does have physical consequences. Suppose, for example, that the two superconductors are characterized by the states $|\phi_L\rangle$ and $|\phi_R\rangle$. The uncoupled situation is then described by the product function $|\phi_L\rangle|\phi_R\rangle$. The coupling arises from the coherent tunneling of a BCS pair from one superconductor to the other. In terms of the individual electrons the tunneling may be represented by a "perturbing" Hamiltonian of the form $T_{kq}c_k^\dagger c_q$, where the state $\mathbf{q}$ is associated with one of the superconductors and $\mathbf{k}$ with the other. In Chapter IV we shall show that if the two superconductors are electrically connected a current flows that is proportional to $|T|^2\sin(\phi_L - \phi_R)$. This dependence on the relative phase was discovered by Josephson and we shall discuss it further in Chapters IV and VII.

The fact that the number of pairs, $N/2$, is the generator for an infinitesimal change in phase means that these are conjugate variables. Thus they satisfy Hamilton's equations. In particular,

$$\hbar(\partial\phi/\partial t) = -\partial \mathcal{H}/\partial(N/2) \equiv -2\mu, \tag{1.36}$$

where μ is the chemical potential. We shall return to the implications of this in Chapter II.

Finally, let us consider what is meant by "long range order." In the case of ferromagnetism this has a simple physical meaning. Namely, the direction of the magnetic moment at any point $\mathbf{r}$ has a component parallel to that at any other point $\mathbf{r}'$. In the case of superconductivity we are again faced with a more complicated situation. Let us begin our discussion by introducing the quantity

$$G_1(\mathbf{r}t, \mathbf{r}'t') = \langle \psi^\dagger(\mathbf{r}'t')\psi(\mathbf{r}t)\rangle \tag{1.37}$$

where $\psi(\mathbf{r}t)$ is just the Heisenberg representation of the field operator we introduced above. The function (1.37) has the physical interpretation of being the amplitude for removing a particle from $\mathbf{r}t$ and replacing one at $\mathbf{r}'t'$ and is called a "propagator." It applies to bosons or fermions. For free electrons (1.37) becomes

---

[19] If we make an infinitesimal change $d\phi$ in the phase of the field operator ψ, then the generator, G, associated with such an infinitesimal phase change $\delta\psi = i\psi(d\phi)$ is defined by $\delta\psi = i[\psi, G](d\phi)$. Since the number operator is given by $N = \int\psi^\dagger(\mathbf{r}')\psi(\mathbf{r}')d\mathbf{r}'$ and since $[\psi(\mathbf{r}')^\dagger, \psi(\mathbf{r})] = \delta(\mathbf{r} - \mathbf{r}')$, it therefore follows that $G = N$.

$$G_1(\mathbf{r}t, \mathbf{r}'t') = \frac{1}{V} \sum_{\mathbf{k}} \exp[i\mathbf{k} \cdot (\mathbf{r} - \mathbf{r}')]\exp[-i\epsilon_{\mathbf{k}}(t - t')]\langle c_{\mathbf{k}}^{\dagger} c_{\mathbf{k}}\rangle. \quad (1.38)$$

Carrying out the sum for a system of electrons at zero temperature ($T = 0$), the equal time propagator

$$G_1(\mathbf{r}t, \mathbf{r}'t) = 3(N/V)(\sin x - x \cos x)/x^3 \quad (1.39)$$

where $x = k_F|\mathbf{r} - \mathbf{r}'|$. This is the familiar oscillation associated with the sharpness of a Fermi surface. Thus G_1 falls off in the Fermi length $1/k_F$.

Now suppose we have a system of noninteracting *bosons* at $T = 0$. Then, as we know from statistical mechanics, all particles are condensed into the $k = 0$ state, i.e., $\langle c_{\mathbf{k}}^{\dagger} c_{\mathbf{k}}\rangle = N\delta_{\mathbf{k},0}$, so

$$G_1(\mathbf{r}t, \mathbf{r}'t) = N/V. \quad (1.40)$$

In this case G_1 remains constant as $\mathbf{r}$ and $\mathbf{r}'$ move apart. Thus we have "long range order" in G_1.

In the superconducting case we must introduce the quantity,

$$G_2(\mathbf{r}_1 t_1, \mathbf{r}_2 t_2; \mathbf{r}_1' t_1', \mathbf{r}_2' t_2') = \langle \psi_{\uparrow}^{\dagger}(\mathbf{r}_2' t_2')\psi_{\downarrow}^{\dagger}(\mathbf{r}_1' t_1')\psi_{\downarrow}(\mathbf{r}_1 t_1)\psi_{\uparrow}(\mathbf{r}_2 t_2)\rangle. \quad (1.41)$$

As we discussed above, a superconductor is characterized by the nonvanishing of $\langle c_{-\mathbf{k}\downarrow} c_{\mathbf{k}\uparrow}\rangle$. This condition is equivalent to $\langle \psi_{\downarrow}(\mathbf{r}t)\psi_{\uparrow}(\mathbf{r}t)\rangle$ being nonzero. The result is that G_2 does not go to zero as the primed and unprimed variables become separated. Thus we speak of "long range order" in G_2. Furthermore, since $c_{-\mathbf{k}\downarrow} c_{\mathbf{k}\uparrow}$ is off-diagonal in the number representation, C. N. Yang has introduced the term off-diagonal long range order (ODLRO) to describe superconductivity.

Let us now consider the evaluation of the order parameter. The simplest form for the interaction $V_{\mathbf{k},\mathbf{k}'}$ is

$$V_{\mathbf{k},\mathbf{k}'} = \begin{cases} -V & |E_{\mathbf{k}}|, |E_{\mathbf{k}'}| < \hbar\omega_D \\ 0 & \text{otherwise} \end{cases} \quad (1.42)$$

where V here refers to a constant, not to be confused with the volume. From Eq. (1.32) we see that this choice of $V_{kk'}$ makes Δ_k independent of k. Furthermore, for a homogeneous system in the absence of a magnetic field the order parameter may also be taken to be real without any loss of generality. The equation for the order parameter (1.32) can then be solved with the result that Δ has the form shown in Fig. I.13a[20,21] Near T_c, $\Delta(T)/\Delta(0) = 1.74 \, \epsilon^{1/2}$. Also shown is the result for the Gorter–Casimir order parameter.

[20] P. Townsend and J. Sutton, *Phys. Rev.* **128**, 591 (1962).
[21] D. E. Moncton, R. J. Birgeneau, L. V. Interrante, and F. Wudl, *Phys. Rev. Lett.* **39**, 507 (1977).

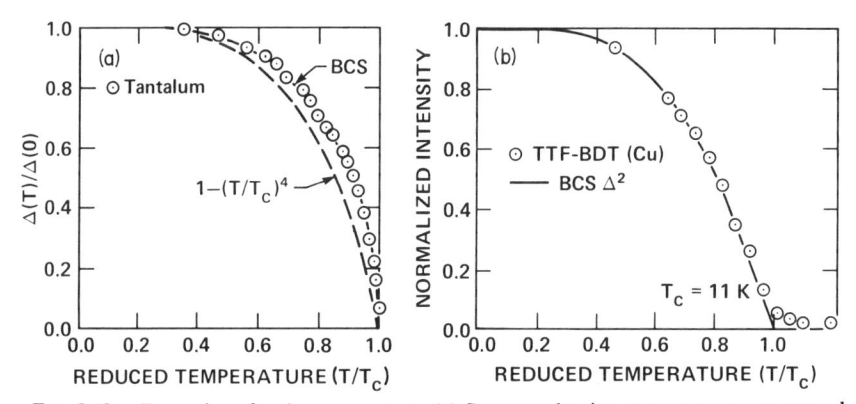

FIG. I.13. Examples of order parameters. (a) Superconducting energy gap as measured by tunneling (after Townsend and Sutton[20]) (b) X-Ray intensity of a new Bragg peak that develops in the quasi one-dimensional Heisenberg antiferromagnet TTF-CuBDT (which stands for a copper complex with tetrathiafulvalene) below 11 K. This transition is due to spin pairing of the spin-$\frac{1}{2}$ TTF$^+$ units. A mean-field spin–lattice dimerization model predicts this intensity should be proportional to the square of the BCS gap function (after Moncton *et al.*[21]).

4. Structural Transitions

As we mentioned in the introductory paragraph to this chapter, Nature offers a variety of transitions between different crystallographic states. Since the basic degrees of freedom in a solid are electronic and vibrational, we may classify these transitions by whether they are driven primarily by electronic or vibrational instabilities. These degrees of freedom are, of course, strongly coupled. In fact, this is the reason that an electronic instability manifests itself in a structural change in the first place.

Since the electronic degrees of freedom in an insulator are different from those in a metal, the studies of their associated phase transitions have taken different paths. However, the underlying principles are equivalent. In an insulator composed of ions with orbitally degenerate ground states it is possible to lower the electronic energy by splitting these degenerate levels apart by means of a lower symmetry distortion. The amplitude of the distortion depends upon the nature of the electronic state and the size of the increase in the elastic energy. Such transitions are referred to as *cooperative Jahn–Teller* transitions.[22] The distorted system is also characterized by the appearance of a spontaneous electric quadrupole moment. The fundamental order parameter, however, is usually taken to be the splitting of the electronic states. Such a transition is illustrated by the second-order transition in PrAlO$_3$ at 151 K. The Pr^{3+} ion has a $^3H_4$

[22] For a comprehensive review, see G. A. Gehring and K. A. Gehring, *Rep. Prog. Phys.* **38**, 1 (1975).

($4f^2$) ground state. Above 151 K this ion finds itself in a site of C_{2v} symmetry, which gives rise, among other levels, to two low-lying singlets. Below 151 K the system gradually distorts eventually becoming tetragonal at about 80 K. The amplitude of this distortion corresponds to the optical phonon order parameter shown in Fig. I.14. The separation of the two low-lying singlets corresponds to the electronic order parameter. The fact that these are proportional to one another supports the identification of this as a cooperative Jahn–Teller transition. In this case the symmetry of this distortion is such that it also couples to an acoustic phonon with the result that the strain order parameter $\epsilon_{zz} - \epsilon_{xx}$ also reflects the transition.

When the phonons themselves are responsible for a structural transition, we can have a *displacive transition*. In such cases the amplitude of the distortion acts as the order parameter. If this distortion is uniform it can lead to a spontaneous electric polarization, i.e., a ferroelectric. An

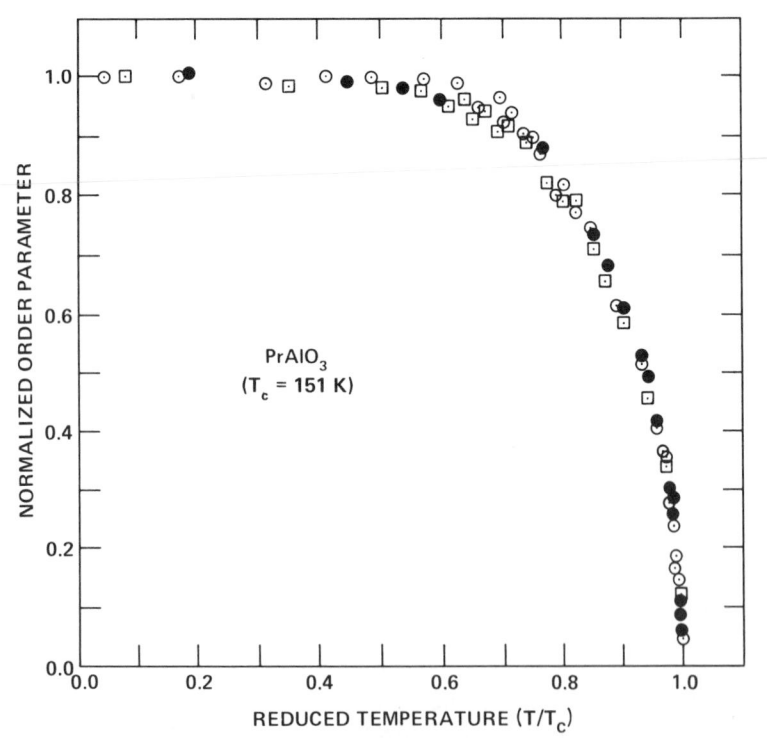

FIG. I.14.   (●) Normalized strain, (□) electronic, and (⊙) optical phonon order parameters as functions of reduced temperature below 151 K in $PrAlO_3$.

example of this is the ferroelectric transition in $BaTiO_3$ at 130 °C. Displacive transitions correspond to a change from one lattice ordering to another, and are the result of the frequency of an optical mode going to zero creating an instability. In the case of $BaTiO_3$ this optical mode is at the zone center. In the case of $SrTiO_3$ the optical mode is at the zone boundary leading to a unit cell doubling but no dipole and therefore no antiferroelectricity. This is illustrated in Fig. I.15.[23] Note that the order parameter in this case is the relative "twist" shown in the figure. The various structural phases occurring in $PrAlO_3$ and $BaTiO_3$ are summarized in Fig. I.16.

Some ferroelectrics are characterized as *order–disorder* ferroelectrics. $NaNO_2$ is such an example. In the disordered phase the NO_2 ions are randomly in either of two configurations, while in the ordered phase they are all in the same configuration. Since the difference between the positions of the oxygen atoms in the two phases is not infinitesimal, the transition is necessarily first order. Order–disorder transitions are not accompanied by "soft modes." In some cases, however, one talks about a soft diffusive mode which is actually a large amplitude thermal hopping between potential wells.

Ammonium chloride, NH_4Cl, also shows an order–disorder transition at 243 K associated with the two possible orientations of the ammonium ion tetrahedra. However, since NH_4^+ does not possess a dipole moment there is no ferroelectricity.

Hydrogen-bonded ferroelectrics such as KH_2PO_4(KDP) undergo a combination of two types of structural transitions. Whereas the protons undergo an order–disorder transition, the metal ions undergo a displacive one from one ordered arrangement to another with a "soft mode."

In metals the electronic degrees of freedom are characterized by a Fermi surface. If additional periodicities are introduced into such a system either by virtue of a lattice distortion or by "waves" in the electron system itself then the Fermi surface and, hence, the electronic energy will be modified.

Whether such a transition occurs depends upon the energies involved. In one dimension, however, Peierls[24] has shown that a conducting chain is unstable with respect to a lattice distortion having a wavelength π/k_F or, equivalently, to the formation of covalently bonded molecules. Diffuse X-ray scattering experiments[25] on the platinum chain complex

[23] R. A. Müller, W. Berlinger, and F. Waldner, *Phys. Rev. Lett.* **21,** 814 (1968).

[24] R. E. Peierls, "Quantum Theory of Solids," p. 108. Oxford Univ. Press (Clarendon), London and New York, 1954.

[25] R. Comes, M. Lambert, H. Launois, and H. R. Zeller, *Phys. Rev. B* **8,** 571 (1973).

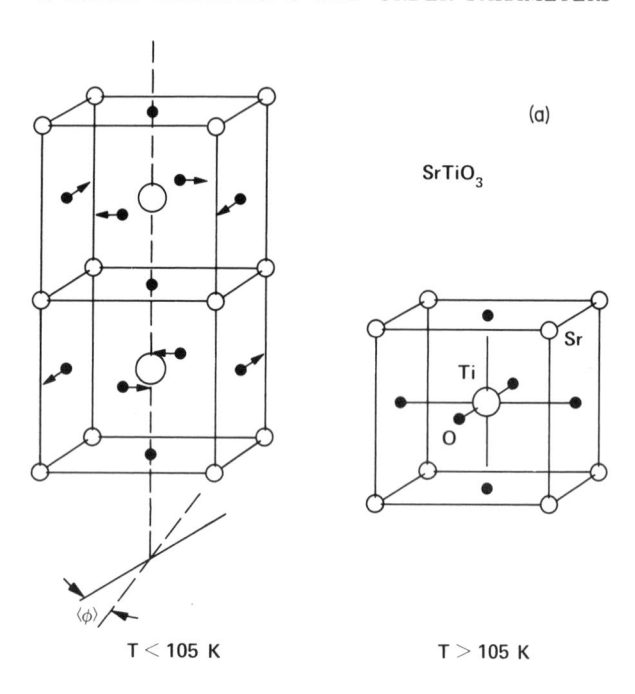

(a)

SrTiO$_3$

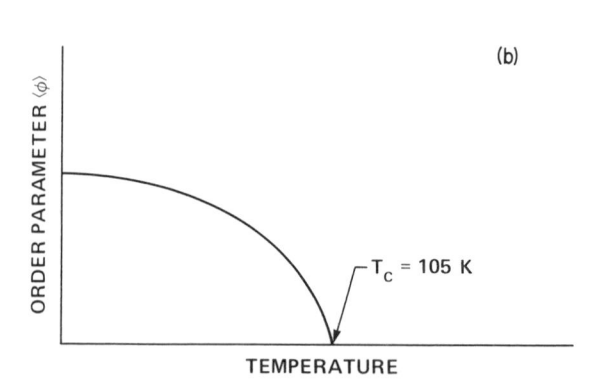

(b)

FIG. I.15.   (a) Soft-mode eigenvector in SrTiO$_3$ defining the order parameter ϕ. The oxygen ions remain in the faces of each cube. (b) Temperature dependence of ϕ as measured by Müller *et al.*[23]

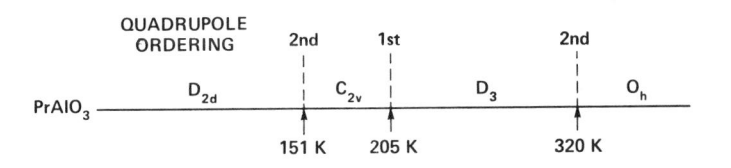

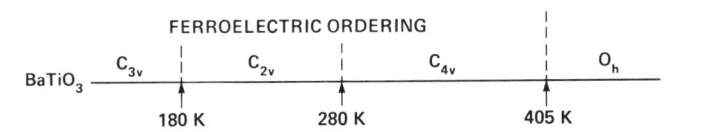

FIG. I.16.   Cation site-symmetries of the various phases of the perovskites $PrAlO_3$ and $BaTiO_3$.

$K_2Pt(CN)_4Br_{0.30} \cdot xH_2O$ suggest it to be an example of such a Peierls distortion.

The Peierls distortion differs from the Jahn–Teller distortion primarily in the nature of the symmetry that is broken. In the Jahn–Teller case this is a point symmetry whereas in the Peierls case it is a translational symmetry.

There is also a magnetic analog of the Peierls transition, called the spin–Peierls transition. In this case a nonconducting antiferromagnetic chain dimerizes and undergoes a second-order transition to a singlet ground state with an energy gap associated with magnetic excitations. An example of such a material is TTF-CuBDT. The theory of this transition has the same mathematical form as the BCS theory. Therefore the gap has the BCS temperature dependence. This has been confirmed by X-ray scattering as shown in Fig. I.13b.

The same (geometrical) considerations that make a Peierls distortion dominant in one dimension also make such instabilities very likely in two dimensions. Electron diffraction studies[26] of the so-called 1T polytype of the layered crystal $TaSe_2$ show a structural transition at $T_0 \simeq 600$ K. This distortion is associated with the existence below T_0 of an incommensurate charge density wave (ICDW). This ICDW is constructed from three symmetry-related $\mathbf{q}$ waves where $\mathbf{q} = 0.285\,\mathbf{a}_0 + \frac{1}{3}\mathbf{c}_0$ where $\mathbf{a}_0$ and $\mathbf{c}_0$ are primary reciprocal lattice vectors. The order parameter associated with this transition may be taken either as the amplitude of the distortion or the amplitude of the charge density. Both are strongly coupled, as in the insulating case. A charge density with wave vector $\mathbf{q}$ is just $\langle c^{\dagger}_{\mathbf{k}+\mathbf{q}} c_{\mathbf{k}} \rangle$. Notice

[26] J. A. Wilson, F. J. Di Salvo, and S. Mahajan, *Phys. Rev. Lett.* **32**, 882 (1974).

that this corresponds to an electron–hole pairing as opposed to the electron–electron pairing in superconductivity. Consequently, the gap equation for a charge density wave state analogous to Eq. (1.27) is

$$\Delta_k = \sum_{k'} V_{kk'} \langle c^\dagger_{k'+q} c_k \rangle. \qquad (1.43)$$

Having mentioned electron–hole pairing in k-space, it is interesting to note that it also occurs in real space. In a semiconductor in which a high density of electrons and holes is produced by optical excitation, the electrons and holes first form excitons which, in germanium, have a binding energy of 4 meV. These excitons then coalesce to form macroscopic (as large as several microns) electron–hole droplets that are metallic. The phase diagram for this unusual situation is shown in Fig. I.17.[27]

5. Liquid Crystals

Although we shall focus primarily on long range order in solids, in this and the following section we shall briefly indicate some ordered states in liquids.

Liquid crystals refer to a large class of organic materials that exhibit an-

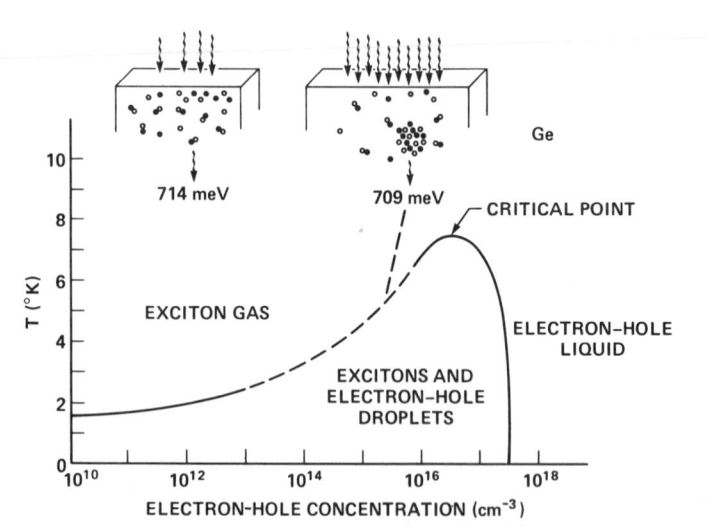

Fig. I.17.   Phase diagram for electrons and holes in pure germanium. The density of excitons is probed by monitoring their infrared luminescence at 714 meV. Electron–hole droplets are associated with luminescence at 709 meV. After Jeffries.[27]

[27] C. D. Jeffries, *Science* **189,** 955 (1975).

isotropy in the liquid state. This anisotropy has its origin in the strongly elongated molecules of which the system is composed. In Fig. 1.18, for example, we illustrate the *nematic* phase of the classic liquid crystal *p*-azoxyanisole. In this phase the centers of mass of the molecules have no long range order. However, there is order in the orientation of the molecules. Thus, although the material flows like a liquid its macroscopic tensor properties are anisotropic. This suggests that we define the order parameter in terms of the angle θ between the long axis of the molecule and the preferred axis. Since the directions "up" and "down" are equivalent, we take the combination

$$s = \langle \tfrac{3}{2}\cos^2\theta - \tfrac{1}{2} \rangle \qquad (1.44)$$

as the order parameter. This parameter may be extracted from various experimental quantities. For example, if $\chi_\parallel$ and $\chi_\perp$ are the (diamagnetic) susceptibilities of the molecule referred to its own axis, and χ_z and χ_x the measured susceptibilities per unit volume, then

$$s = \frac{\chi_z - \chi_x}{N(\chi_\parallel - \chi_\perp)} \qquad (1.45)$$

where N is the number of molecules per unit volume.

If the molecule has a "handedness," or chirality, then the structure develops a helical phase called *cholesteric*. A cholesteric is very similar to a nematic. The centers of mass of the molecules are not ordered, but their orientation lies along a preferred axis that now spirals through the mate-

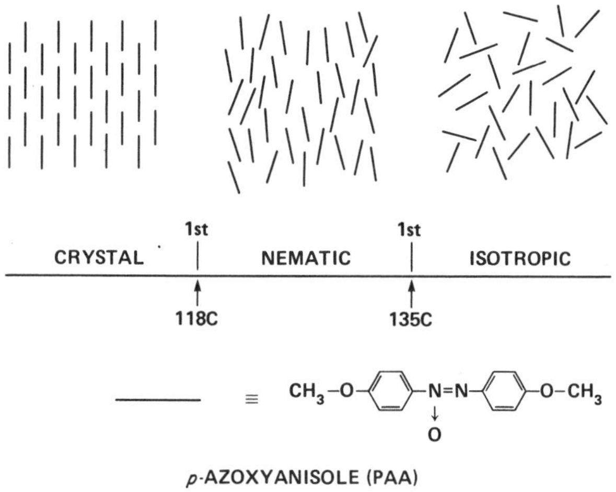

FIG. I.18.   Schematic illustration of the molecular ordering in the nematic liquid crystal PAA.

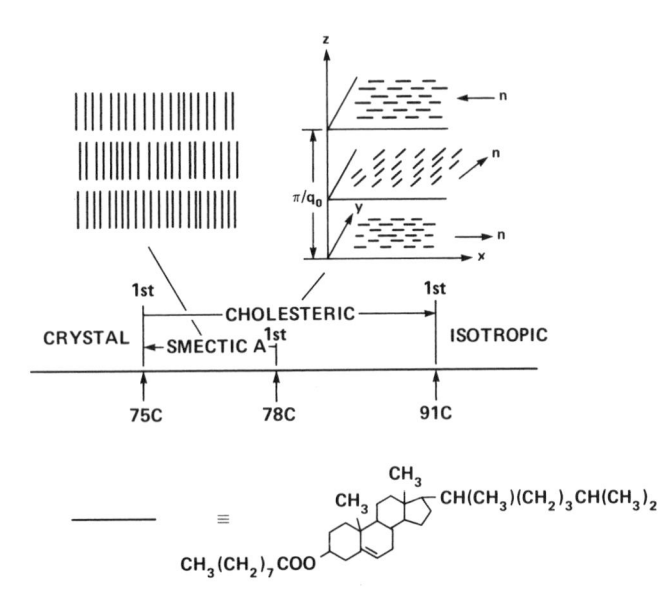

CHOLESTERYL NONANOATE

FIG. I.19   Schematic illustration of the molecular ordering in the cholesteric and smectic A phases of cholesteryl nonanoate.

rial as illustrated in Fig. I.19. The wavelength of this spiral is of the order of 3000 Å. This leads to dramatic optical properties. In fact, all the liquid crystal phases may be made to exhibit optical properties that make them attractive for display purposes.

Notice in Fig. I.19 that when the cholesteric phase of cholesteryl non-anoate is cooled it passes through a so-called *smectic A* phase. This phase has a layer structure with a thickness approximately equal to the length of the molecule. Within each layer the centers of mass are not ordered, i.e., they behave like a two-dimensional liquid.

There are an enormous number of liquid crystals and the list continues to grow. Most of the phase transitions are first order. The particular liquid crystal, *p*-cyanobenzylidene-amino-*p*-*n*-octyloxybenzene (CBAOB), is interesting because it undergoes a *second-order* transition from the nematic to the smectic A with decreasing temperature. McMillan[28] has suggested that we view this as the appearance of a density wave in the direction of the preferred axis. The order parameter is then given by

$$\sigma = \langle \cos(\mathbf{r} \cdot \mathbf{q})(\tfrac{3}{2}\cos^2 \theta - \tfrac{1}{2}) \rangle \tag{1.46}$$

[28] W. L. McMillan, *Phys. Rev. A* **4**, 1238 (1971).

where **q** lies in the direction of the preferred axis and has the value 2π divided by the observed interplanar distance.

6. Superfluid Helium

As we mentioned earlier the zero-point motion of helium prevents solidification. However, both isotopes of helium exhibit "electronic" ordering.

a. Superfluid ^4He: The boiling point of ^{4}He at atmospheric pressure is 4.2 K. Below this temperature the liquid continues to bubble until $T_\lambda = 2.17$ K at which point it suddenly becomes clear and quiet. Below this point ^{4}He is a superfluid. The bubbling ceases because of the presence of second sound that provides an efficient means for dissipating thermal fluctuations. ^{4}He obeys Bose statistics and the appearance of superfluidity is a manifestation of Bose condensation. If the number of particles in the $k = 0$ mode is very large, this component of the boson field operator $\psi(\mathbf{r})$ may be replaced by its average value, ξ_0. Thus

$$\psi(\mathbf{r}) = \xi_0 + \frac{1}{\sqrt{V}} \sum_{k \neq 0} e^{i\mathbf{k}\cdot\mathbf{r}} a_\mathbf{k} \tag{1.47}$$

where ξ_0 plays the role of the order parameter. It has both magnitude and phase, $\xi_0 = \sqrt{n_0}\, e^{i\phi}$. If ϕ varies slowly in space $\phi(\mathbf{r}) = \phi_0 + \nabla\phi \cdot \mathbf{r}$. The wave function is then proportional to $\exp(im\mathbf{v}_s \cdot \mathbf{r})$ corresponding to a macroscopic motion with a flow velocity $\mathbf{v}_s = \nabla\phi/m$. Thus the existence of a condensate implies superfluidity where the superfluid velocity is given by the gradient of the phase of the order parameter.

b. Superfluid ^3He: Below 2.7 mK ^{3}He, which is a Fermion, also enters a superfluid state. The phase diagram is shown in Fig. I.20.[29] It is thought that in both the A and B phases the ^{3}He atoms form Cooper pairs. Since the ^{3}He atoms are in an electronic spin-singlet ground state, the nuclear spin $\frac{1}{2}$ is the appropriate spin variable. The fact that the magnetic susceptibility of the A phase remains essentially that of the normal phase as opposed to the superconducting case where the susceptibility drops dramatically suggests that the Cooper pairs are nuclear spin *triplets* with $I = 1$ in which the $m_I = 0$ component is not present. Since the susceptibility of the B phase is *less* than that of the normal phase, it is presumed that *all* three magnetic substates contribute. Thus, the order parameter takes the form of a symmetric 2×2 matrix in spin space,

$$\Delta_{\alpha\beta}(\mathbf{k}) \sim \langle c_{\mathbf{k}\alpha} c_{-\mathbf{k}\beta} \rangle. \tag{1.48}$$

[29] A. J. Leggett, *Rev. Mod. Phys.* **47**, 331 (1975); J. C. Wheatley, *ibid*, p. 415.

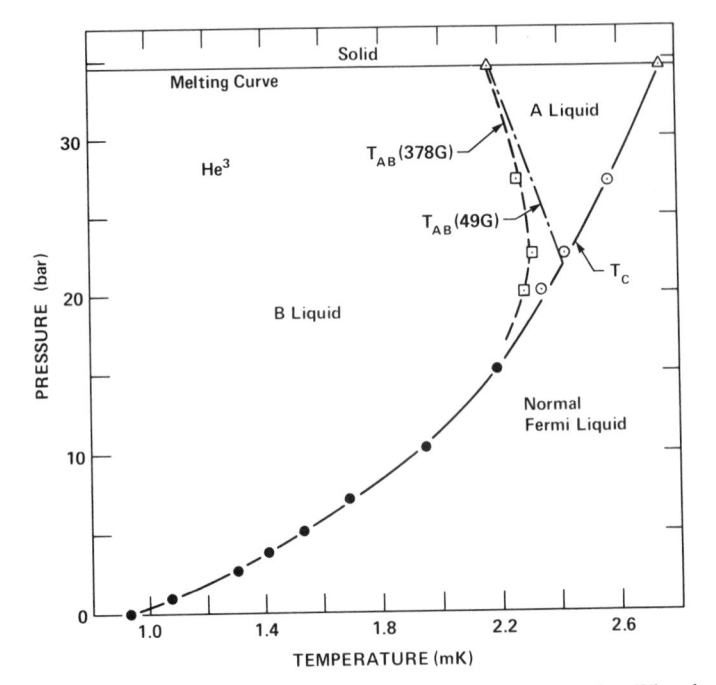

FIG. I.20. Phase diagram of ³He in the low-milliKelvin region. After Wheatley.[29]

Since the relevant momenta lie on the Fermi surface, $\Delta_{\alpha\beta}(\mathbf{k})$ will depend only on the direction of $\mathbf{k}_F$. Thus $\Delta_{\alpha\beta}(\mathbf{k}_F)$ may be expanded in spherical harmonics. However, since the spin state is symmetrical, the Pauli principle will only allow the odd spherical harmonics, the simplest being $L = 1$. Thus, the superfluid Cooper pair has a complex structure involving two unit angular momenta, $\mathbf{L}$ and $\mathbf{I}$. As a result, the static and dynamic magnetic properties of superfluid ³He are very interesting. For example, Paulson and Wheatley[30] have shown that in ³He-A the orbital momentum $\mathbf{L}$ orders in a ferromagnetic sense. We refer the reader to the review articles by Leggett[29] and Wheatley.[29]

D. BROKEN SYMMETRY

The term "broken symmetry" refers to the situation in which the ground state of a system does not have the full symmetry possessed by the Hamiltonian $\mathcal{H}$ used to describe the system. We have already seen several examples of this. It means that the canonical ensemble,

[30] D. N. Paulson and J. C. Wheatley, *Phys. Rev. Lett.* **40**, 557 (1978).

$\exp(-\beta\mathcal{H})$, is not appropriate in the ordered state. Rather, one must use a "restricted ensemble." It is always a challenge to physicists to identify this restricted ensemble.

1. Order of Transition

The manner in which the symmetry is broken finally enables us to define the order of the transition without running into Ehrenfest's difficulties. If the order parameter vanishes discontinuously then the transition is said to be *first order*, while if it vanishes continuously it is *second order*. In this latter case the symmetries on the two sides of the transition are related. In particular, it can be shown that the symmetry group on the "lower" (lower symmetry) side of the transition must be a subgroup of the "higher" side, while in a first-order transition, the two symmetries need not have any relation to one another.

2. Goldstone's Theorem

When the symmetry group that is broken is *continuous* (such as translational or rotational invariance) a new excitation may appear, whose frequency goes to zero at long wavelength. This phenomenon was first noted theoretically by Goldstone[31] and sometimes is elevated to the status of a "theorem." Consider a ferromagnet. The ground state is characterized by a nonzero magnetization in some direction we define as the z-direction. The fact that this direction is arbitrary implies that it costs no energy to rotate this spin configuration as a whole. This corresponds to an infinitely long wavelength excitation. This also means that it costs very little energy to produce a long but finite wavelength "twist." This is easily seen in the Heisenberg Hamiltonian (1.19) where the energy is proportional to the cosine of the angle between adjacent spins. Since this relative angle is the only relevent degree of freedom, we obtain one single branch of elementary excitations called spin waves, or *magnons*.

In the case of a crystal it costs no energy to translate the crystal uniformly in any direction. If, however, a finite wavelength displacement is produced the energy will, in general, depend upon the three possible directions in which the displacement can be made. Thus we obtain three "Goldstone modes" commonly known as acoustic *phonons*.

In the low-frequency regime where the wavelengths become very long we are in a regime where a hydrodynamic description applies. In this sense Goldstone modes are also hydrodynamic modes. This should not surprise us, for hydrodynamics is based on the existence of conserved

[31] J. Goldstone, *Nuovo Cimento* **19**, 154 (1961).

quantities and, as we saw above, the generator of the symmetry that is broken is just such a conserved quantity. We shall discuss the hydrodynamic theory of these modes in Chapter II.

Having brought up hydrodynamics, its perhaps worth saying a little more about phonons. In a normal fluid, hydrodynamic theory shows that the fluctuation spectrum consists of a transverse mass diffusion "mode," a longitudinal heat diffusion "mode," and a longitudinal sound mode. The longitudinal fluctuations give rise to the Rayleigh and Brillouin peaks, respectively, in light scattering. The presence of a longitudinal sound mode in the liquid phase does not imply a broken symmetry. Goldstone's theorem merely states that when a continuous symmetry is broken we expect long wavelength modes with $\omega \to 0$ as $k \to 0$. It does not exclude such modes in the symmetric phase, should they be there for some other reason. In the case of longitudinal sound waves, they can arise from the conservation of matter as well as the breaking of translational invariance. The same differences apply to *second sound*. In liquid helium second sound results from the breaking of gauge symmetry, while in a solid it can result from the conservation of phonon momentum that occurs when umklapp scattering becomes very small.[32]

Let us now consider superconductivity. Since gauge symmetry is continuous we might expect Goldstone modes. That such modes do not occur in this case follows from a caveat regarding the applicability of this theorem.

Lange[33] has discussed Goldstone's theorem with respect to superconductors and ferromagnets. In both cases the argument is based on the conservation law for the generator of the relevent transformation. For a superconductor this generator is the number operator $N = \int d^3 r n(\mathbf{r})$. The change in the order parameter associated with a change in the phase is then proportional to $[N, \psi]$. Particle conservation requires

$$\int d^3 r (\partial/\partial t) \langle [n(\mathbf{r}, t), \psi(\mathbf{r}')] \rangle = -\int d^3 r \nabla \cdot \langle [\mathbf{j}(\mathbf{r}, t), \psi(\mathbf{r}')] \rangle \quad (1.49)$$

---

[32] The usual discussion of heat flow in a solid proceeds by combining the definition of heat flux, $\mathbf{Q} = -K \nabla T$, where K is the thermal conductivity, with the continuity relation, $C \, \partial T / \partial t = -\nabla \cdot \mathbf{Q}$, where C is the heat capacity to obtain a diffusion equation. The heat flux relation implies that the heat flow responds instantly to the temperature gradient. But we know that because of the finite phonon–phonon collision time it takes a finite time to establish a heat flux. Only the so-called umklapp collision processes contribute to the thermal conductivity. Therefore we write $\mathbf{Q} = -K \nabla T - \tau_u \, \partial \mathbf{Q} / \partial t$ where τ_u is the umklapp collision time. If $\partial T / \partial t \gg T / \tau_u$, then the temperature obeys a wave equation.

[33] R. Lange, *Phys. Rev. Lett.* **14**, 3 (1965); *Phys. Rev.* **146**, 301 (1966).

where

$$n(t) \equiv \exp(i\mathcal{H}t/\hbar)n \exp(-i\mathcal{H}t/\hbar). \qquad (1.50)$$

The fact that we have such a conservation law is based on the existence of a continuous symmetry group. In fact, it is a general result (Noether's theorem) that the presence of continuous symmetry transformations lead to conservation relations. Applying the divergence theorem

$$\int d^3r(\partial/\partial t)\langle[n(\mathbf{r}, t), \psi(\mathbf{r}')]\rangle = -\int d\mathbf{S} \cdot \langle[\mathbf{j}(\mathbf{r}, t), \psi(\mathbf{r}')]\rangle. \qquad (1.51)$$

Let us introduce the spatial Fourier transform

$$L(\mathbf{k}, t) = \int d^3r \exp[-i\mathbf{k} \cdot (\mathbf{r} - \mathbf{r}')]\langle[n(\mathbf{r}, t), \psi(\mathbf{r}')]\rangle. \qquad (1.52)$$

Then we recognize the left-hand side of the preceding equation as the limit as $k \to 0$ of the spatial Fourier transform of the integrand; thus

$$(\partial/\partial\tau) \lim_{k\to 0} L(\mathbf{k}, t) = \text{surface integral.} \qquad (1.53)$$

The essence of Goldstone's theorem is that *if the surface integral is zero,* then $L(\mathbf{k}, t)$ must be independent of time which means that the limit as $\mathbf{k} \to 0$ of its time Fourier transform must be proportional to $\delta(\omega)$. The time dependence of the commutator on the left-hand side of Eq. (1.49) is obtained by introducing a complete set of intermediate states between $n(\mathbf{r}, t)$ and $\psi(\mathbf{r}')$. The resulting dependence has the form $\exp[-i\epsilon(\mathbf{k})t/\hbar]$ where $\epsilon(\mathbf{k})$ are the excited state eigenvalues. Thus, the time Fourier transform of $L(\mathbf{k}, t)$ is proportional to $\delta[\omega - \epsilon(\mathbf{k})]$. Combining this with our result we see that in the limit as $\mathbf{k} \to 0$ $\epsilon(\mathbf{k})$ must equal 0. That is, there must be excitations whose frequency goes to zero at long wavelength.

It turns out that if the unscreened interactions are of long range the surface integral will not vanish. In particular, the Coulomb interaction is long range and, as we know, leads to plasmons in metals that have a finite frequency at $\mathbf{k} = 0$. For this same reason there are no Goldstone bosons associated with the appearance of superconductivity. It is interesting to note in this respect that if we "remove" the long range nature of the Coulomb interaction by restricting the electrons to a plane, then the plasmon frequencies do indeed go to zero at zero wave vector.

Notice that Goldstone's "theorem" only deals with the situation at $T = 0$. As one approaches a second-order transition from above, the system will fluctuate into the ordered state. The spatial extent of these fluctuations is characterized by a "coherence length" ξ which we shall discuss in the next chapter. The time that this fluctuation persists will be of the order of $\Delta G/\hbar$ where ΔG is the difference in the free energy between the ordered and disordered states. During this time the ordered "droplet" will support collective modes with wavelengths shorter than ξ. Thus the observation[34] of spin waves above the Curie temperature, for example, is not a violation of Goldstone's "theorem."

3. Soft Modes[35]

This is perhaps an appropriate place to say something about "soft modes" that are not particularly related to Goldstone bosons. Goldstone's "theorem" is concerned with the restrictions that symmetry places on the possible states connected by a phase transition. It does not say anything about the mechanism responsible for the transition. In certain cases this mechanism is due to a decrease or "softening" of a collective mode frequency as we mentioned in the case of the displacive transition in $BaTiO_3$. How this works is particularly easy to see in the case of ferroelectricity. The Lyddane–Sacks–Teller relation relates the low and high (but still below optical) frequency dielectric constants to the longitudinal optical (LO) and transverse optical (TO) phonon frequencies according to

$$\frac{\epsilon_0(T)}{\epsilon_\infty} = \frac{\omega_{LO}{}^2}{\omega_{TO}(T)^2}. \tag{1.54}$$

If the TO phonon "goes soft," i.e., $\omega_{TO} \to 0$, then $\epsilon_0(T) \to \infty$ which heralds the onset of ferroelectricity. Another example of this behavior is found in the incipient ferroelectric $KTaO_3$ as shown in Fig. I.21a. If the phonon modes go soft at the Brillouin zone boundary we obtain a structural transition of the type illustrated by $SrTiO_3$ in Fig. I.15. The phonon spectrum is shown in Fig. I.21b.[36] Although the soft mode in $SrTiO_3$ splits in the lower symmetry phase, there are no *new* modes appearing. In this case, the symmetry group that is broken is *finite* so we would not expect new modes.

In many cases the soft mode is coupled to another mode in the system with the result that the latter also becomes soft. $BaTiO_3$ and $PrAlO_3$ are examples.

In $PrAlO_3$ the proximity of the Pr^{3+} ions makes it possible for electronic excitations to propagate from one ion to another giving rise to an exciton. It turns out that it is the softening of the exciton associated with the

[34] H. A. Mook, J. W. Lynn, and R. M. Nicklow, *Phys. Rev. Lett.* **30**, 556 (1973).

[35] For a general review of soft modes, see J. F. Scott, *Rev. Mod, Phys.* **46**, 83 (1974).

[36] G. Shirane, *Rev. Mod. Phys.* **46**, 437 (1974).

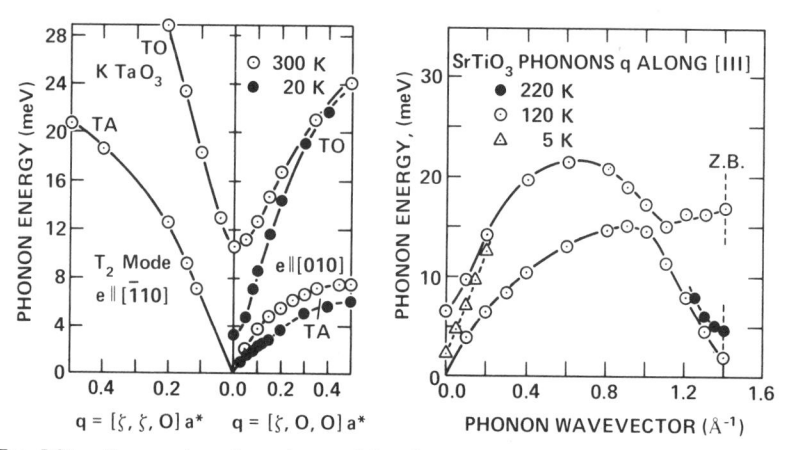

FIG. I.21. Temperature dependence of the phonon modes in (a) KTaO₃ and (b) SrTiO₃. In KTaO₃ the soft mode is at the zone center while in SrTiO₃ it is at the zone boundary. In SrTiO₃ the zone center mode also "softens" leading to incipient ferroelectricity. After Shirane.[36]

lowest-lying electronic transition in $PrAlO_3$ that initiates the structural transition at 151 K. However, as we saw above, the electronic degrees of freedom are coupled to the lattice. In particular this exciton couples to an acoustic phonon eventually driving its long wavelength velocity to zero as shown in Fig. I.22.[37]

In $BaTiO_3$ we have an optical phonon also coupling to an acoustic phonon through the piezoelectric interaction. When the transition is accompanied by a soft *acoustic* phonon the new structure will show a macroscopic strain. Thus, $BaTiO_3$ is referred to as an intrinsic ferroelectric and an extrinsic ferroelastic.

In addition to the soft phonon, $SrTiO_3$ also shows a *central peak* in the neutron scattering spectrum as far as 65 K above T_c. As $T \to T_c$ the weight of this central peak grows relative to that of the phonon peak. This is illustrated in Fig. I.23.[38] Similar behavior has also been observed in a number of other perovskites. There have been various attempts to ascribe this behavior to intrinsic mechanisms. However, none of these models reproduce the observed characteristic that the width of the central peak is very narrow compared to the phonon width even at temperatures well above T_c where the weight of the central peak is small compared to that of

[37] R. J. Birgeneau, J. K. Kjems, G. Shirane, and L. G. Van Uitert, *Phys. Rev. B* **10**, 2512 (1974).

[38] S. M. Shapiro, J. D. Axe, G. Shirane, and T. Riste, *Phys. Rev. B* **6**, 4332 (1972).

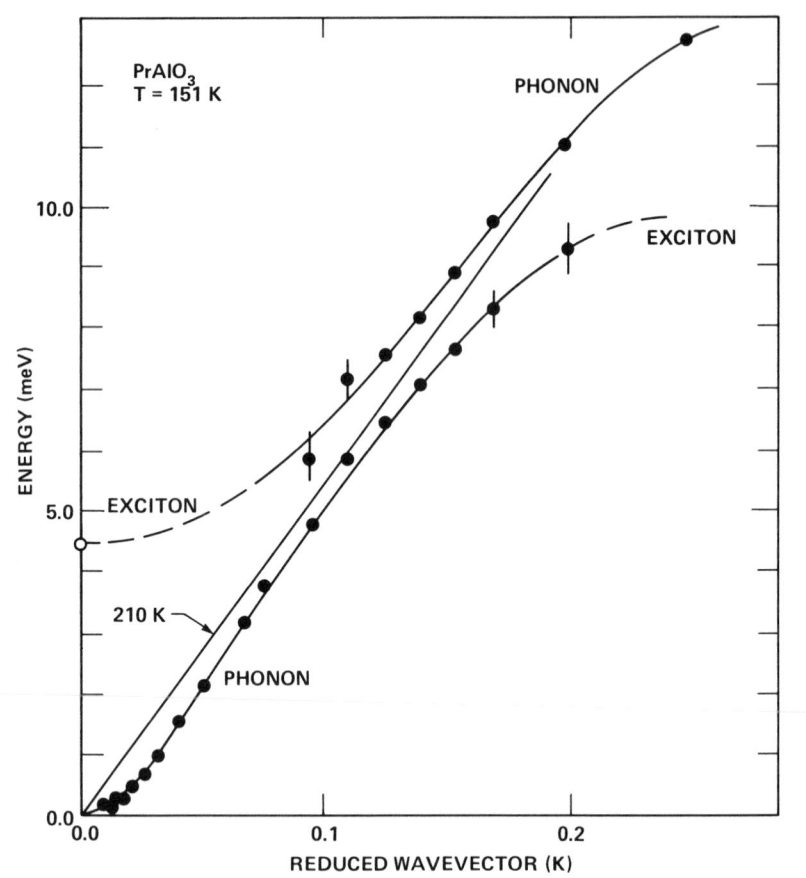

FIG. I.22. Dispersion relations for the quadrupole excitons and the [101] transverse acoustic phonons in PrAlO$_3$. The open circle at 4.4 meV is the $k = 0$ Raman value. After Birgeneau et al.[37]

the phonon peak. Halperin and Varma[39] have shown that a small concentration of defect cells which couple linearly to the order parameter, but relax on a slow time scale between different equivalent orientations, may account for the narrow central peak.

E. Long Range Order and Dimensionality

A standard approach used by physicists in dealing with difficult problems is to simplify them by generally considering lower dimensional

[39] B. I. Halperin and C. M. Varma, *Phys. Rev. B* **14,** 4030 (1976).

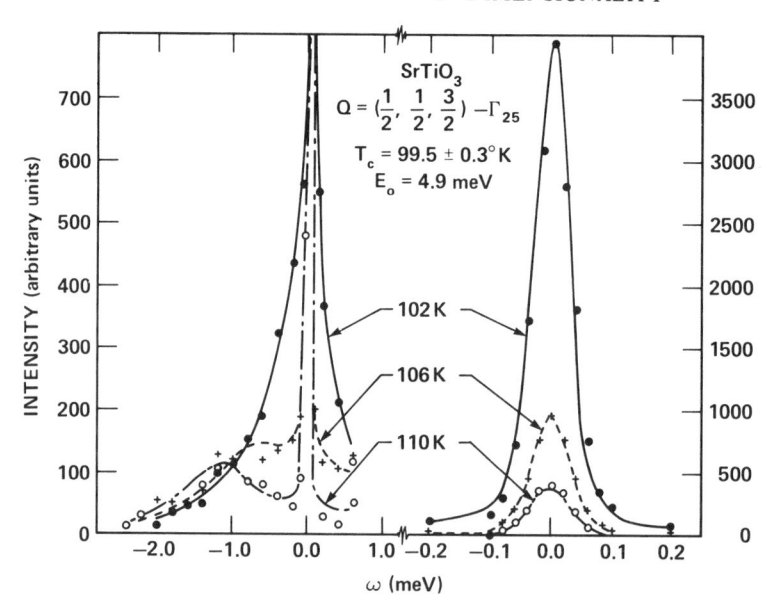

FIG. I.23. Scattered neutron intensity of $SrTiO_3$ at several temperatures. The left-hand side indicates the softening of the phonon while the right-hand side emphasizes the diverging central peak. After Shapiro et al.[38]

systems. We say "generally" because Wilson has introduced a scheme for calculating quantities near a critical point that involves an expansion in dimensionality about some dimension d^* that we shall discuss more fully in Chapter II.

Although certain real systems exhibit one- and two-dimensional behavior, such studies are important theoretically in their own right because of the general and exact results that can be derived. One such example that is commonly referred to is the following.

THEOREM: *Long-range order cannot occur in a one-dimensional system whose elements interact through finite-range forces.*[40]

This result had been appreciated for a long time. As a simple illustration consider a system whose elementary excitations are bosons with an energy-momentum relationship of the form either $\hbar\omega_k \propto \hbar ck$ or Dk^2. Phonons are an example of the former while ferromagnetic spinwaves the latter. From our previous discussion we recognize these as the Goldstone bosons associated with continuous symmetry operations. The total

---

[40] L. Landau and E. Lifshitz, "Statistical Physics," Sec. 149. Pergamon, Oxford, 1958.

number of quanta at temperature T is then

$$\sum_k n_k = \frac{1}{(2\pi)^D} \int_0^k d^D k n_k \qquad (1.55)$$

where D is the dimensionality. At high temperatures $n_k = [\exp(\hbar\omega_k/k_B T) - 1]^{-1} \sim k_B T/\hbar\omega_k$ and the integral above becomes

$$T \int_0^k \frac{d^D k}{k \text{ or } k^2} \qquad (1.56)$$

which diverges at small k for dimensionality $D = 1$. Since these excitations correspond to fluctuations in the order parameter, this divergence suggests that the original assumption of long range order may be incorrect.

These arguments can be put on a firmer theoretical foundation. In 1966, for example, Hohenberg[41] showed that Bose superfluidity and Cooper pairing in a Fermi system are inconsistent with certain sum rules in one and two dimensions. Similarly, Mermin and Wagner[42] showed that a one- or two-dimensional isotropic, spin S, Heisenberg system with finite range exchange can be neither ferromagnetic nor antiferromagnetic. It is interesting to note that the BCS theory that we discussed above does predict superconductivity in any dimension, including one dimension. This contradiction is due to the fact that this is a mean field theory and such theories favor phase transitions.

The validity of these theorems in two dimensions depends upon the existence of a continuous symmetry. Thus, for example, Griffiths[43] has shown that the Ising model, which does *not* possess a continuous symmetry, *is* ferromagnetic in two dimensions.

Notice that the theorem deals with *long range order,* not *phase transitions* as such. The term "phase transition" refers to a qualitative change in the properties of a system associated with a discontinuity in the thermodynamic equation of state. One might argue that this need not be accompanied by long range order. For example, one could have a situation in which the susceptibility diverges at a different temperature from that at which the spontaneous magnetization appears.

A more important point, however, is that a sharp phase transition into a state of long range order is not necessary for the existence of dynamical phenomena characteristic of the ordered state. For example, it has been suggested that it should be possible to establish a persistent current in an

[41] P. C. Hohenberg, *Phys. Rev.* **158**, 383 (1967).

[42] N. Mermin and H. Wagner, *Phys. Rev. Lett.* **17**, 1133 (1966).

[43] R. B. Griffiths, *Phys. Rev.* **136**, A437 (1964).

organic chain whose decay time is finite but too long to measure. We shall see how this might arise in Chapter II.

The size of the system, in particular the number of particles N, is also an important aspect of this theorem. Landau and Lifshitz consider a linear system comprised of segments of two different phases. If the energy associated with a boundary is u and if there are n such boundaries, then the free energy is

$$F = nu - k_B T \ln \binom{N}{n} \qquad (1.57)$$

where $\binom{N}{n}$ is the binomial coefficient. Minimizing this with respect to n gives an average segment size of the order of $l \exp(\beta u)$ where $l = L/N$ is a characteristic microscopic length. In the thermodynamic limit ($N \to \infty$) $l \to 0$ and the segment size goes to zero indicating no long range order. If, however, one allows the system to have a small, but finite, cross section σ ($\sqrt{\sigma} \ll$ correlation length), then Imry and Scalapino[44] have shown that the single phase regions can be quite large, giving one, in fact, extremely sharp pseudotransitions.

Another point to note about the nonexistence of long range order in one dimension theorem is that it only holds for *finite* temperatures. The ground state problem is another matter and has its own set of "theorems." One of these is due to Lieb and Mattis[45] and states:

THEOREM: *For any one-dimensional many-electron system with a Hamiltonian of the form*

$$\mathcal{H} = -\frac{1}{2} \sum_i^N \nabla_i^2 + V(x_1, \cdots x_N) \qquad (1.58)$$

where V *is any spin-independent potential and with boundary conditions which are local and linear (i.e., not periodic) then the lowest eigenvalue associated with spin* S *is less than or equal to the lowest eigenvalue associated with spin* S + 1.

This means that the ground state of such a one-dimensional system will be nonmagnetic. Although real systems are never really one dimensional, there are many models that are either mathematically equivalent to one-dimensional systems or take this form as an approximation. In such

[44] Y. Imry and D. Scalapino, *Phys. Rev. A* **9**, 1672 (1974).
[45] E. Lieb and D. Mattis, *Phys. Rev.* **125**, 164 (1962).

cases it is important to have a rigorous theorem for testing the validity of the approximations used in solving such problems.

It is worth mentioning at this point a one-dimensional result obtained by Lieb and Wu[46] for a particular, but very important, system—that of a narrow band of electrons interacting only when they encounter one another on the same site. The Hamiltonian has the form

$$\mathcal{H} = t \sum_{i,j} c_{i\sigma}^{\dagger} c_{j\sigma} + U \sum_{i} c_{i\uparrow}^{\dagger} c_{i\uparrow} c_{i\downarrow}^{\dagger} c_{i\downarrow} \qquad (1.59)$$

where t is a hopping integral and U is the Coulomb repulsion. Lieb and Wu show that the ground state for the half-filled band is insulating for any nonzero U, and conducting for $U = 0$. This means that if one varies t by, say, changing the intratomic spacing one does *not* obtain a metal-to-insulator transition.

[46] E. Lieb and F. Y. Wu, *Phys. Rev. Lett.* **20**, 1445 (1968).

II. Phenomenological Theories

The fact that solids undergo phase transitions that result in dramatic changes in their macroscopic properties has, of course, been known for a long time. For an equally long time man has attempted to understand these changes. As we mentioned in Chapter I, however, the success of any theory is dependent upon the appropriate choice of the quantities in terms of which the theory is developed. The thermodynamic description of phase transitions, for example, had to await the recognition by Clausius of the importance of the quantity entropy.

In this chapter we explore the consequences of a phenomenological description of phase transitions based on the concept of the order parameter introduced in Chapter I. In its simplest form this theory gives the "classical," or mean field, behavior of a system. Recently, a great deal of progress has been made at improving this theory. Our purpose here is to quantify some of the concepts introduced in Chapter I and present the additional concept of a coherence length.

A. LANDAU THEORY

As we noted in Chapter I, the appearance of an order parameter implies a lowering of the symmetry of the system. This observation implies that the ground state of a system may have lower symmetry than that of its Hamiltonian. In 1937 Landau[1] developed a theory of second-order phase transitions based on symmetry considerations.

Landau begins by considering some property ρ of the system that is directly affected by the phase change. In a structural transition this is the charge density of the system while in a magnetic transition it is the spin density. Above the transition this property, call it ρ_0, is invariant under a set of symmetry operations that define a group G_0. Below the transition this property ρ_1 is now invariant under the operations of a group G_1, which is a subgroup of G_0. Landau then expands the difference $\delta\rho = \rho_0 - \rho_1$ in terms of the basis functions $\phi_i^{(n)}$ of the irreducible representations of G_0:

[1] The English translations of Landau's papers may be found in "Collected Papers of L. D. Landau" by D. Ter Haar. Gordon & Breach, New York, 1965.

$$\delta\rho = \sum_n{}' \sum_{i=1} c_i^{(n)} \phi_i^{(n)} \tag{2.1}$$

where n labels of irreducible representation and i denotes the basis function. The prime indicates that we exclude the identity representation since it must be included in both ρ_0 and ρ_1. The coefficients $c_i^{(n)}$ must all vanish at the transition temperature. Furthermore, if $\delta\rho$ is to be invariant under the operations of G_1, which is a subgroup of G_0, then the $c_i^{(n)}$ must transform like the $\phi_i^{(n)}$.

Since the $c_i^{(n)}$ characterize the system in equilibrium below the transition, their values are determined by minimizing the free energy. At a second-order transition the $c_i^{(n)}$ take arbitrarily small values. Therefore, the free energy may be expanded in powers of the $c_i^{(n)}$. The appropriate free energy to consider is the one whose parameters we control experimentally, namely, temperature and pressure. From our earlier discussion this would be the Gibbs free energy $G(T, P, c_i^{(n)})$. In order to apply the theory to a system, such as a magnetic one, in which the order parameter can be coupled to an external field, we must set this field to zero. Otherwise the phase transition is "smeared out." We shall return to the question of what happens in the presence of an external field.

Since the Gibbs free energy is continuous at the transition, it may be written as $G_0(T, P)$, the free energy in the disordered phase, plus the expansion in $c_i^{(n)}$. Each term in this expansion must contain only combinations of the $c_i^{(n)}$ which are invariant under the operations of G_0. Since we have excluded the identity representation, there is no invariant linear in $c_i^{(n)}$. That is, if G did contain a term linear in $c_i^{(n)}$ then G would have the lower symmetry of $c_i^{(n)}$. Furthermore, it can be shown from group theory that for each representation $\Gamma^{(n)}$ of G_0 there is only one *quadratic* invariant, namely, $\sum_i (c_i^{(n)})^2$. Thus

$$G(T, P, c_i^{(n)}) = G_0(T, P) + \sum_n{}' A^{(n)}(T, P) \sum_i (c_i^{(n)})^2 + \cdots . \tag{2.2}$$

Minimizing the free energy at the transition point itself must give $c_i^{(n)} = 0$. This will only be so if the $A^{(n)}$ are nonnegative. If the $A^{(n)}$ are positive at the transition point, then they will also be positive just below the transition point. But minimizing the free energy then gives $c_i^{(n)} = 0$, which means no transition has occurred. Therefore the $A^{(n)}$ must, in fact, be zero at the transition, i.e., $A^{(n)}(T_c, P_c) = 0$. If the coefficient associated with another irreducible representation also changed sign, then $A^{(n')}(T_c, P_c) = 0$, and we would have two equations for T_c and P_c giving a critical *point*. Since this is not generally the case, the quadratic term in Eq. (2.2) is simply $A^{(n)}(T, P)\sum_i(c_i^{(n)})^2$. The simplest form for $A(T, P)$ that changes sign at T_c is

$$A(T, P) = a(P)(T - T_c). \qquad (2.3)$$

Let us define an order parameter ψ with components $\psi_i = c_i^{(n)}$. Thus, the number of independent components of the order parameter is equal to the dimensionality of the representation according to which it transforms. Then

$$G(T, P) = G_0(T, P) + A(T, P)\psi^2 + \cdots. \qquad (2.4)$$

Let us now consider higher order terms in the expansion. For simplicity, let us consider a one-component order parameter. If the direct product $\Gamma^{(n)} \times \Gamma^{(n)} \times \Gamma^{(n)}$ contains the identity representation of G_1, then a cubic term, $B\psi^3$, will exist. If $B(P_c, T_c) = 0$, then the transition will occur at a single point, while if $B(P_c, T_c) \neq 0$, the transition will be first order. In the latter case one must apply the theory cautiously, for in a first-order transition $c_i^{(n)}$ changes discontinuously, whereas the whole expansion concept depends on $c_i^{(n)}$ being small. Since the direct product $\Gamma^{(n)} \times \Gamma^{(n)} \times \Gamma^{(n)} \times \Gamma^{(n)}$ does contain the identity representation, there will always be a quartic term in the expansion, $C\psi^4$. If the coefficient C is positive the transition is second order; whereas if it is negative the transition is first order. We approximate this by $C(P, T_c) \equiv C$. Thus, we finally have the expansion

$$G(T, P, \psi) = G_0(T, P) + a(P)(T - T_c)\psi^2 + C\psi^4. \qquad (2.5)$$

Let us consider the consequences of a free energy of the form Eq. (2.5).

1. Order Parameter

Minimizing the free energy with respect to ψ gives

$$\partial G/\partial\psi = 2a(T - T_c)\psi + 4C\psi^3 = 0. \qquad (2.6)$$

From this we see that the order parameter has the form

$$\psi = (a/2C)^{1/2}(T_c - T)^{1/2}. \qquad (2.7)$$

2. Entropy

$$S = -(\partial G/\partial T) = S_0 - a\psi^2 - 2a(T - T_c)\psi(\partial\psi/\partial T) - 4C\psi^3(\partial\psi/\partial T). \qquad (2.8)$$

The last two terms cancel by virtue of $\partial G/\partial\psi = 0$. Therefore, since $\psi^2 \sim (T_c - T)$, the entropy decreases linearly with temperature below the transition.

3. Specific Heat

$$C_p = T \frac{\partial S}{\partial T}\bigg|_p = \begin{cases} C_0(T) & T > T_c \\ C_0(T) + \dfrac{a^2 T}{2C} & T < T_c. \end{cases} \tag{2.9}$$

Thus the Landau theory predicts a specific heat discontinuity of $a^2 T_c / 2C$.

Experimentally it is found that this theory gives a fairly good description of superconductors. The order parameter, for example, does appear to fit a behavior proportional to $(T_c - T)^{1/2}$ near T_c and there is, as we have seen, a discontinuity in the specific heat. Ferromagnets, on the other hand, are not accurately described by this theory. The magnetization varies more like $(T_c - T)^{1/3}$ in many cases and the specific heat often shows a λ-anomaly. The origin of this difference lies in their different coherence lengths as we shall see below.

4. Equation of State

Let us now consider how one extends the Landau theory to determine the equation of state. Such an equation relates the order parameter to its conjugate field. Let us denote this field by η. The lowest order coupling will just involve the product of this field with the order parameter ψ. If the presence of this external field increases the value of the order parameter, then the interaction enters the total free energy with a minus sign, $G - \psi\eta$. Minimizing this with respect to ψ gives

$$\partial G / \partial \psi = \eta. \tag{2.10}$$

Therefore, along the critical isotherm ($T = T_c$), Eq. (2.6) gives

$$\psi = (1/2C)^{1/3} \eta^{1/3}. \tag{2.11}$$

Notice that ψ^2 versus η/ψ is a straight line with a slope $1/2C$. We shall return to this when we discuss the magnetic susceptibility in Chapter V.

5. Susceptibility

For $T > T_c$, in the presence of η, $\partial G / \partial \psi \simeq 2a(T - T_c)\psi$. Therefore, the susceptibility, $\chi = \psi/\eta$, becomes

$$\chi = [2a(T - T_c)]^{-1}. \tag{2.12}$$

Below the transition we must retain the ψ^4 term in the free energy. The result is that χ diverges as $|T - T_c|^{-1}$, but with an additional factor of 2.

Collecting these results we find that the critical exponents associated with the Landau theory defined by Eq. (1.13) are

$$\alpha = \alpha' = 0 \text{ (discontinuity)}$$
$$\beta = \tfrac{1}{2} \qquad\qquad (2.13)$$
$$\gamma = \gamma' = 1.$$

These are often referred to as the *classical* critical exponents. They are also identical to those one would obtain from a *mean field* theory.

6. Examples

a. Antiferromagnetic Ordering in MnSe₂: The Landau theory tells us that in the immediate vicinity of a second-order transition the spin density of the ordered phase transforms as a basis function for a single irreducible representation of the symmetry group of the disordered phase. This imposes certain restrictions on the magnetic structures that can be reached through a second-order transition. Consider[2] $MnSe_2$, which becomes antiferromagnetic below $T_N = 75$ K. At 4.2 K the spin structure, as determined from neutron scattering, has the form shown in Fig. II.1 (top). This has the analytical representation $\rho = (\tfrac{4}{3})\rho_1 - (\tfrac{1}{3})\rho_2$ where

$$\rho_1 = \mu_0 \cos(2\pi/a)z \, \cos(2\pi/3a)x$$
$$\rho_2 = \mu_0 \cos(2\pi/a)z \, \cos(2\pi/a)x. \qquad (2.14)$$

Both ρ_1 and ρ_2 transform as a basis function for an irreducible representation of the paramagnetic symmetry group. However, the fact that *two* irreducible representations are required is not consistent with Landau's theory. Dimmock proposes that immediately below T_N the spin density is characterized by ρ_1 [Fig. II.1 (bottom)] and that the contribution from ρ_2 gradually increases as we move away from the transition. The fact that the ρ_2 representation is contained in the decomposition of the direct cube of the ρ_1 representation suggests that ρ_2 is arising from the third-order invariant in the free energy.

b. The A-15 Martensitic Transformations: Since Landau's theory is based on group theory it carries with it the same power for making general, model-independent statements. Let us consider the A-15 compounds such as V_3Si and Nb_3Sn. These particular materials are of great interest for their relatively high superconducting transition temperatures (17 K and 18 K, respectively). They exhibit transformations from cubic to tetragonal structure at temperatures slightly above their superconducting transitions. These structural transitions have been described as "martensitic" in the sense that there is no atomic diffusion involved. However, unlike most martensitic transitions which are strongly first order, these

[2] J. Dimmock, *Phys. Rev.* **130**, 1337 (1963).

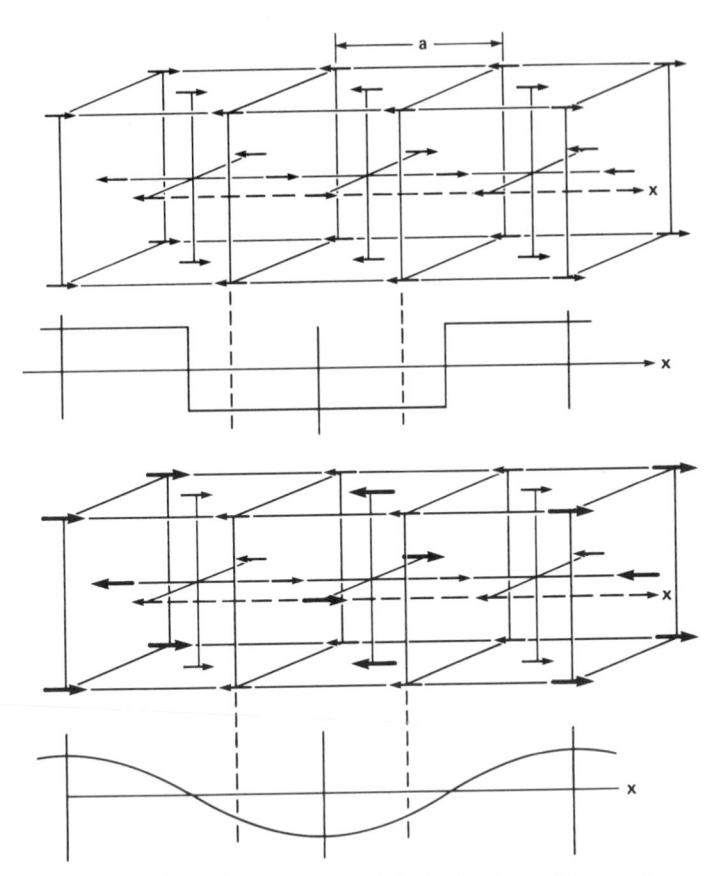

FIG. II.1. Spin configurations of $MnSe_2$. Only the Mn ions of the pyrite structure are shown. The top configuration is that deduced from neutron scattering at 4.2 K. The square wave indicates the directions of the spins in the front face. As explained in the text, this configuration is inconsistent with a second-order phase transition. Dimmock therefore suggested that the lower configuration, with a sinusoidal spin structure, appears at the transition and evolves continuously to the upper configuration with decreasing temperature.

transitions appear to be second order. That is, the characteristic first-order features discussed in Chapter I had not been unambiguously observed. The Landau theory then enables us to make a statement about the order parameter. Since the transition from a cubic structure to a tetragonal one means that the ratio of the lattice parameters, c/a, is no longer 1 this suggests that $\psi_1 = c/a\text{-}1$ might be the appropriate order parameter.

Anderson and Blount,[3] however, argue that if the transition is truly sec-

[3] P. W. Anderson and E. I. Blount, *Phys. Rev. Lett.* **14**, 217 (1965).

ond order this cannot be. Their point is that ψ_1 is one of two symmetry-equivalent strains

$$\psi_1 = (1/\sqrt{6})(2\epsilon_{zz} - \epsilon_{xx} - \epsilon_{yy})$$

$$\psi_2 = (1/\sqrt{2})(\epsilon_{xx} - \epsilon_{yy})$$

and that a third-order invariant of these does exist, namely $(\psi_1^2 - 3\psi_2^2)\psi_1$. Therefore, the free energy contains the term $B(T)(\psi_1^2 - 3\psi_2^2)\psi_1$. For a second-order transition the coefficient of the second-order invariant, $A(T)$ as well as $B(T)$, must vanish simultaneously which Anderson and Blount consider "infinitely improbable." This means that either ψ_1 is *not* the order parameter, or the transition is indeed first order. Anderson and Blount suggested a "hidden" order parameter. However, subsequent measurements have shown that the elastic constant, $c_{11} - c_{12}$, which is just the coefficient $A(T)$, does indeed vary as $(T - T_c)$. Thus a Landau expansion based on strain as the order parameter is appropriate, which means that the transition is weakly first order.

Problem II.1
Prove that on the low-temperature side of a second-order soft mode structural transition there is always a totally symmetric zone center phonon regardless of its soft mode progenitor above T_c. Such a mode is Raman active. Therefore, this result means that Raman scattering may always be used to study the dynamics of such a transition below T_c.

Problem II.2
The proposed phase diagram for the ferroelectric KH_2PO_4 is shown in Fig. II.2. Let us represent the free energy of such a material as

$$G = a(T - T_0)P^2 + BP^4 + CP^6$$

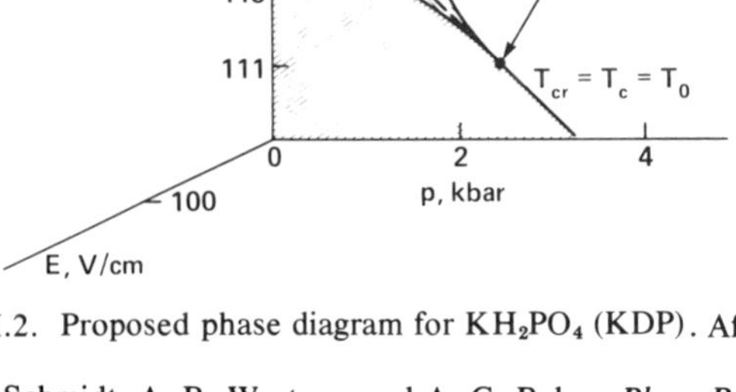

FIG. II.2. Proposed phase diagram for KH_2PO_4 (KDP). After V. H. Schmidt *et al.*[4]

[4] V. H. Schmidt, A. B. Western, and A. G. Baker, *Phys. Rev. Lett.* **37**, 839 (1976).

where P here refers to the polarization. Take, a, $C > 0$ while B changes sign with pressure, p. The electric field is given by $E = \partial G/\partial P$.

    a. Derive expressions for the first-order transition line (T_c) in the $E = 0$ plane in terms of the Landau coefficients.

    b. Derive expressions for the coordinates of the "wing" critical line (T_{cr}). Note that in this case stability also requires $\partial^3 G/\partial P^3 = 0$.

B. GINZBURG–LANDAU THEORY[5]

The Landau theory described above does not allow for the possible *spatial* variation of the order parameter. This creates problems in calculating the critical fields for superconductors with various geometries and motivated Ginzburg and Landau to generalize the Landau theory in 1950. This generalization took the form of assuming that the order parameter ψ was a complex and slowly varying function of position $\psi(\mathbf{r})$ so that the free energy now became a *functional* of $\psi(\mathbf{r})$ and $\psi(\mathbf{r})^*$. Another way of saying this is that the order parameter $\psi(\mathbf{r})$ depends upon the order parameters at all other points, $\psi(\mathbf{r}')$. Such a nonlocal correction contributes a term to the free energy density proportional to $|\nabla\psi(\mathbf{r})|^2$. The coefficient of proportionality depends upon the microscopic system being considered.

1. Superconductors

In extending the Landau theory to superconductivity Ginzburg and Landau argued that the order parameter $\psi(\mathbf{r})$ plays the role of the superconducting wave function and, therefore, its square should be normalized to the density of superconducting electrons,

$$|\psi(\mathbf{r})|^2 = n_s^*(\mathbf{r}).\tag{2.15}$$

In the Schrödinger equation the spatial variation of the wave function corresponds to the kinetic energy term, $(-\hbar^2/2m^*)\nabla^2\psi(\mathbf{r})$. If this is to arise from a minimization of the free energy, the free energy density must contain the term $(\hbar^2/2m^*)|\nabla\psi(\mathbf{r})|^2$. In this way Ginzburg and Landau arrived at an expression for the coefficient of the $|\nabla\psi(\mathbf{r})|^2$ term. However, the effective mass, m^* and n_s^*, were treated as effective quantities. Thus the total free energy density assumed by Ginzburg and Landau for the superconducting state takes the form

$$\mathcal{G}_s(\mathbf{r}) = \mathcal{G}_0(\mathbf{r}) + A(T)|\psi(\mathbf{r})|^2 + \tfrac{1}{2}C|\psi(\mathbf{r})|^4 + (\hbar^2/2m^*)|\nabla\psi(\mathbf{r})|^2.\tag{2.16}$$

[5] The English translation of the original Landau–Ginzburg paper may be found in ref. 1; for a comprehensive review of this subject, see M. Cyrot, *Rep. Prog. Phys.* **36**, 103 (1973).

$\psi(\mathbf{r})$ is obtained by requiring that the total free energy,

$$G = \int \mathscr{G}(\mathbf{r}) \, d\mathbf{r}, \tag{2.17}$$

be a minimum with respect to variations of $\psi(\mathbf{r})$ and $\psi(\mathbf{r})^*$. Setting the functional derivative $\delta G/\delta\psi^*(\mathbf{r}) = 0$ and adopting the boundary condition $\hat{n} \cdot \nabla\psi = 0$, gives

$$A(T)\psi + C|\psi|^2\psi - (\hbar^2/2m^*)\nabla^2\psi = 0. \tag{2.18}$$

This particular boundary condition assures that no current flows across the surface. However, the less restrictive condition $\hat{n} \cdot \nabla\psi = \psi/b$ where b is an "extrapolation" length also satisfies this condition. The reader may wonder why we do not choose the simpler boundary condition, $\psi = 0$, which also results in no supercurrent leaving the sample. The reason is that it can be shown[6] that the solution of the G–L equation with this boundary condition predicts that a film would not be superconducting unless the thickness were greater than a certain critical value. (Approximately the coherence length defined below.) Since superconductivity is observed in very thin films this boundary condition is not appropriate.

   a. Penetration Depth: The G–L free energy density, Eq. (2.16), is easily generalized to include the presence of a microscopic magnetic field $\mathbf{h} = \nabla \times \mathbf{A}$ by adding the vector potential $\mathbf{A}$ just as in quantum mechanics,

$$-i\hbar\nabla \rightarrow -i\hbar\nabla + (e^*A/c). \tag{2.10}$$

It was not until the observation of flux quantization that the effective charge e^* could be identified as $2e$. It is an interesting historical point to note that experimental data did exist at that time which would have led Ginzburg and Landau to the pairing concept had they attempted a quantitative comparison between their theory and the data. Equation (2.19) is based on the assumption that the order parameter behaves like a wave function under a gauge transformation. We shall see below that the free energy of a ferromagnet also contains a term of the form $|\nabla\psi|^2$. However, in this case ψ is not a wave function. Therefore in the presence of a field the transformation [Eq. (2.19)] does not apply. We must also add the field energy density, $h^2/8\pi$, to $\mathscr{G}_s(\mathbf{r})$. Notice that since $\langle \mathbf{h} \rangle = \mathbf{B}$ the magnetic contribution to the free energy is not in terms of the variable we usually control, namely, the external field H. In this sense $\mathscr{G}(\mathbf{r})$ would be a Helmholtz free energy. We may transform to the Gibbs free energy by making

   [6] See, for example, N. R. Werthamer, *in* "Superconductivity" (R. D. Parks, ed.), Vol. 1, pp. 326–327. Dekker, New York, 1969.

a Legendre transformation,

$$\mathscr{G}(\mathbf{r}) \rightarrow \mathscr{G}(\mathbf{r}) - (1/4\pi)\mathbf{h} \cdot \mathbf{H}. \qquad (2.20)$$

We shall hereafter assume that $\mathscr{G}(\mathbf{r})$ is the appropriate Gibbs free energy. Minimizing this total free energy with respect to variations in $\mathbf{A}(\mathbf{r})$ with the boundary condition that the tangential component of the magnetic field be continuous across the surface, i.e., $\hat{n} \times (\mathbf{h} - \mathbf{H}) = 0$, gives

$$\mathbf{j} = \frac{c}{4\pi} \nabla \times \mathbf{h} = - \frac{e^*\hbar}{i2m^*} (\psi^* \nabla \psi - \psi \nabla \psi^*) - \frac{(e^*)^2}{m^*c} |\psi|^2 \mathbf{A}. \qquad (2.21)$$

Writing $\psi = |\psi| \exp(i\phi)$ this leads to a current

$$\mathbf{j} = \frac{e^*\hbar}{m^*} |\psi|^2 \left(\nabla\phi - \frac{e^*}{\hbar c} \mathbf{A} \right). \qquad (2.22)$$

Let us assume for the moment that $\psi(\mathbf{r})$ is uniform, i.e., $\psi(\mathbf{r}) = |\psi_\infty|$. Then Eq. (2.22) reduces to $\mathbf{j} = -n_s^*(e^*)^2\mathbf{A}/m^*c$. Taking the curl of this current,

$$\mathbf{h} = -c\nabla \times \left(\frac{m^*}{n_s^*(e^*)^2} \mathbf{j} \right), \qquad (2.23)$$

which is the constitutive relation initially proposed by F. London and H. London in 1935 to explain the Meissner effect. Equation (2.23) is actually valid even if the phase is spatially varying as long as $|\psi|$ is constant. Combining this with the Maxwell equation $\nabla \times \mathbf{h} = (4\pi/c)\mathbf{j}$ leads to

$$\nabla^2 \mathbf{h} = [1/\lambda(T)^2]\mathbf{h} \qquad (2.24)$$

where

$$\lambda(T) = \left(\frac{m^*c^2}{4\pi n_s^*(e^*)^2} \right)^{1/2} \qquad (2.25)$$

is the characteristic depth to which a magnetic field will penetrate a superconductor. If $m^* = 2m$, $n_s^* = n/2$, and $e^* = 2e$, this reduces to the constant

$$\lambda_L(0) = \left(\frac{mc^2}{4\pi ne^2} \right)^{1/2}. \qquad (2.26)$$

From the pairing concept it is easy to understand why we take $m^* = 2m$ and $e^* = 2e$. But why do all the electrons appear to take part in the screening? This is particularly puzzling since only those pairs in a thin slice at the Fermi surface affect the total energy. Specifically, if we divide the total energy density difference between the normal state and the superconducting state (the so-called condensation energy density, $N(E_F)\Delta^2/2$)

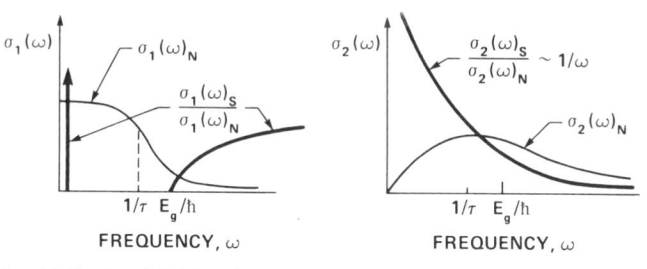

FIG. II.3.   (a) Real and (b) imaginary parts of the conductivity for a normal metal (N) in the Drude approximation and a superconductor (S).

by the energy required to break a pair, 2Δ, we find that the average number of pairs is of the order of $(\Delta/E_F)n$ which is $\ll n$. To see why n enters the penetration depth, consider the real part of the frequency-dependent conductivity, $\sigma_1(\omega)$. If the electron scattering in the normal metal is characterized by a relaxation time τ the normal conductivity has the classical Drude form illustrated in Fig. II.3, i.e., $\sigma_1(\omega) = (ne^2\tau/m)/(1 + \omega^2\tau^2)$. The superconductor, on the other hand, is characterized by a delta function in $\sigma_1(\omega)$ at $\omega = 0$. This reflects the fact that a current-carrying state can be constructed without requiring excitations across the gap (as one would require in a semiconductor). Tinkham and Ferrell[7] have shown that the oscillator strength in this delta function, which governs the penetration depth, has come from the region $\omega < E_g/\hbar$ where $\sigma_1(\omega)$ is now zero. If $E_g > \hbar/\tau$, then $\int_0^{E_g} \sigma_1(\omega)_N d\omega$, which measures the oscillator strength, is proportional to n, the density of electrons entering the Drude expression. Thus, as long as $E_g > \hbar/\tau$ the density of electrons involved in the Meissner screening is the same as that appearing in the conductivity of the normal state. (If $E_g < \hbar/\tau$, then the oscillator strength in the delta function is reduced which reduces the effective number of electrons entering the expression for the penetration depth.)

One might now ask why all the electrons appear to take part in *normal* conduction. One might argue that n must appear in $\sigma_1(\omega)$ in order to satisfy the general sum rule $\int_0^\infty \sigma_1(\omega)\, d\omega = \pi n e^2/2m$. However, this merely tells us that by going to high enough frequencies we can excite all the electrons. It does not tell us about the dc conductivity. The fact is that all the electrons do not take part in the dc conductivity. A derivation of this conductivity based on a Boltzmann equation shows that the conductivity is proportional to an integral of the velocity (as well as the relaxation time) over the Fermi surface. In particular, $\mathbf{j} = \int d^3k\, \mathbf{v}_k \tau \mathbf{k} \cdot \nabla_k f$ where f is the electron distribution function. In lowest order this is just the Fermi–

[7] M. Tinkham and R. A. Ferrell, *Phys. Rev. Lett.* **2**, 331 (1959).

Dirac function, and its gradient converts the volume integral to a surface integral. The integral of the velocity over the surface is proportional to the volume enclosed by the Fermi surface, k_F^3, which is, in turn, proportional to n. Thus, n appears not because there are n electrons involved in the transport, but because they determine the velocity of those few that are. Thus, it is not surprising that pairing, which only occurs near the Fermi surface, has an effect on the conductivity that appears to involve all the electrons. Let us now consider what happens when $\psi(\mathbf{r})$ is *not* uniform.

b. Coherence Length: One of the attractive features of the G–L theory is the natural appearance of a characteristic "coherence length." The need for a coherence length was arrived at independently by Pippard[7a] on the basis of the observation that the penetration depth increased appreciably when $\hbar/\tau > E_g$ as discussed above. This led Pippard to introduce a non-local version of the current density–vector potential relation in which the current at $\mathbf{r}$ is governed by the vector potential at $\mathbf{r}'$ out to a distance $|\mathbf{r}' - \mathbf{r}| \sim \xi_p$ where

$$1/\xi_p = (1/\xi_0) + (1/l). \tag{2.27}$$

Here l is the electron mean free path and ξ_0 is the "coherence length." Using the uncertainty principle, Pippard argued that $\xi_0 \sim \hbar v_F/k_B T_c$. Except for a numerical constant, this is the result obtained from the microscopic theory [see Eq. (2.46)]. The mean free path is related to the normal state resistivity ρ_n by

$$l = m v_F / n e^2 \rho_n. \tag{2.28}$$

Let us now consider the G–L length. For simplicity let us assume that ψ varies only in, say, the z-direction. Then Eq. (2.18) becomes

$$-(\hbar^2/2m^*)(d^2\psi/dz^2) + A(T)\psi + C|\psi|^2\psi = 0. \tag{2.29}$$

If ψ did *not* vary in space this would have the trivial solution

$$|\psi_\infty|^2 = -[A(T)/C]. \tag{2.30}$$

The difference in free energies is

$$\mathscr{G}_s - \mathscr{G}_n = A(T)|\psi_\infty|^2 + \tfrac{1}{2}C|\psi_\infty|^4 = -[A(T)^2/2C]. \tag{2.31}$$

We shall now use a thermodynamic argument to show that $A(T)$ must be negative in the ordered state.

[7a] A. B. Pippard, *Proc. Roy. Soc. (London)* **A126,** 547 (1953).

From the definition of the Gibbs free energy we have

$$\mathscr{G}(T, H) - \mathscr{G}(T, 0) = -\frac{1}{4\pi}\int_0^H B(H')dH'. \tag{2.32}$$

If we neglect the small Pauli susceptibility of a normal metal and take the magnetization $M(H') = 0$, then $B = H$ and

$$\mathscr{G}_\text{n}(T, H) - \mathscr{G}_\text{n}(T, 0) = -(H^2/8\pi). \tag{2.33}$$

On the other hand, since a superconductor is a perfect diamagnet, $B(H') = 0$. Therefore,

$$\mathscr{G}_\text{s}(T, H) - \mathscr{G}_\text{s}(T, 0) = 0. \tag{2.34}$$

Under conditions of constant temperature, pressure, and field the Gibbs free energies of the normal and superconducting phases must be equal along the phase boundary where the field has the critical value H_c (see Fig. V.5). Thus

$$\mathscr{G}_\text{s}(T, H_c) = \mathscr{G}_\text{n}(T, H_c) \tag{2.35}$$

which leads to

$$\mathscr{G}_\text{s}(T, 0) - \mathscr{G}_\text{n}(T, 0) = -[H_c(T)^2/8\pi]. \tag{2.36}$$

This shows that a negative condensation energy, $-H_c{}^2/8\pi$, is associated with the superconducting state. By comparing Eq. (2.36) with Eq. (2.31) we see that C must be greater than zero, the same necessary condition we found in the previous section for a second-order transition. Thus if $|\psi_\infty|^2$ is to be an acceptable solution then, from Eq. (2.30), $A(T)$ must be less than zero.

As in the previous section we take $A(T)$ to be $a(T - T_c)$. Notice that from Eqs. (2.15) and (2.30) this predicts that $n_s(T)*$ vanishes linearly with temperature at the transition point.

If we introduce the variable

$$f = \frac{\psi}{|\psi_\infty|}, \tag{2.37}$$

then the equation for f becomes

$$-\frac{\hbar^2}{2m^*|A(T)|}\frac{d^2f}{dz^2} - f + f^3 = 0. \tag{2.38}$$

The coefficient of the second derivative must have the dimensions of a length squared. This is the Ginzburg–Landau *coherence length*

$$\xi(T) = \left(\frac{\hbar^2}{2m^*A(T)}\right)^{1/2} = \left(\frac{\hbar^2}{2m^*aT_c}\right)^{1/2}\left(\frac{T_c}{T_c - T}\right)^{1/2} = \xi(0)\left(\frac{T_c}{T_c - T}\right)^{1/2}. \tag{2.39}$$

Notice that the coherence length diverges as $(T_c - T)^{-1/2}$. In general, we ascribe the critical exponent ν to the coherence length, i.e.,

$$\xi \propto \begin{cases} \epsilon^{-\nu} & T \to T_c{}^+ \\ (-\epsilon)^{\nu'} & T \to T_c{}^-. \end{cases} \tag{2.40}$$

Thus the mean field or classical value of ν and ν' is $\frac{1}{2}$. Physically, above T_c ξ corresponds to the length over which the order parameter exists, while below T_c ξ corresponds to the length over which the order parameter varies from its equilibrium value. Notice that this is physically quite different from the Pippard length which characterizes the range of nonlocal electromagnetic effects.

Anisotropy may be introduced into the theory through an effective mass tensor. Equation (2.18), for example, becomes

$$A(T)\psi + C|\psi|^2\psi + \frac{1}{2}(-i\hbar\mathbf{\nabla}) \cdot \left(\frac{1}{m^*}\right) \cdot (-i\hbar\mathbf{\nabla})\psi = 0. \qquad (2.41)$$

This leads to an anisotropic coherence length. This is important in layered superconductors as we shall mention in Chapter VIII.

Since $|\psi_\infty|^2 = n_s^* = -A(T)/C$ the penetration depth, Eq. (2.25) may be written

$$\lambda(T) = \left(\frac{m^*c^2C}{4\pi(e^*)^2 aT_c}\right)^{1/2} \left(\frac{T_c}{T_c - T}\right)^{1/2} = \lambda(0)\left(\frac{T_c}{T_c - T}\right)^{1/2}. \qquad (2.42)$$

By dividing the penetration length by the coherence length we remove the divergence in temperature and obtain the *Ginzburg–Landau parameter*,

$$\kappa = \frac{\lambda(T)}{\xi(T)} = \frac{m^*c}{\hbar e^*}\left(\frac{C}{2\pi}\right)^{1/2}. \qquad (2.43)$$

For many years superconductors were classified as "soft" (e.g., Pb) and "hard" (e.g., all the superconducting transition metals) by experimentalists in recognition of an imprecise relationship between their mechanical, magnetic, and calorimetric properties. In terms of the parameters of the Ginzburg–Landau theory a well-defined classification emerges.

Type I superconductors are those where the penetration depth λ, is less than the coherence length, ξ. In an applied field H, any boundary between normal and superconducting regions has a positive energy associated with it (high-energy region) because in the boundary region over a distance $\approx(\lambda - \xi)$ the positive energy necessary to exclude the magnetic field is not compensated by the condensation energy (Fig. II.4a). Therefore, if a field is applied to a long thin sample (so shaped to eliminate the complications of demagnetizing fields), it will be excluded at all fields below the critical field. The boundary will be at the surface only where, for a depth characterized by λ, shielding currents will flow so as to exactly cancel the applied field inside the sample.

For the *type II* superconductor, where $\xi < \lambda$, the boundary region is

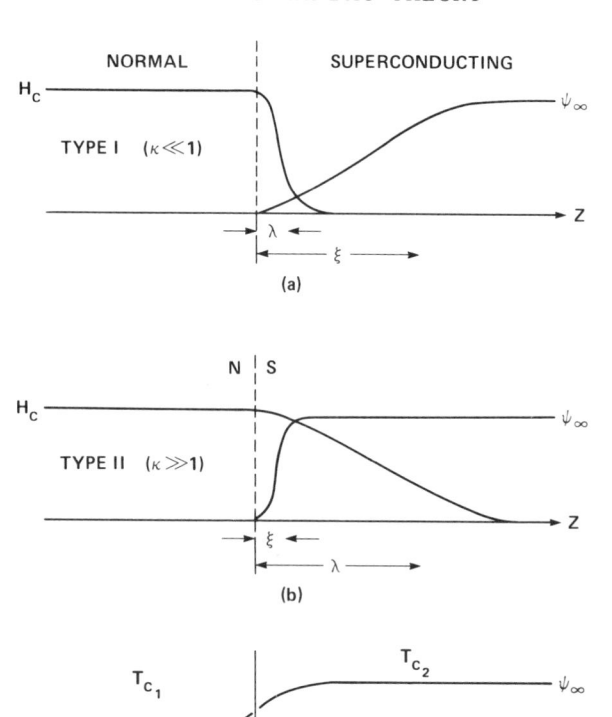

FIG. II.4. Variation of the order parameter at the interface between a superconducting region and, in (a) and (b), a region driven normal by the presence of a critical field. In (c) the normal side is a different metal with a lower transition temperature, i.e., $T_{c_1} < T_{c_2}$. In this case the superconducting wave function penetrates into the normal metal a distance given by the "extrapolation" length, b.

one of negative surface energy. As seen in Fig. II.4b the magnetic field has penetrated into the region where the condensation into the superconducting state has already taken place. We shall consider type II superconductors in much more detail in Chapter VIII.

In Fig. II.4c we illustrate what happens at the boundary between a normal metal (this could be another superconductor *above* its transition temperature) and a superconductor. The boundary condition $\hat{n} \cdot \nabla\psi = \psi/b$, which is appropriate in this case, allows Cooper pairs to leak into the normal metal. This is referred to as the "proximity effect." The probability of finding a Cooper pair in the normal metal falls off with some characteristic length, b which is of the order of $\hbar v_F/2\pi k_B T$, where v_F is the Fermi velocity in the normal metal.

c. Relation to Microscopic Theory: In 1959 Gorkov[8] derived the Ginzburg–Landau equations from the microscopic BCS theory thereby relating the Ginzburg–Landau order parameter to the pair potential, $\Delta(\mathbf{r})$,

$$\psi(\mathbf{r}) = \left(\frac{7\zeta(3)n}{8(\pi k_B T_c)^2}\right)^{1/2} \Delta(\mathbf{r}). \tag{2.44}$$

where n is the total electron density and $\zeta(3)$ is the Riemann zeta function, 1.202. In the absence of a field Gorkov's result for the Ginzburg–Landau free energy takes the form

$$G = \int d\mathbf{r}[\alpha(T)|\Delta(\mathbf{r})|^2 + (\beta/2)|\Delta(\mathbf{r})|^4 + N(0)(0.74\xi_0)^2|\nabla\Delta(\mathbf{r})|^2] \tag{2.45}$$

where

$$\alpha(T) = N(0)(T - T_c)/T_c$$

$$\beta = 0.098N(0)/(k_B T_c)^2 \tag{2.46}$$

$$\xi_0 = 0.180(\hbar v_F/k_B T_c).$$

Here $N(0)$ is the density of electron states in the normal metal, $N(0) = (mk_F/2\pi^2\hbar^2)$ and ξ_0 is the BCS coherence length.

By comparing Gorkov's resulting free energy with that of Ginzburg and Landau, Eq. (2.16), we obtain values for the G–L parameters:

$$a = \frac{6\pi^2 k_B^2 T_c}{7\zeta(3)\epsilon_F} \qquad C = \frac{6\pi^2 k_B^2 T_c^2}{7\zeta(3)\epsilon_F n}. \tag{2.47}$$

Therefore, the penetration depth, coherence length, and the Ginzburg–Landau parameter become

$$\lambda(T)_{\text{clean}} = \lambda_L(0)[2(1 - T/T_c)]^{-1/2}$$

$$\xi(T)_{\text{clean}} = 0.74\xi_0(1 - T/T_c)^{-1/2} \tag{2.48}$$

$$\kappa_{\text{clean}} = 0.96\lambda_L(0)/\xi_0.$$

We have used the subscript "clean" to indicate that these are the results for a system in which the coherence length ξ_0 is smaller than the electron mean free path l. If the reverse is true, i.e., $l \ll \xi_0$, we refer to the sample as "dirty." The Gorkov derivation of the G–L equations may also be carried out for the dirty case. The results are

$$\lambda(T)_{\text{dirty}} = 0.615\lambda_L(0)(\xi_0/l)^{1/2}(1 - T/T_c)^{1/2}$$

$$\xi(T)_{\text{dirty}} = 0.85(\xi_0 l)^{1/2}(1 - T/T_c)^{-1/2} \tag{2.49}$$

$$\kappa_{\text{dirty}} = 0.725\lambda_L(0)/l.$$

[8] L. P. Gorkov, *Sov. Phys.—JETP (Engl. Transl.)* **9**, 1364 (1959).

The difference between the coherence length and the Pippard length [Eq. (2.27)] is particularly evident in the dirty case. One of the interesting features of the κ_{dirty} is that it can be expressed completely in terms of normal state parameters. The mean free path, for example, is related to the normal state resistivity according to Eq. (2.28) while the density of electrons is related to the coefficient, γ, of that contribution to the electronic specific heat that is linear in temperature according to

$$n/E_F = 2N(0)/3 = 2\gamma/\pi^2 k_B^2. \tag{2.50}$$

Thus,

$$\kappa_{\text{dirty}} = (ce/2k_B T\pi^{3/2})\gamma^{1/2}\rho_n. \tag{2.51}$$

The BCS coherence length ξ_0 is an intrinsic length introduced by the microscopic theory. It is the distance over which a paired state maintains its identity. It bears no obvious relationship to the macroscopic Ginzburg–Landau coherence length $\xi(T)$ which is a measure of the shortest distance over which the order parameter can be varied. The equilibrium density of superconducting pairs and their phase are established in a distance $\xi(T)$ from a normal-superconducting interface.

d. Microbridges: It is interesting to consider what happens when the sample dimensions become comparable to the coherence length. In Fig. II.5[9] we illustrate two structures which are part of a class of structures known as superconducting microbridges. If the thickness, d, is less than the width, w, and if w is made small compared with the perpendicular magnetic penetration length and the coherence length, then the superconducting properties can only vary along the length. Thus, we have, in effect, a one-dimensional superconductor from the Ginzburg–Landau point of view. (If the length, L, is also short compared with the coherence length, the microbridge exhibits the Josephson effect, which we shall discuss in Chapter V.) Microbridges are of interest because their geometry provides

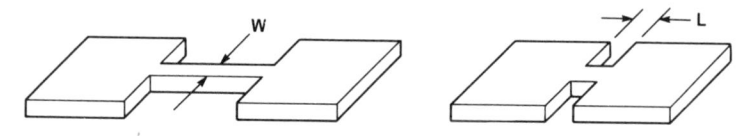

UNIFORM THICKNESS (DAYEM) BRIDGES

FIG. II.5. Typical microbridge geometries. These are often referred to as "Dayem" bridges after A. H. Dayem who first developed them for studying the effects of microwave fields on such "weak links."[9]

[9] P. W. Anderson and A. H. Dayem, *Phys. Rev. Lett.* **13**, 195 (1964); A. H. Dayem and J. J. Wiegand, *Phys. Rev.* **155**, 419 (1967).

simplifications in studying the resistive process in a superconductor when the critical current is exceeded. For our purposes, they provide a nice example of the use of the Ginzburg–Landau theory.

In one dimension, and in the absence of applied fields, the Ginzburg–Landau equations (2.38) and 2.22) become

$$-\xi^2(d^2f/dz^2) - f + f^3 = 0 \tag{2.52}$$

$$j = \frac{e^*\hbar|\psi_\infty|^2|f|^2\nabla\phi}{m^*}. \tag{2.53}$$

Let us introduce the normalized coordinate z/ξ. Equation (2.53) then becomes

$$|f|^2\phi' = \frac{j}{e^*\hbar|\psi_\infty|^2\xi/m^*} = J. \tag{2.54}$$

Since there is no accumulation of charge, the equation of continuity requires $\partial j/\partial z = 0$. This means that J, as defined by Eq. (2.54) is a constant. Using the continuity relation together with the fact that J is a constant enables us to decouple the two G–L equations and obtain one equation for the one-dimensional microbridge:

$$-|f| + |f|^3 - |f|'' - \frac{J^2}{|f|^3} = 0. \tag{2.55}$$

Once $|f|$ is known the total "phase winding," i.e., 2π times the number of twists along the bridge, is given by

$$\Delta\phi = J \int_{-l/2}^{l/2} \frac{dz}{|f|^2} \tag{2.56}$$

where $l = L/\xi$. As the current density through the microbridge increases, the phase winding per unit length also increases as illustrated in Fig. II.6.

Problem II.3
    Consider a *long* microbridge (i.e., a superconducting filament). Assume that the amplitude of ψ is uniform along the length so that the f'' term in Eq. (2.55) may be neglected.
    a. Solve for the current J as a function of the total phase winding per unit length, $\Delta\phi/l$.
    b. How does the maximum current depend upon $\epsilon = (T_c - T)/T_c$?

The uniform solution assumed in Problem II.3 is not, of course, the only solution to Eq. (2.55). There are a whole class of unstable solutions (corresponding to maxima in the Ginzburg–Landau free energy) first obtained by Langer and Ambegaokar.[10] One of these is illustrated at

[10] J. Langer and V. Ambegaokar, *Phys. Rev.* **164**, 498 (1967).

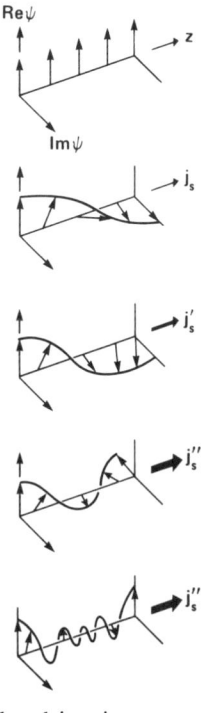

FIG. II.6. Variation of the real and imaginary parts of the order parameter along a microbridge as a function of the current density through the bridge. Notice that the lower two configurations correspond to the same current density.

the bottom of Fig. II.6. Since these solutions must satisfy $|f|^2\phi' =$ constant, a variation in amplitude must be compensated by a variation in the phase winding for a given current density. The importance of such solutions lies in the fact that if $|f|$ goes to zero at some point along the filament, the phase coherence is broken allowing the phase to slip by $\pm 2\pi$, thereby adding or subtracting one turn from the helix. According to Eq. (1.36) such a change in phase must be accompanied by a difference in chemical potential, i.e., a voltage. Therefore, such phase slip events lead to dissipation in the superconducting state. Langer and Ambegaokar found the activation energy for such a fluctuation to be $\sqrt{2}\, H_c(T)^2 A\xi(T)/3\pi$, where A is the cross-sectional area of the filament. This leads to an astronomically small probability for a phase slip unless one is within a millidegree of T_c. This has an interesting implication for "superconductivity" in one dimension as we mentioned in Chapter I. Suppose we could, somehow, establish a state with a particular phase winding, i.e., a particular current density. The decay of this current must occur through phase slip events. This may be so unlikely that we effectively have a persistent current, even though we may not have had a phase transition.

We conclude this section by pointing out that the correlation length in superfluid ^{3}He is also of the order of $\hbar v_F/k_B T_c$ as expected in the clean limit [Eq. (2.46)]. Although the gap in this case is 1000 times smaller than a typical superconducting gap, the correlation length is ≈ 150 Å, only slightly smaller than that of a superconductor, because v_F now refers to the massive helium atom.

2. Ferromagnets

Slow variations in the direction of the magnetization also give rise to a contribution to the energy density proportional to $|\nabla M|^2$. Actually, since M is a vector, one can construct three quadratic forms: $(\nabla \cdot M)^2$, $(\nabla \times M)^2$, and $|\nabla M|^2 \equiv (\nabla M_x)^2 + (\nabla M_y)^2 + (\nabla M_z)^2$. However, since the Hamiltonian of the system is invariant under rotations in spin space, we require that the inhomogeneous term also have this symmetry. This restricts us to the $|\nabla M|^2$ term. This contribution to the energy density is written

$$A_{ex}|\nabla M|^2/M_0^2 \tag{2.57}$$

where we have added the subscript to distinguish the symbol from the Landau coefficient of Eq. (2.3).

Just as in the case of superconductivity, we may express the Landau coefficients in terms of the parameters of a microscopic model. In particular, the mean field susceptibility associated with the Heisenberg exchange interaction is

$$\chi = C_S/(T - T_c) \tag{2.58}$$

where the Curie constant C_S is given by

$$C_S = \frac{Ng^2\mu_B^2 S(S + 1)}{3k_B}. \tag{2.59}$$

Comparing this result with Eq. (2.12) we see that the Landau coefficient $a = \frac{1}{2}C_S$. Similarly, the Heisenberg model also gives

$$A_{ex} = \frac{zS^2Jd^2}{6V_0} = \frac{Sk_BT_cd^2}{4(S + 1)V_0} \tag{2.60}$$

where z is the number of neighboring spins S at a distance d, J the interatomic exchange integral, and V_0 the atomic volume.

Problem II.4

The Heisenberg exchange interaction between a spin i and its neighbors j is

$$\mathcal{H} = -2 \sum_j J_{ij} \mathbf{S}_i \cdot \mathbf{S}_j.$$

If u_i and u_j are unit vectors in the directions of the spins with direction cosines α_{ix}, α_{iy}, etc., use expansions of the form

$$\alpha_{ix}\alpha_{jx} = \alpha_{ix}(\alpha_{ix} + \mathbf{r}_{ij} \cdot \nabla\alpha_{ix} + \tfrac{1}{2}(\mathbf{r}_{ij} \cdot \nabla)^2\alpha_{ix} + \cdots)$$

to derive Eq. (2.60).

The ferromagnetic coherence length then becomes

$$\xi(T) = \left(\frac{A_{ex}}{A(T)}\right)^{1/2} = d\left(\frac{T_c}{6(T_c - T)}\right)^{1/2} \qquad (2.61)$$

showing that the range of the exchange interaction, d, takes the place of the BCS length ξ_0. We shall see later that this is the origin of the difference in the heat capacity of superconductors and ferromagnets that we noted in Chapter I. At $T = 0$, Eq. (2.61) predicts a coherence length less than the interspin spacing which merely indicates that the theory does not apply here.

Photoemission provides an interesting way of observing the magnetic coherence length. Just as in the superconducting case illustrated in Fig. II.4c, we expect the magnetic order parameter to decrease at a surface with the characteristic length $\xi(T)$. As the temperature increases towards T_c the coherence length will eventually exceed the escape depth. The electrons will then all come from a volume in which the magnetization is reduced. The escape depth of photoelectrons depends upon the frequency of the exciting light. By changing the photon energy the point at which this crossover occurs shifts. This behavior is shown in Fig. II.7.[11]

3. Charge Density Waves

As we mentioned in Chapter I, the compound $TaSe_2$ shows a first-order transition at 600 K which has been identified as being associated with charge density wave formation. $TaSe_2$ is a member of the transition metal dichalcogenide family. These compounds have a layered, or two-dimensional structure. The group IVa dichalcogenides such as TiS_2, ZrS_2, etc. have the CdI_2 structure, the so-called 1T structure, in which the metal ion is octahedrally coordinated; the Va dichalcogenides such as

---

[11] S. F. Alvarado, M. Erbudak, F. Meier, and H. C. Siegmann, *Phys. Rev. Lett.* **39**, 219 (1977).

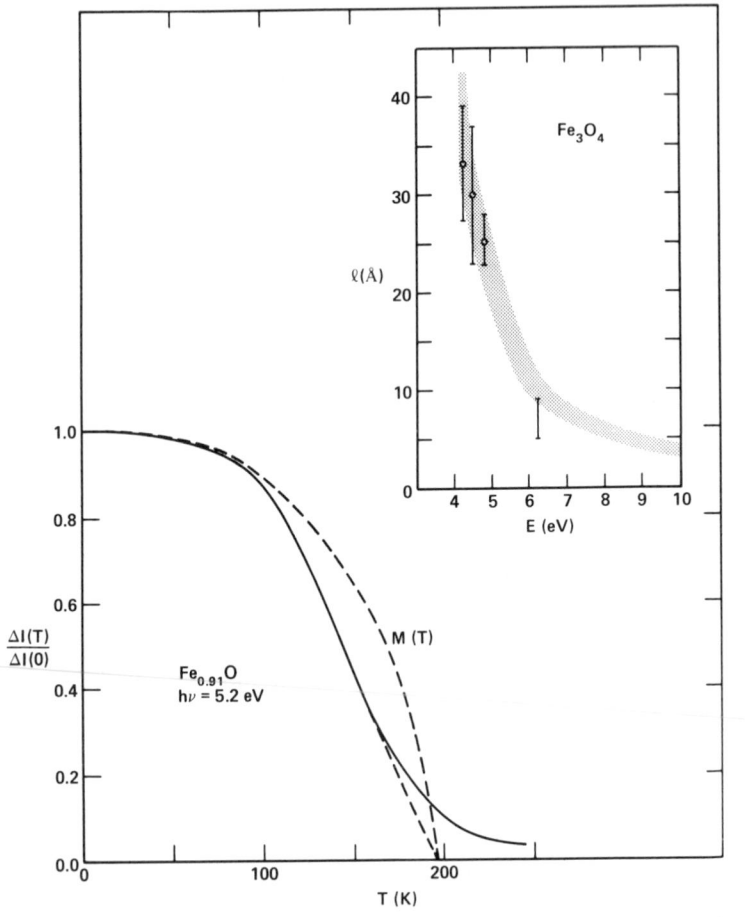

FIG. II.7. Comparison of the temperature dependence of the normalized change in photocurrent with the bulk magnetization. The photoelectron escape depth for Fe_3O_4 is shown in the insert. The corresponding values for $Fe_{0.91}O$ should be the same since the quantum yields and optical properties of these materials are very similar. After Alvarado et al.[11]

NbS_2, TaS_2, etc. have both the 1T structure and the 2H polytype in which the metal ion is trigonal-prismatically coordinated; the group VIa dichalcogenides have only this 2H structure. These structures are indicated at the bottom of Fig. (II.8). These compounds have been of interest because of the possibility of inserting (intercalating) organic molecules such as amines and pyridine between the layers to enhance their two-dimensional character. More recently, the interest has centered on the charge density waves. The electrical properties of the sulfides and selenides is summarized in Fig. II.8.

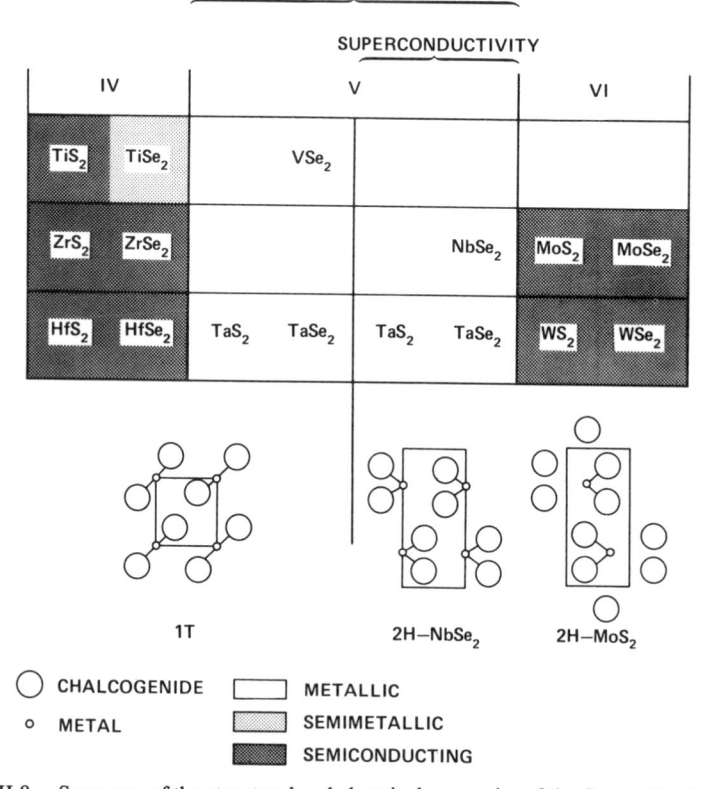

FIG. II.8. Summary of the structural and electrical properties of the Group IV, V, and VI transition metal dichalcogenides.

Charge density waves manifest themselves as superlattices in high-energy electron diffraction patterns. $1T$-TaS_2 shows three CDW phases, $1T_1$, $1T_2$, and $1T_3$. In all cases the main CdI_2 structure reflections are surrounded by groups of extra reflections with threefold symmetry. However, in $1T_1$ and $1T_2$ the magnitudes of the three vectors defining these extra reflections are an irrational fraction of the undistorted structure. We refer to this as a triple incommensurate charge density wave (ICDW). In $1T_3$ these vectors become commensurate defining a perfect $\sqrt{13} \times \sqrt{13}$ superlattice in the plane. The atomic displacements of this phase are shown in Fig. II.9.[12] The various phases are summarized in Fig. II.10. In all cases we see that as the temperature decreases the system passes through the sequence: normal $\rightarrow$ ICDW $\rightarrow$ CCDW, the latter being separated by first-order transitions.

[12] P. Fazekus and E. Tosatti, *Proc. Int. Conf. Phys. Semicond., 13th,* Int. Conf., Rome, 1976 (F. G. Fumi, ed.), p. 415.

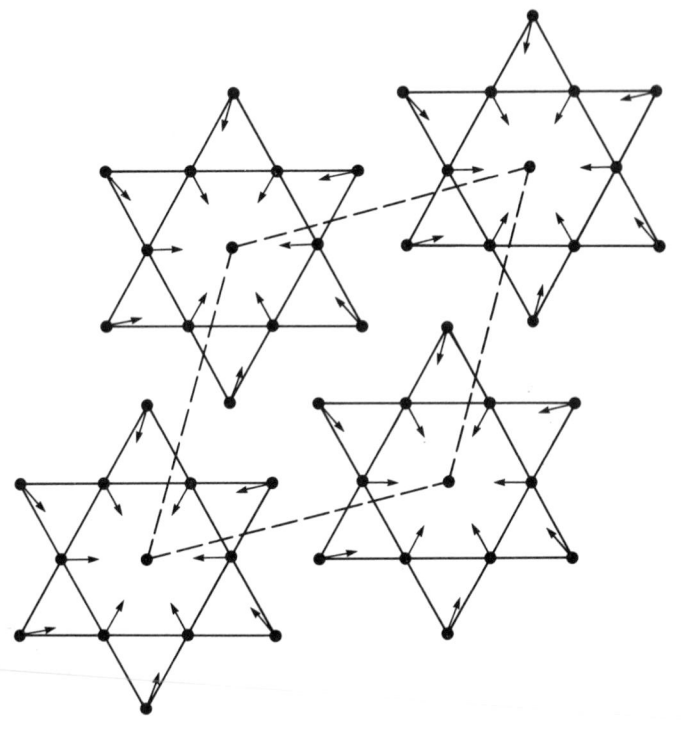

FIG. II.9. Atomic displacements associated with the commensurate charge density wave in 1T-TaS$_2$. The side of the parallelogram is $\sqrt{13}$ times the high-temperature lattice constant, a. After Fazekus and Tosatti.[12]

Since these are "weak" first-order transitions, McMillan[13] has employed the Ginzburg–Landau theory to describe this behavior. Recognizing the fact that the experiments show three charge density waves that have components which are 120° apart in the planes perpendicular to the c-axis, McMillan introduces a six-component order parameter consisting of three complex functions, $\psi_i(r)$. These are related to the conduction electron charge density by

$$\rho(\mathbf{r}) = \rho_0(\mathbf{r})[1 + \alpha(\mathbf{r})] \tag{2.62}$$

where $\rho_0(\mathbf{r})$ is the charge density in the normal state and

$$\alpha(\mathbf{r}) = \mathrm{Re}\left[\sum_i^3 \psi_i(\mathbf{r})\right] = \mathrm{Re}[\psi_1(\mathbf{r}) + \psi_2(\mathbf{r}) + \psi_3(\mathbf{r})]. \tag{2.63}$$

[13] W. L. McMillan, *Phys. Rev. B* **12**, 1187 (1975).

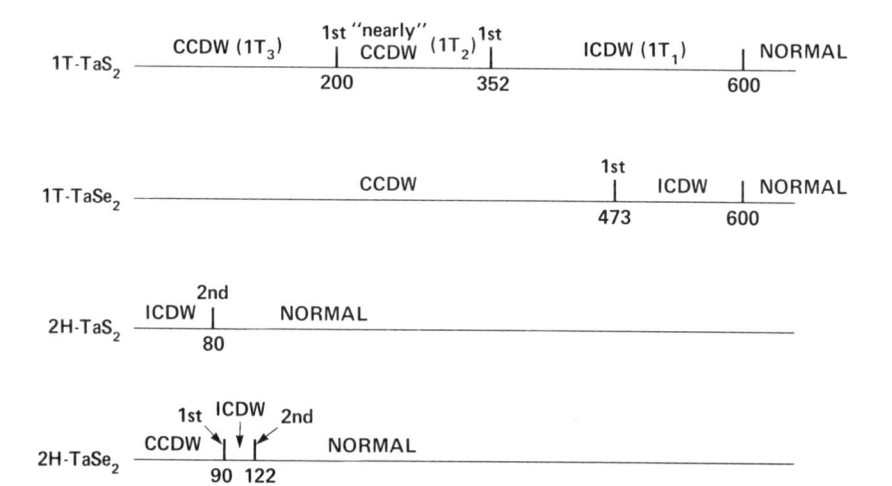

FIG. II.10. Summary of the charge density wave states in the tantalum dichalcogenides.

We must now construct the free energy density. At this point one must rely on intuition and the experimental facts. For example, one might begin with the simplest form for the "Landau part" of the free energy density: $a\alpha^2 + b\alpha^3 + c\alpha^4$ where the cubic term must be included to describe the first-order transition. This, however, does not permit a distinction between single and triple charge density waves. Therefore, McMillan adds cross-terms of the form $d(|\psi_1\psi_2|^2 + |\psi_2\psi_3|^2 + |\psi_3\psi_1|^2)$. He shows that if $d > 9c/4$ the triple CDW will be more stable than a single CDW.

The gradient terms are chosen such that the free energy is a minimum when the wave vectors of the CDW's correspond to those observed:

$$e \sum_{i=1}^{3} |(\mathbf{q}_i \cdot \nabla - i q_i^2)\psi_i|^2 + f \sum_{i=1}^{3} |\mathbf{q}_i \times \nabla\psi_i|^2. \qquad (2.64)$$

Here $|\mathbf{q}_i| = 2\pi/\lambda$ where λ is the wavelength of the *incommensurate* CDW and the directions lie in the plane separated by 120°. The coefficients e and f are referred to as the "elastic" constants for the CDW. As we shall see in Chapter IV, these particular $\mathbf{q}_i$ are related to the geometry of the Fermi surface.

All the coefficients a, b, etc. must have the periodicity of the lattice, i.e.,

$$a(\mathbf{r}) = a_0 + a_1 \sum_i \exp(i\mathbf{K}_i \cdot \mathbf{r}) \qquad (2.65)$$

where $\mathbf{K}_i$ is one of the six shortest reciprocal lattice vectors of the hexagonal planar lattice.

The coherence length is given by the square root of the ratio of the coefficient of the gradient term in the free energy to that of the quadratic term. Thus, assuming an order parameter of the form $\psi_i \sim \exp[i(\mathbf{q}_i + \delta\mathbf{q}) \cdot \mathbf{r}]$ with $\delta\mathbf{q} \| \mathbf{q}_i$, we obtain

$$\xi = \left(\frac{2e_0\mathbf{q}_i^2}{|a_0|} \right)^{1/2}. \qquad (2.66)$$

Unfortunately, a theory does not yet exist which relates these parameters to microscopic quantities.

If we now substitute $\psi_i = \phi_0 \exp(i\mathbf{q}_i \cdot \mathbf{r})$ into these expressions and integrate over d^2r the free energy of one layer becomes

$$F = \tfrac{3}{2}\alpha_0\phi_0^2 + \tfrac{2}{3}b_0\phi_0^3 + \tfrac{3}{8}(15c_0 - 8d_0)\phi_0^4. \qquad (2.67)$$

The cubic term survives the integration because $\Sigma_i\mathbf{q}_i = 0$. This term can have either sign depending on the relative phases of the three waves. Experimentally it is found that the charge densities of each of the three CDW's add up. This suggests that we take $b_0 < 0$. The coefficient a_0 is taken to have the familiar form $a_0 = a'_0(T - T^*)$.

Minimizing F with respect to ϕ_0 gives the transition temperature

$$T_{N \to ICDW} = T^* + b_0^2/a'_0(15c_0 - 8d_0) \qquad (2.68)$$

with a change in entropy

$$\Delta S = 3a'_0 b_0/(15c_0 - 8d_0) \qquad (2.69)$$

indicating a first-order transition, although a "weak" one if b_0 is small. Notice that the presence of the cubic term also shifts the transition temperature away from T^*.

In 2H-TaSe$_2$ the commensurate wave vectors $\mathbf{q}_i = \tfrac{1}{3}\mathbf{K}_i$. Therefore, we write $\psi_i = \phi_0 \exp(i\mathbf{K}_i \cdot r/3)$. Since the gradient terms vanish for the *incommensurate* wavelength, we now obtain an *elastic* contribution to the free energy: $3e_0q_1^2(q_1 - K_1/3)^2\phi_0^2$. In the commensurate case we also have an additional contribution from the periodic part of $b(\mathbf{r})$. In particular, since $b(\mathbf{r}) = b_0 + b_1\Sigma_i \exp(i\mathbf{K}_i \cdot \mathbf{r})$, the integral over $b(r)\psi_i^3(r)$ for *each i* will be nonzero. This corresponds to an *umklapp* contribution to the cubic term. Thus, we find that forcing the wave to be commensurate costs elastic energy, but enables the system to gain umklapp energy, assuming $b_1 < 0$. The resulting ICDW $\to$ CCDW transition then occurs at a lower temperature than the N $\to$ ICDW transition and is first order.

Since the electronic charge density in the normal state is uniform, the appearance of CDW's (even incommensurate ones) indicates a broken symmetry. To investigate modes arising from fluctuations in the order parameter, we write $\psi_i(\mathbf{r}) = \phi(\mathbf{r}) \exp(i\mathbf{p}_i \cdot \mathbf{r})$ where $\mathbf{p}_i = \mathbf{q}_i$ for the incommensurate phase and $\mathbf{p}_i = \mathbf{K}_i/3$ for the commensurate phase, and expand the coefficient in a Fourier series:

$$\phi(\mathbf{r}) = \phi_0 + \sum_q \phi_q e^{i\mathbf{q}\cdot\mathbf{r}}. \tag{2.70}$$

Substituting this expansion into the free energy and minimizing with respect to ϕ_0 and ϕ_q shows that ϕ_q and ϕ_{-q}^* are coupled together. If we write $\phi_q = |\phi_q| \exp i\theta_q$, then the linear combinations $\phi_q \pm \phi_{-q}^*$ are proportional to $|\phi_q| \cos \theta_q$ and $|\phi_q| \sin \theta_q$, respectively. If θ_q is small then these two modes correspond to fluctuations in the amplitude and phase of the order parameter. McMillan finds that for a single ICDW the energy of the phase fluctuation is

$$\epsilon_q^- = e_0(\mathbf{q}_i \cdot \mathbf{q})^2 + f_0(\mathbf{q}_1 \times \mathbf{q})^2. \tag{2.71}$$

Notice that this vanishes as $q \to 0$. Overhauser[14] was the first to suggest such an excitation and called it a *phason*. In the commensurate phase the phason corresponds to a soft optic phonon with a finite energy gap. These modes are illustrated in Fig. II.11. Of course, q_1 now becomes a new reciprocal lattice vector so the phason dispersion curve may be "folded back" about the point $q_1/2$ which places it at the original origin, thereby making it accessible to light scattering.

With these concepts we can qualitatively understand the appearance of the CCDW and ICDW phases. At low temperature the umklapp coupling stabilizes the commensurate phase. As the temperature increases the entropy gain associated with the phasons of the incommensurate phase eventually gives this phase a lower free energy. Finally, the entropy of the electron–hole excitations drive the system "normal."

4. Fluctuations

As mentioned above, the Landau and the Ginzburg–Landau descriptions work quite well for superconductors. The reason for this is that fluctuations, which these theories neglect, are very small in superconductors. To demonstrate this let us calculate the quantity $\langle \Delta(0)\Delta(\mathbf{r}) \rangle$ in the normal metal above T_c. This is a measure of how the order parameter fluctuates. Let us introduce the Fourier transform

$$\Delta(\mathbf{r}) = \sum_k \Delta_k e^{-i\mathbf{k}\cdot\mathbf{r}} \tag{2.72}$$

[14] A. W. Overhauser, *Phys. Rev. B* **3**, 3173 (1971).

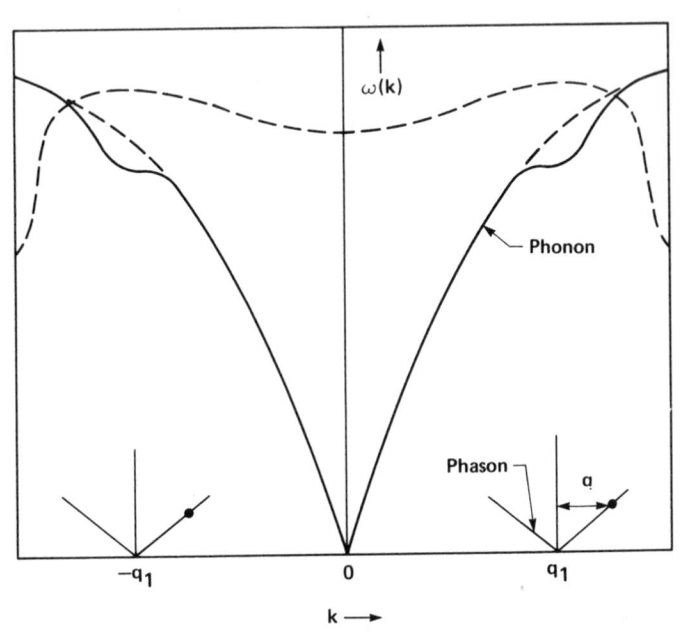

FIG. II.11.   Excitation spectrum associated with an incommensurate charge density wave state.

and assume that the fluctuations at different wave vectors are uncorrelated, i.e., $\langle \Delta_k \Delta_{k'} \rangle = \langle |\Delta_k|^2 \rangle$ for $k' = -k$ and zero otherwise. Then

$$\langle \Delta(0)\Delta(r) \rangle = \sum_k c^{ik \cdot r} \langle |\Delta_k|^2 \rangle. \tag{2.73}$$

If we assume that the metal is maintained at a constant temperature and pressure, then the probability that some parameter of the system has the value, say, x is proportional to $\exp[-G(x)/k_B T]$ where $G(x)$ is the Gibbs free energy. Thus

$$\langle |\Delta_k|^2 \rangle = \frac{\int d\Delta_k e^{-G/k_B T} |\Delta_k|^2}{\int d\Delta_k e^{-G/k_B T}} \tag{2.74}$$

where the integrals extend from 0 to infinity. Neglecting the fourth-order term in Eq. (2.45), the free energy for a clean superconductor has the Fourier representation

$$G = VN(0)(0.74\xi_0)^2 \sum_k [\xi(T)^{-2} + k^2]|\Delta_k|^2 \tag{2.75}$$

where $\xi(T)$ is the "clean" Ginzburg–Landau coherence length given by Eq. (2.48), and V is the volume of the sample. Carrying out the integrations in Eq. (2.74),

$$\langle |\Delta_k|^2 \rangle = \frac{1}{2VN(0)(0.74\xi_0)^2} \frac{k_B T}{[\xi(T)^{-2} + k^2]}. \tag{2.76}$$

Notice that if this is substituted into Eq. (2.75),

$$G = \sum_k \left(\frac{k_B T}{2} \right) \tag{2.77}$$

which is just the equipartition result.

Substituting Eq. (2.76) into Eq. (2.73) and converting the sum over k into an integral gives

$$\langle \Delta(0)\Delta(r) \rangle = \frac{k_B T}{8\pi N(0)(0.74\xi_0)^2} \frac{e^{-r/\xi(T)}}{r}. \tag{2.78}$$

In a dirty superconductor, the factor $(0.74\xi_0)^2$ is replaced by $(0.85)^2\xi_0 l$ and $\xi(T)$ takes the dirty value given by Eq. (2.49). Equation (2.78) shows that $\xi(T)$ also governs the spatial fluctuation in the number of superconducting electrons, n_s. More importantly, however, Eq. (2.78) forms the basis for a discussion of the validity of the Ginzburg–Landau theory.

5. Ginzburg Criterion

The Ginzburg–Landau expansion for the free energy restricts its validity to small values of the order parameter and thus to regions not to far from T_c. However, Ginzburg[15] has also pointed out that for the Landau theory to be valid fluctuations in the order parameter averaged over a volume Ω_ξ determined by the coherence length must be small in comparison with the order parameter itself. This places a *lower* limit on the size of the order parameter. The condition is

$$\overline{\langle \Delta(0)\Delta(r) \rangle}^{\Omega_\xi} \ll \langle \Delta(0) \rangle^2. \tag{2.79}$$

The order parameter as given by Eq. (2.7) may be expressed in terms of the quantities introduced in Eq. (2.46) as

$$\langle \Delta(0) \rangle = \left(\frac{N(0)|\epsilon|}{\beta} \right)^{1/2} \tag{2.80}$$

where $\epsilon = (T - T_c)/T_c$. Since the discontinuity in the specific heat, Eq.

[15] V. L. Ginzburg, Sov. Phys.—Solid State (Engl. Transl.) 2, 1824 (1960).

(2.9), is $\Delta C_p = N(0)^2/2\beta T_c$, the Ginzburg criterion (2.79) for the validity of the Landau theory may be written

$$|\epsilon| \gg \epsilon_c \equiv \left(\frac{0.03k_B}{\Delta C_p \, \xi_0^3}\right)^2. \tag{2.81}$$

The factor of ξ_0^6 in the denominator indicates that as the range of the correlation length becomes longer Landau theory becomes better. For an infinite range interaction we expect to obtain Landau behavior all the way into the critical point. Kittel and Shore[16] have shown that this is indeed the case for a Heisenberg Hamiltonian of the form $-2J\mathbf{S}_i \cdot \mathbf{S}_j$. In the case of dipolar interactions, however, which are also long range, the angular dependence can lead to non-Landau behavior.

For a superconductor such as tin, $\xi_0 \simeq 2300$ Å while $\Delta C_p = 0.8 \times 10^4$ ergs/cm³ deg, giving $\epsilon_c = 10^{-4}$, which explains why the Landau theory works so well. Notice that this does not imply that fluctuation effects will be unobservable. It merely means that they are well described by the mean field theory sketched in Section 4 above.

The Ginzburg criterion also applies to ferromagnets. In this case, however, the correlation length is the distance between interacting spins which is of the order of 2 Å. Thus, for iron where $\Delta C_p = 3 \times 10^7$ erg/cm³ deg, $\epsilon_c = 10^{-2}$.

The criterion (2.81) is based on a three-dimensional argument. It is interesting to consider what happens when we generalize to d-dimensions. The d-dimensional Fourier transform in going from Eq. (2.76) to Eq. (2.78) followed by a d-dimensional spatial average Ω_ξ gives

$$\overline{\langle \Delta(0)\Delta(r)\rangle}^{\Omega_\xi} \sim \frac{\xi(T)^2}{\xi(T)^d} \sim \epsilon^{(d-2)/2}. \tag{2.82}$$

The order parameter itself varies as $\epsilon^{1/2}$. Therefore the Ginzburg criterion gives

$$\text{constant } \epsilon^{(d-2)/2} \ll \epsilon^1. \tag{2.83}$$

Notice that when $d \geq d^* = 4$ this condition is satisfied for arbitrarily small ϵ. Thus for dimensions greater than 4, the Landau mean field theory is exact! d^* is called the *marginal dimensionality*. The smaller d becomes relative to d^*, the larger become the deviations from mean field theory. This is illustrated in Fig. II.12 which shows the values of the critical exponents associated with the susceptibility (γ) and the order parameter (β).

The usefulness of the concept of marginal dimensionality, however, lies

[16] C. Kittel and H. Shore, *Phys. Rev. A* **138**, 1165 (1965).

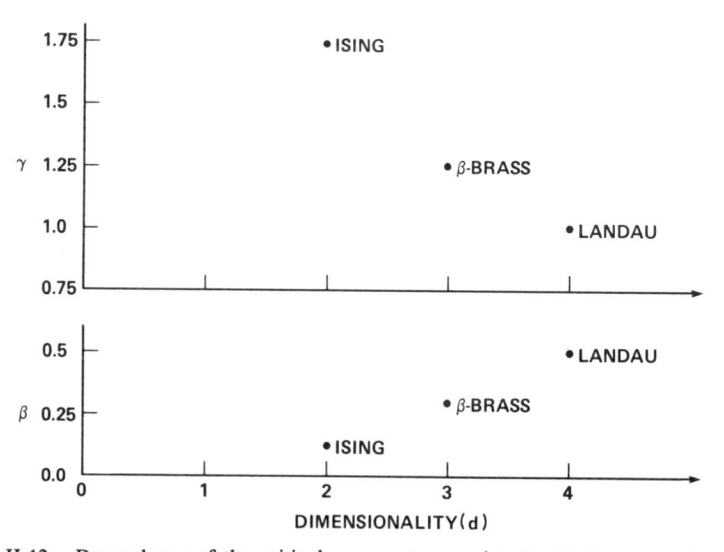

FIG. II.12. Dependence of the critical exponents associated with the susceptibility (γ) and the magnetization (β) upon dimensionality.

in the fact that, depending upon the nature of the interactions, d^* may become 3 of less.[17] When $d^* = d$ we say the system exhibits behavior of marginal dimensionality. It turns out that fluctuation effects appear only as logarithmic corrections to the mean field results. This occurs, for example, in uniaxial, dipolar-coupled ferromagnets such as $LiTbF_4$. If one calculates the generalized susceptibility [i.e., the analog of Eq. (1.15) for this system] one finds

$$\chi(q)^{-1} = 1 + \xi^2 q^2 + g\xi^2 q_z^2/q^2 \qquad (2.84)$$

where g is a constant. Rewriting the last term as $(q_z\xi^2)^2/(q\xi)^2$ shows that the longitudinal correlation range is "superdiverging." That is, it varies as the *square* of the diverging transverse correlation range ξ. The correlation volume over which the fluctuations are to be averaged is therefore ξ to the power *four* rather than the cube of ξ. The marginal dimensionality then becomes $d^* = 3$.

If the fluctuations in q-space are confined to a line, d^* becomes 2. In this case the system shows mean field behavior. The structural transition in $PrAlO_3$ at 151 K, which we described in Chapter I, is such an example. In Fig. II.13[17] we show the order parameter in the vicinity of T_c for the four systems mentioned in this discussion. The point to note is that near

[17] See, for example, J. Als-Nielsen and R. J. Birgeneau, *Am. J. Phys.* **45**, 554 (1977).

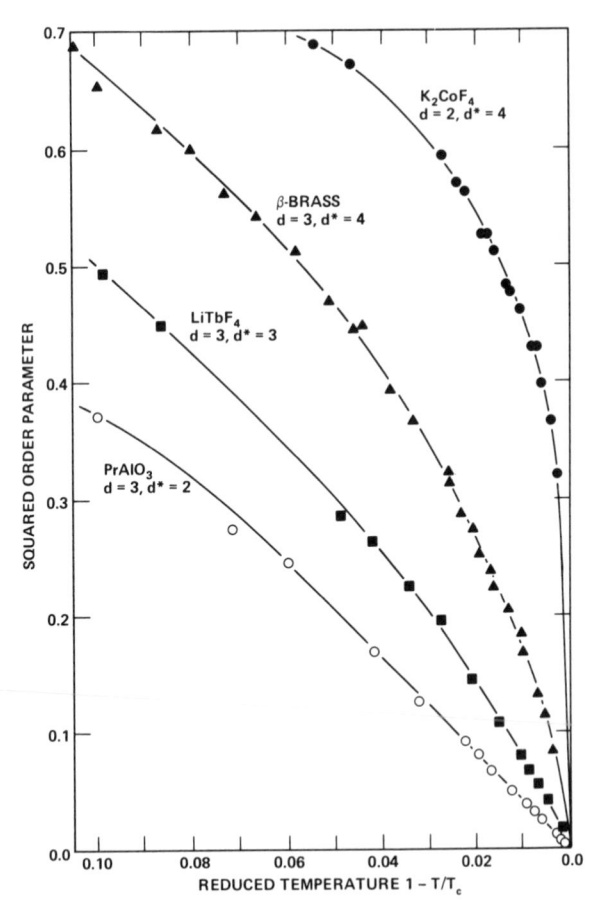

FIG. II.13. Squared order parameter as determined from neutron scattering for four different systems. After Als-Nielsen and Birgeneau.[17]

T_c the data for $PrAlO_3$ vary *linearly* with $1 - T/T_c$ in agreement with the Landau result. The other materials show a stronger dependence on $1 - T/T_c$.

C. BEYOND GINZBURG–LANDAU

In the late 1950's Domb and his co-workers employed series expansion methods to investigate the critical behavior of magnetic systems. These results stimulated very precise measurements of various physical properties in the vicinity of the critical point. The way in which the sublattice magnetization in MnF_2 goes to zero at the Néel temperature, for example, was measured to an accuracy of 5 millidegrees.[18] Experiments such as these provided a quantitative measurement of the breakdown of the Landau, or mean field, theories.

The deficiency of the Landau or Ginzburg–Landau theory lies in the fact that it deals with essentially one value of the order parameter, ψ. A rigorous statistical theory would deal with all values of ψ, weighing each by an appropriate Boltzmann factor. More specifically, the value of the order parameter is the proper statistical average over a region of size $l \ll \xi$, the correlation length. Thus the Landau theory does include fluctuations with wavelengths $\lambda < l$, but does not statistically treat fluctuations with wavelengths $\lambda > l$. Since the coherence length becomes very long near the critical point, we expect such long wavelength fluctuations to be very important in this region. Therefore, to treat this problem properly we should continuously "renormalize" the region l over which the order parameter is averaged. This means that the coefficients in the G–L free energy will depend on l.

1. Scaling

In an effort to replace the difficult problem of a large correlation length by one with a small correlation length Widom[19] and Kadanoff[20] developed the concept of "scaling" which is the phenomenological predecessor to a powerful theoretical technique known as the renormalization group. We can illustrate the concept of scaling by considering a d-dimensional lattice of spins with lattice constant a.

For simplicity suppose the Hamiltonian of this system has the Ising form

$$\mathcal{H} = -\mathcal{J} \sum_{i,j} s_i s_j - H \sum_i s_i. \tag{2.85}$$

We now divide this lattice into blocks of spins, each having a side of length $L \ll \xi$. The total spin of the Ith block is

$$S_I' = \sum_{i \in I} s_i. \tag{2.86}$$

[18] P. Heller and G. Benedek, *Phys. Rev. Lett.* **8**, 428 (1962).

[19] B. Widom, *J. Chem. Phys.* **43**, 3892 and 3898 (1965).

[20] L. Kadanoff, *Physics* **2**, 263 (1966); L. P. Kadanoff, W. Götze, D. Hamblen, R. Hecht, E. A. S. Lewis, V. V. Palciauskas, M. Rayl, J. Swift, D. Aspens, and J. Kane, *Rev. Mod. Phys.* **39**, 395 (1967).

Since these spins are strongly correlated we rescale the total spin, $S_1' = ZS_1$ so that S_1 takes only the values ± 1. For complete spin alignment Z would equal L^d. The Hamiltonian appropriate for these block spins is then

$$\mathcal{H}_L = -\mathcal{J}_L \sum_{I,J} S_I S_J - H_L \sum_I S_I \qquad (2.87)$$

where $\mathcal{J}_L$ is a new coupling constant and $H_L = ZH$.

Near the critical point the free energy per spin will contain a singular part $g(\epsilon, H)$ where $\epsilon = (T - T_c)/T_c$. Since we have forced the block spins to have the same properties as the original spins, the block Hamiltonian has exactly the same form as the original Hamiltonian. The free energy per spin will then have the same functional form except for a volume factor of L^d, i.e.,

$$g(\epsilon_L, H_L) = L^d g(\epsilon, H). \qquad (2.88)$$

We do not know how $\mathcal{J}_L$ relates to $\mathcal{J}$. However, when the original system goes critical at $\epsilon \to 0$ the block spin system should also go critical. Thus ϵ_L should be proportional to ϵ. Furthermore, increasing the block size L requires larger correlation lengths which effectively moves us away from the critical point. Thus we expect ϵ_L should be a function of L, say, L^x where x is a positive number. Similarly, we write Z as L^y which allows for incomplete alignment. Thus, Eq. (2.88) becomes

$$g(L^x \epsilon, L^y H) = L^d g(\epsilon, H). \qquad (2.89)$$

Any function satisfying such a relation is known as a *generalized homogeneous function*. Thus, scaling theory is equivalent to the hypothesis that the free energy is a generalized homogeneous function.

To appreciate the implications of Eq. (2.89) let us differentiate both sides with respect to the field. This gives

$$L^y M(L^x \epsilon, L^y H) = L^d M(\epsilon, H). \qquad (2.90)$$

Since this must be true for any $L \ll \xi$, we choose $L = (-1/\epsilon)^{1/x}$. In the limit $H \to 0$, Eq. (2.90) then becomes

$$(-1/\epsilon)^{(y-d)/x} M(-1, 0) = M(\epsilon, 0). \qquad (2.91)$$

But as $\epsilon \to 0^-$, $M(\epsilon, 0) \sim (-\epsilon)^\beta$ so that $(d - y)/x = \beta$. By taking similar derivatives we can relate the various critical exponents to the two scaling parameters x and y. Thus, $x = 1/\nu = d/(2 - \alpha)$. In this way many of the inequalities mentioned in Chapter I become equalities. When the experimentally measured exponents satisfy these equalities we say that they "exhibit scaling." Notice that Eq. (2.90) may be written

$$\frac{M(\epsilon, H)}{|\epsilon|^\beta} = M\left(\frac{\epsilon}{|\epsilon|}, \frac{H}{|\epsilon|^{\beta\delta}}\right)$$

where $\delta = y/(d - y)$. This suggests that if we define a scaled magnetization $m = |\epsilon|^{-\beta} M$ and a scaled field $\mathcal{H} = |\epsilon|^{-\beta\delta} H$ then $\mathcal{H}$ versus m should be the same for all temperatures. This consequence of the scaling hypothesis has been beautifully demonstrated for $CrBr_3$ by Ho and Litster.[21]

2. Renormalization Group

In 1971 K. Wilson[22] established the mathematical foundation of scaling. The problem as we saw above is the relation between $\mathcal{J}_L$ and $\mathcal{J}$. In general we may have several exchange interactions: near neighbor, next-near neighbor, etc. We shall therefore write this set as a vector $\mathcal{J}$. Consider the partition function for N spins on a d-dimensional lattice,

$$Z(\mathcal{J}, N) = \sum_{\{S_i\}} \exp[\mathcal{H}(\mathcal{J}, \{S_i\}, N)], \tag{2.92}$$

where $\{S_i\}$ refers to a given spin configuration. We now assume that we can divide the N spins into two groups: NL^{-d} block spins, S_I', and a remaining group of internal spins, σ_I, such that when the sum over the internal spin configurations is performed in the partition function we are left with a Hamiltonian of the same form as we had originally. Thus,

$$Z(\mathcal{J}, N) = \sum_{\{S_I'\sigma_I\}} \exp[\mathcal{H}(\mathcal{J}, \{S_I', \sigma_I\}, N)]$$

$$= \sum_{\{S_I\}} \exp[\mathcal{H}(\mathcal{J}_L, \{S_I\}, NL^{-d})] \tag{2.93}$$

$$= Z(\mathcal{J}_L, NL^{-d})$$

where $\mathcal{J}_L$ is some function of $\mathcal{J}$. Let us describe this relation by a transformation operator T, i.e., $\mathcal{J}_L = T(\mathcal{J})$. We can repeat this block construction,

[21] J. T. Ho and J. D. Litster, *Phys. Rev. Lett.* **22**, 603 (1969).

[22] K. Wilson, *Phys. Rev. B* **4**, 3174 (1971). For a good introductory discussion, see Finn Ravndal, "Scaling and Renormalization Groups," Lecture Notes. Nordisk Institut for Teoretisk Atomfysik (NORDITA), Copenhagen, 1975. H. J. Maris and L. P. Kadanoff [*Am. J. Phys.* **46**, 652 (1978)] have also described the renormalization group approach at an undergraduate level.

obtaining new coupling constants,

$$\mathcal{J}_{(n+1)L} = T(\mathcal{J}_{nL}).\qquad(2.94)$$

These transformations are what are referred to as the "renormalization group." As Ravndal points out,[22] these transformations do not, in fact, form a group in the mathematical sense—a more appropriate name would be "effective coupling theory."

One may eventually reach a point called the "fixed point" defined by

$$\mathcal{J}^* = T(\mathcal{J}^*).\qquad(2.95)$$

As an illustration of what is involved in the calculation of a fixed point, consider a triangular two-dimensional lattice of Ising spins. As our block spins we take the three spins on the corners of a triangle so $L = \sqrt{3}$. In order that the block spins take on only two values like the original Ising spins, we define them by

$$S_I = \text{sign}(S_1{}^I + S_2{}^I + S_3{}^I)\qquad(2.96)$$

where $S_i{}^I$ is the ith spin within the Ith block. The set of internal block spins σ_I can take on four different configurations for every value of S_I. The condition that determines the coupling constant transformation is then

$$\exp[\mathcal{H}(S_I)] = \sum_{\{\sigma_i\}} \exp[\mathcal{H}(S_I, \sigma_I)].\qquad(2.97)$$

Niemeijer and van Leeuwen[23] have formulated the summation over $\{\sigma_I\}$ in Eq. (2.97) in terms of a perturbation series by writing

$$\mathcal{H} = \mathcal{H}_0 + V$$

where

$$\mathcal{H}_0 = \mathcal{J} \sum_I \sum_{\substack{i \in I \\ j \in I}} S_i S_j\qquad(2.98)$$

and

$$V = \mathcal{J} \sum_{IJ} \sum_{\substack{i \in I \\ j \in J}} S_i S_j.$$

The details may be found in the article by Ravndal. The result is that to first order in V the block spin Hamiltonian satisfying Eq. (2.93) has the

[23] Th. Niemeijer and J. M. J. van Leeuwen, *Phys. Rev. Lett.* **31,** 1411 (1973).

same form as the original Hamiltonian:

$$\mathcal{H}(S_I) = \mathcal{J}_L \sum_{I,J} S_I S_J$$

with

$$\mathcal{J}_L = 2 \mathcal{J} \frac{e^{3\mathcal{J}} + e^{-\mathcal{J}}}{e^{3\mathcal{J}} + 3e^{-\mathcal{J}}}. \tag{2.99}$$

This is the desired transformation. From Eq. (2.95) we see that it has a trivial fixed point at $\mathcal{J}^* = 0$ and a nontrivial fixed point for $\mathcal{J}^* = (\frac{1}{4}) \ln(1 + 2\sqrt{2}) = 0.34$.

The importance of the fixed point lies in the fact that the physics of the critical point is governed by the transformation near this point. We cannot go into the details here, but Wilson has shown that if one linearizes the transformation around the fixed point the eigenvalues of the resulting transformation matrix are directly related to the critical exponents. In the two-dimensional Ising example above, the transformation matrix is simply a scalar so there is only one eigenvalue λ of the linearized transformation, namely $(\partial \mathcal{J}_L / \partial \mathcal{J})|_{\mathcal{J}=\mathcal{J}*} = 1.62$. This is related to the scaling parameter x introduced in Eq. (2.89) by $x = \ln\lambda / \ln L = 0.87$. The exact Onsager solution gives $x = 1$. The parameter y is obtained by adding a field to the Hamiltonian (2.98) and finding how it transforms when block spins are introduced.

If the repeated application of the renormalization-group transformations does *not* lead to a fixed point, this means there is no critical point for the system, at least as a function of temperature. Therefore, one expects the transition to be discontinuous, i.e., first order. Mukamel, Krinsky, and Bak[24] have used this result to predict first-order transitions in a number of magnetic systems. They begin by identifying the number of components making up the order parameter as we discussed in Chapter I. They then construct Ginzburg–Landau Hamiltonians to fourth order in the order parameters which are invariant under the symmetry operations of the high-symmetry group. The critical phenomena associated with each of these Hamiltonians is then studied using the renormalization group technique. In this way they "explain" the first-order antiferromagnetic transitions in Cr, Eu, UO_2, and MnO.

In the context of our discussion of the deficiency of the Ginzburg–Landau theory, the renormalization group approach leads to differential

[24] D. Mukamel and S. Krinsky, *Phys. Rev. B* **13**, 5065 and 5078 (1976); P. Bak and D. Mukamel, *ibid.* p. 5086.

equations for how the coefficients in the G–L free energy depend upon l. In particular, if we write the free energy as

$$G = \int d^3r[r_0|\psi(\mathbf{r})|^2 + u_0|\psi(\mathbf{r})|^4 + |\nabla\psi(\mathbf{r})|^2], \qquad (2.100)$$

then renormalization theory requires that r_0 and u_0 satisfy

$$\frac{dr_0(l, T)}{dl} = -3l^{\epsilon-1}r_0(l, T)u_0(l, T) \qquad (2.101)$$

$$\frac{du_0(l, T)}{dl} = -9l^{\epsilon-1}u_0(l, T)^2$$

where we have adopted the conventional notation $\epsilon = 4 - d$, d being the dimensionality. A term linear with u_0 has been neglected in the first equation since it does not depend on T near T_c. The solutions for $r_0(l, T)$ and $u_0(l, T)$ for T near T_c are

$$r_0(l, T) \propto l^{-\epsilon/3}(T - T_c)$$
$$u_0(l, T) \propto l^{-\epsilon}. \qquad (2.102)$$

Notice that $r_0(l, T)$ and $u_0(l, T)$ are independent of l in the fourth dimension. This is a consequence of the interplay between the strength of a fluctuation, given by the $(\nabla\psi)^2$ term in the free energy, and the volume l^d over which the fluctuation acts as we saw above. Using these renormalized coefficients in the Ginzburg–Landau theory the coherence length, Eq. (2.39) becomes

$$\xi \propto \frac{1}{\sqrt{r_0}} \propto \xi^{\epsilon/6}(T - T_c)^{-1/2}. \qquad (2.103)$$

Since $\xi \sim ((T - T_c)/T_c)^{-\nu}$ we find that the critical exponent ν now becomes

$$\nu = \frac{1}{2 - \epsilon/3} \qquad (2.104)$$

which for three dimensions ($\epsilon = 1$) gives $\nu = 0.6$, which is very close to the experimental values observed in magnetic systems.

This demonstrates how renormalization theory complements scaling theory by actually providing values for the critical exponents.

D. HYDRODYNAMICS

Another theoretical approach very similar to those described above, in the sense that it relies only on conservation laws, symmetry properties, and expansions about equilibrium, is the hydrodynamic approach. The advantage of this approach is that it includes time-dependent as well as spatial variations. Time-dependent effects must be considered if one wants to understand dynamic phenomena such as light scattering and neutron scattering. Hydrodynamics also enables us to determine the nature of the Goldstone boson spectrum.

The first step in setting up a hydrodynamic theory is to identify those quantities which are conserved. The underlying assumption is that one can find a regime in which all nonconserved quantities have relaxed to their local equilibrium values and therefore need not be considered. We must also consider quantities which have no "restoring forces." As we shall see in the example below, one of these quantities turns out to be the gradient of the phase of the order parameter. Having once identified these quantities their time derivatives are related to the divergence of some current. From the solutions of these equations we can construct various dynamic correlation functions.

As an illustration of how the hydrodynamic arguments go, let us consider[25] a system described by the anisotropic Heisenberg Hamiltonian

$$\mathcal{H}_0 = -2 \sum_j \sum_{j>i} [J_z S_i^z S_j^z + J_\perp (S_i^x S_j^x + S_i^y S_j^y)]. \tag{2.105}$$

If $J_\perp$ is positive and greater than J_z, then the ground state will be that of ferromagnetic alignment in the $x - y$ plane as illustrated in Fig. II.14. The order parameter appropriate to this "planar" ferromagnet is $m_x + im_y = m_\perp \exp(i\phi)$. If an external field H_z is applied along the z-axis, then the magnetization will develop a component along this direction. Since the total Hamiltonian, $\mathcal{H}_0 - H_z \Sigma_i S_i^z$, is invariant under a rotation of the spins about the z-axis, the z-component of the magnetic moment, M_z,

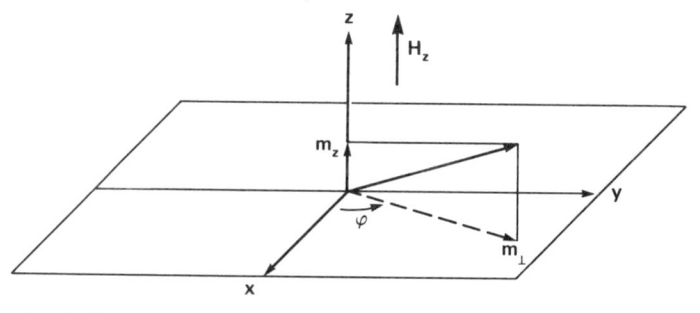

FIG. II.14.   Coordinates used for discussion of the planar ferromagnet.

[25] B. I. Halperin and P. C. Hohenberg, *Phys. Rev.* **188**, 898 (1969).

is a constant of motion. The internal energy, $E = \langle \mathcal{H}_0 \rangle$, is also a conserved quantity. We now assume that we can define local values for the magnetization $m_z(\mathbf{r})$, $m_\perp(\mathbf{r}) \exp[i\phi(\mathbf{r})]$, and the energy density, $\epsilon(\mathbf{r})$. Since the correlation length, ξ, is a measure of the range of fluctuations, if these local quantities are to have meaning, they must be spatially uniform over a distance greater than ξ. Thus, if this spatial variation is characterized by a wavevector q, the hydrodynamic region is $q\xi \ll 1$. This is illustrated in Fig. II.15.[26] In this regime the magnitude of the transverse magnetization is determined by $m_z(\mathbf{r})$ and $\epsilon(\mathbf{r})$ in the same way that the total transverse moment $m_\perp$ is related to E: $m_\perp(\mathbf{r}) = [\epsilon(\mathbf{r}), m_z(\mathbf{r})]$. Thus, the state of the system is determined by $\epsilon(\mathbf{r})$, $m_z(\mathbf{r})$, and $\phi(\mathbf{r})$. This means that the time derivatives $\partial\epsilon/\partial t$, $\partial m_z/\partial t$, and $\partial\phi/\partial t$ are functionals of these three quantities. Since $\int\epsilon(\mathbf{r}) \, d^3r = E$ is constant with time, the derivative $\partial\epsilon/\partial t$ may be expressed as the divergence of some vector, i.e.,

$$\partial\epsilon(\mathbf{r})/\partial t = -\nabla \cdot \mathbf{j}_\epsilon(\mathbf{r}) \tag{2.106}$$

where $\mathbf{j}_\epsilon(\mathbf{r})$ plays the role of an energy current. Similarly,

$$\partial m_z(\mathbf{r})/\partial t = -\nabla \cdot \mathbf{j}_m(\mathbf{r}). \tag{2.107}$$

The phase $\phi(\mathbf{r})$ is not a conserved quantity. However, from Fig. II.14 we see that $\partial\phi/\partial t$ is just the frequency at which the transverse magnetization

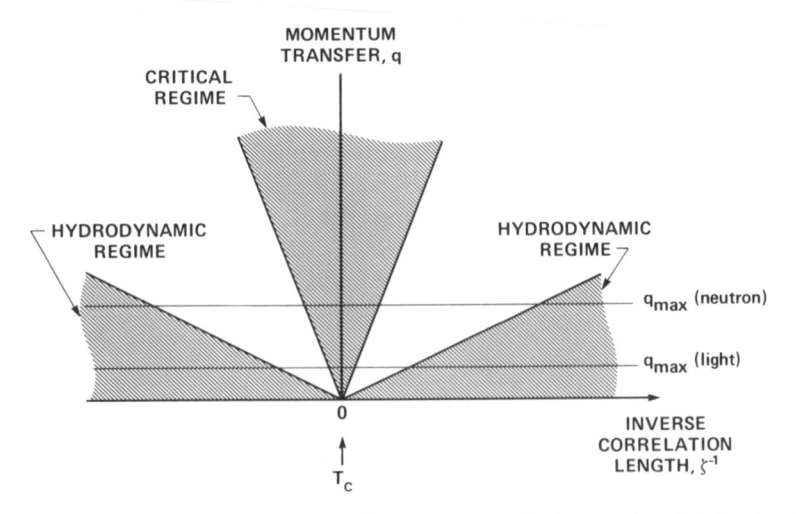

FIG. II.15.   The microscopic domain of wave vector and coherence length indicating the hydrodynamic and critical regions. Notice that neutrons have wavelengths which make them more appropriate for probing the critical region. After Halperin and Hohenberg.[26]

[26] B. I. Halperin and P. C. Hohenberg, *Phys. Rev.* **177**, 952 (1969).

precesses about the z-axis. The thermal equilibrium value of M_z in the presence of an applied field H_z is obtained by minimizing the total energy $E - H_z M_z$, i.e., by the relation

$$h(E, M_z) \equiv \partial E/\partial M_z = H_z. \qquad (2.108)$$

In this state of *true* equilibrium $\partial\phi/\partial t = 0$. Let us now introduce a small local variation in H_z. Since M_z is a constant of the motion, it will not be affected by this variation in H_z. But the phase will now have a time dependence proportional to the deviation from equilibrium, i.e.,

$$\partial\phi/\partial t = h(E, M_z) - H_z. \qquad (2.109)$$

For example, if $M_z = 0$ and we apply a field H_0 in the z-direction $M_\perp$ will precess at the Larmor frequency given by H_0. Equation (2.109) describes the time rate of change of the phase of the order parameter. In the context of superconductivity Eq. (2.109) is just the Josephson equation. It is convenient to introduce a new variable defined by $\mathbf{v} = \nabla\phi(\mathbf{r})$. Its equation of motion in zero field is then

$$\partial\mathbf{v}(\mathbf{r})/\partial t = \nabla h[\epsilon(\mathbf{r}), m_z(\mathbf{r})]. \qquad (2.110)$$

Equations (2.106), (2.107), and (2.110) constitute the basic hydrodynamic equations for this system. We must now develop constitutive relations between $\mathbf{j}_\epsilon$, $\mathbf{j}_m$, h and the functions $\epsilon(\mathbf{r})$, $m_z(\mathbf{r})$, and $\mathbf{v}(\mathbf{r})$.

The susceptibility χ is defined by $M_z = \chi H_z$ where M_z is the *true* equilibrium value. Therefore $H_z = h$ and

$$h(\epsilon(\mathbf{r}), m_z(\mathbf{r})) = \chi^{-1} m_z(\mathbf{r}). \qquad (2.111)$$

The determination of the constitutive relations for $\mathbf{j}_\epsilon$ and $\mathbf{j}_m$ is more involved. Basically, one argues that since $\mathbf{j}_\epsilon$ and $\mathbf{j}_m$ are vectors they are proportional to $\mathbf{v}$ plus terms involving various space derivatives of ϵ, m_z, and $\mathbf{v}$. We refer the reader to the paper of Halperin and Hohenberg[25] for the details. The results are

$$\begin{aligned}
\mathbf{j}_\epsilon(\mathbf{r}) &= -\rho_s h\mathbf{v}(\mathbf{r}) - \kappa\nabla T \\
\mathbf{j}_m(\mathbf{r}) &= -\rho_s \mathbf{v}(\mathbf{r}) - \Gamma\nabla h
\end{aligned} \qquad (2.112)$$

where ρ_s is a parameter called the "stiffness" constant, κ is the thermal conductivity, and Γ is a spin transport coefficient. These quantities all have microscopic expressions, but for our purpose will merely be regarded as phenomenological parameters. Since $\nabla T = C^{-1}\nabla\epsilon$ where C is the specific heat we may finally write our three fundamental hydrodynamic equations as

$$\partial m_z/\partial t = \rho_s\nabla\cdot\mathbf{v} + \Gamma\chi^{-1}\nabla^2 m_z \qquad (2.113)$$

$$\partial\epsilon/\partial t = \kappa C^{-1}\nabla^2\epsilon \tag{2.114}$$

$$\partial v/\partial t = \chi^{-1}\nabla m_z. \tag{2.115}$$

Assuming plane wave solutions of these equations yields a spin wave mode with frequency (in zero field)

$$\omega_m(\mathbf{k}) = ck - i\tfrac{1}{2}\Gamma\chi^{-1}k^2 \tag{2.116}$$

where $c = (\rho_s/\chi)^{1/2}$, and a heat diffusion mode with frequency

$$\omega_\epsilon(\mathbf{k}) = -i\omega C^{-1}k^2. \tag{2.117}$$

The spin wave frequency arises from the combination of Eqs. (2.113) and (2.115). The latter equation arises from fluctuations in the phase of the order parameter. Thus, we may think of the *Goldstone bosons as being related to fluctuations in the phase of the order parameter*. In the paramagnetic state, the phase is undefined so the terms involving v vanish. This leaves us with just a diffusion equation for m_z.

If $\rho_s = 0$, then the frequency of the mode given by Eq. (2.116) is purely imaginary.

The fact that the equations for $m_z(\mathbf{r})$ and $\epsilon(\mathbf{r})$ contain Laplacians which lead to diffusive behavior is a consequence of the fact that the integrals over space of these quantities are conserved. [The spatial integral over ∇^2 may be converted to an integral over a surface that may be chosen where $m_z(\mathbf{r})$ is zero, i.e., outside the sample.] If ϵ and m_z were not conserved, then we would have terms directly proportional to $m_z(\mathbf{r})$ and $\epsilon(\mathbf{r})$ which would lead to relaxation.

This hydrodynamic approach may be applied to a variety of systems to determine the nature of their excitation spectra. In Table II.1 we list some systems with their order parameters and Goldstone bosons. Here N is the staggered magnetization.

TABLE II.1
SUMMARY OF THE ORDER PARAMETERS ASSOCIATED WITH
VARIOUS MAGNETIC SYSTEMS

System	Order parameter	Conserved?	Excitations
Planar FM	$M_\perp e^{i\phi}$	No	$\omega_k \propto k$
Planar AFM	$N_\perp e^{i\phi}$	No	$\omega_k \propto k$
Isotropic FM	$M_0\hat{m}$	Yes	$\omega_k \propto k^2$
Isotropic AFM	$N_0\hat{n}$	No	$\omega_k \propto k$
Uniaxial FM	M_z	Yes	None
Uniaxial AFM	N_z	No	None

E. Time-Dependent Ginzburg–Landau Theory

Although the hydrodynamic approach outlined in the previous section successfully predicts the existence of spin wave excitations and spin diffusion, we do not expect it to apply near a critical point as Fig. II.15 clearly illustrates. It does not tell us, for example, how a nonconserved order parameter actually decays toward equilibrium. This is an important question, for it is widely observed that it takes an increasingly longer time for a system to reach equilibrium as one approaches the critical point.

There have been a number of efforts made to develop a theory of such dynamic critical phenomena. One class of such efforts are based on a time-dependent formulation of the Ginzburg–Landau equations. It turns out that a rigorous time dependent generalization of the Ginzburg–Landau theory for superconductors is valid only in the so-called "gapless" regime. Nevertheless, a time-dependent form can be developed by marrying hydrodynamics to the Ginzburg–Landau theory that serves as a model for critical dynamics. Consider hydrodynamic Eq. (2.113) for the *conserved* quantity m_z when the external field H_z is zero:

$$\partial m_z/\partial t = \rho_S \nabla \cdot \mathbf{v} + \Gamma \chi^{-1} \nabla^2 m_z. \tag{2.118}$$

The quantity $\chi^{-1} m_z$ is the "internal" magnetic field h that was defined as $\partial \epsilon/\partial m_z$. A time-dependent Ginzburg–Landau (TDGL) equation is obtained by replacing h by the functional derivative $\delta G/\delta \psi$ where G is given by Eq. (2.100), and neglecting the term proportional to ρ_s. The neglect of the ρ_s term means that the theory will not contain propagating solutions. One also adds a statistically defined noise source, $\eta(\mathbf{x}, t)$ to insure that the system relaxes to an equilibrium distribution proportional to $\exp[-G/k_B T]$. The TDGL equation then becomes

$$\partial \psi/\partial t = \Gamma \nabla^2 (\delta G/\delta \psi) + \eta. \tag{2.119}$$

At long wavelengths this tells us that the Fourier components of the order parameter relax by a diffusion process with a characteristic frequency

$$\omega_k = \Gamma \chi_\psi^{-1}(k, \xi) k^2 \tag{2.120}$$

where $\chi_\psi(k, \xi)$ is the static susceptibility as a function of the wave vector k and the correlation length ξ. Since the susceptibility diverges as $k \to 0$ and $T \to T_c$ then $\omega_k \to 0$, which is referred to as *critical slowing down*.

In the case of a superconductor Eq. (2.119) gives

$$\partial \psi/\partial t = -\Gamma[1 + \xi^2(T)\nabla^2]\psi \tag{2.121}$$

where microscopic considerations show that $\Gamma = 8k_B(T - T_c)/\pi\hbar$. We shall use this later in our discussion of the resistivity of superconductors.

If the order parameter is *not* conserved then the TDGL equation becomes

$$\partial\psi/\partial t = -\Gamma(\delta G/\delta\psi) + \eta. \tag{2.122}$$

When $u_0 = 0$ in Eq. (2.100) then the Fourier components of the order parameter relax exponentially with a rate

$$\omega_k = \Gamma\chi_\psi^{-1}(k, \xi). \tag{2.123}$$

When $u_0 \neq 0$ the Fourier components do not in general relax in a simple exponential form. However, a characteristic frequency may be introduced in analogy with Eq. (2.123).

$$\omega_k = \Gamma(k, \xi)\chi_\psi^{-1}(k, \xi) \tag{2.124}$$

which defines the generalized transport coefficient $\Gamma(k, \xi)$. Renormalization theory has been used to explore these and other models in an attempt to understand critical dynamics. It would be beyond the scope of this text to pursue this any further. Our purpose has merely been to indicate to the reader what *time-dependent* Ginzburg–Landau theory involves.

III. The Superconducting Transition Temperature

In the previous chapter we discussed phase transitions from a general point of view without any specific reference to microscopic mechanisms. We did indicate that the Ginzburg–Landau parameters for superconductors and ferromagnets could be obtained from microscopic theory, but the details were buried in the transition temperatures. In this and the next chapter we shall review the physical considerations that enter the superconducting and Curie temperatures. Such a venture into the microscopic theory is necessary if we are to compare theory with experiment. It is also necessary in order to understand how other physical properties affect the long range order.

In Chapter I we derived the gap equation, Eq. (1.32),

$$\Delta_{\mathbf{k}} = - \sum_{\mathbf{k}'} V_{\mathbf{k}\mathbf{k}'}\Delta_{\mathbf{k}'} \tanh(E_{\mathbf{k}'}/2k_{\mathrm{B}}T)/2E_{\mathbf{k}'}. \tag{3.1}$$

If, following BCS, we make the approximation indicated in Eq. (1.42), then $\Delta_{\mathbf{k}} = \Delta_{\mathbf{k}'} = \Delta$ and the gap equation reduces to

$$\frac{1}{V} = \frac{1}{2} \sum_{\mathbf{k}} \frac{\tanh(E_{\mathbf{k}}/2k_{\mathrm{B}}T)}{E_{\mathbf{k}}}. \tag{3.2}$$

(Notice that if $V > 0$ then $\Delta = -C\Delta$ where C is a positive quantity. The only solution in this case is $\Delta = 0$.) At the transition temperature $E_{\mathbf{k}} = \epsilon_{\mathbf{k}}$. The sum over k may be converted into an integral by introducing the density of states, $N(\epsilon)$:

$$\sum_{\mathbf{k}} \rightarrow \int_{-\hbar\omega_{\mathrm{D}}}^{\hbar\omega_{\mathrm{D}}} N(\epsilon)\,d\epsilon \tag{3.3}$$

where ω_{D} is the Debye frequency. Assuming the density of states does not vary over this interval, Eq. (3.2) becomes

$$\frac{1}{N(E_{\mathrm{F}})V} = \frac{1}{2} \int_{-\hbar\omega_{\mathrm{D}}}^{\hbar\omega_{\mathrm{D}}} \frac{\tanh(\epsilon/2k_{\mathrm{B}}T_{\mathrm{C}})}{\epsilon}\,d\epsilon$$

$$= \int_{0}^{\hbar\omega_{\mathrm{D}}/2k_{\mathrm{B}}T_{\mathrm{C}}} \frac{\tanh x}{x}\,dx = \ln\left(\frac{2\gamma}{\pi}\frac{\hbar\omega_{\mathrm{D}}}{k_{\mathrm{B}}T_{\mathrm{C}}}\right) \tag{3.4}$$

where γ is Euler's constant. Thus,

$$k_{\mathrm{B}}T_{\mathrm{BCS}} = 1.13\ \hbar\omega_{\mathrm{D}}e^{-1/N(E_{\mathrm{F}})V}.$$

(3.5)

At $T = 0$, $\tanh(E_k/2k_{\mathrm{B}}T)$ in Eq. (3.2) becomes 1. Since $E_k = (\epsilon_k{}^2 + \Delta^2)^{1/2}$ we obtain $2\Delta = 3.5k_{\mathrm{B}}T_{\mathrm{c}}$.

What are the contributions to $V_{kk'}$? What makes it attractive? These are some of the questions we shall attempt to answer in this chapter.

A. ELECTRON–PHONON MECHANISMS

Although superconductivity was discovered by Kamerlingh Onnes in 1911, it was not until the work of Bardeen, Cooper, and Schrieffer[1] in 1957, that an understanding of this phenomenon was finally achieved. The BCS theory is based on the physical idea that when an electron moves through a lattice of ions these ions respond in an attempt to screen the electron. Because the inertia of an ion is so much greater than that of an electron, the electron leaves behind an "overscreened" region which can *attract* a second electron.

The role of the electron–phonon interaction in superconductivity had been suspect for many years. In fact, Kamerlingh Onnes himself thought of this as evidenced by this quote from a paper with Tuyn[2] in 1922:

> The object of the investigation was to establish the vanishing point of [the resistivity of] Pb more accurately, as well as to trace a possible difference in the vanishing point of Pb and uranium Pb. Regarding a difference of vanishing point temperature for isotopes it seemed not impossible that the occurrence of the superconductivity might be influenced by the mass of the nucleus.

However, his gas thermometer was only good to an accuracy of 0.01 K. This magnitude is almost exactly the shift in T_{c} which we estimate would be expected from the "uranium Pb," i.e., ^{206}Pb. Thus, Onnes barely missed discovering the isotope effect.

The first detailed theory based on the electron–phonon interaction was developed by Fröhlich[3] in 1950. He calculated the phonon correction to the self-energy of the electrons. This quantity is diagonal in occupation number space. Therefore, it does not say anything about off-diagonal order, which, as we saw in Chapter I, is the characteristic of superconductivity. Nevertheless, Fröhlich's theory did predict that the superconducting transition temperature should be inversely proportional to the iso-

---

[1] J. Bardeen, L. N. Cooper, and J. R. Schrieffer, *Phys. Rev.* **108**, 1175 (1957).

[2] H. Kamerlingh Onnes and W. Tuyn, *Leiden Commun.* No. 160 (1922).

[3] H. Fröhlich, *Phys. Rev.* **79**, 845 (1950).

topic mass. This behavior was, in fact, established experimentally the same year.[4] But the microscopic origin of this isotope effect is more subtle than Fröhlich's theory would suggest. Both experimental groups measured the isotopic dependence of T_c in Hg for which $T_c \sim M^{-1/2}$. It is interesting to consider how the history of superconductivity might have been delayed had these measurements been made on Zr or Ru for which $T_c \sim M^0$!

The observation of the isotope effect clearly indicated that the electron–phonon interaction was playing a major role in superconductivity. In Fröhlich's theory the electron–phonon interaction was characterized by an unknown coupling constant. It, therefore, became necessary to develop a detailed theory for this interaction. Both Fröhlich[5] and Bardeen and Pines[6] consider what form the effective electron–electron interaction would take in a metal when electron–phonon interactions are added to the electron–electron interactions. Fröhlich applied his results to a one-dimensional system,[7] about which we shall have more to say later, while Bardeen's work evolved into the BCS theory. Let us now consider the origin of the phonon-mediated electron–electron interactions.

1. Energy Renormalization

The electron(e)–nuclear(n) Hamiltonian is

$$\mathcal{H} = T_e + V_{en} + V_{ee} + T_n + V_{nn}. \tag{3.6}$$

The physics of this Hamiltonian is obviously very rich, and powerful techniques have been developed to deal with this. Later in this chapter we shall touch upon some of these, but for the moment we shall merely introduce some definitions and indicate the origin of the attractive electron–electron interaction.

Since $\mathcal{H}$ involves *all* the electrons, it is convenient first to separate out the core electrons. This is done by recognizing that the dominant effect of the core electrons is to introduce "orthogonalization wiggles" in the conduction electron wave function near the nucleus. As a result the pseudo-conduction electrons see an average potential in this region which is relatively weak. Thus we can take account of the core electrons by introducing ionic pseudopotentials in V_{en}. If we assume that the resulting

[4] E. Maxwell, *Phys. Rev.* **78**, 477 (1950); C. Reynolds, B. Serin, W. Wright, and L. Nesbitt, *ibid.* p. 487.

[5] H. Fröhlich, *Proc. R. Soc. London, Ser. A* **215**, 291 (1952).

[6] J. Bardeen and D. Pines, *Phys. Rev.* **99**, 1140 (1955).

[7] H. Fröhlich, *Proc. R. Soc. London, Ser. A* **223**, 296 (1954).

screened ionic potential does not strongly perturb the pseudoconduction electrons, the electron field operator may be expanded in plane waves.

If we now expand the positions of the ions about their equilibrium positions,

$$\mathbf{R}_\alpha = \mathbf{R}_\alpha{}^0 + \mathbf{u}_\alpha, \tag{3.7}$$

and introduce phonon operators $b_{\mathbf{q}\lambda}$ with polarization vectors $\hat{\boldsymbol{\epsilon}}_{\mathbf{q}\lambda}$ according to

$$\mathbf{u}_\alpha = \sum_{\mathbf{q}\lambda} \frac{1}{(2NM\omega_{\mathbf{q}\lambda})^{1/2}} \hat{\boldsymbol{\epsilon}}_{\mathbf{q}\lambda}(b_{\mathbf{q}\lambda} + b^\dagger_{-\mathbf{q}\lambda})e^{i\mathbf{q}\cdot\mathbf{R}_\alpha}, \tag{3.8}$$

the Hamiltonian becomes

$$
\begin{aligned}
\mathcal{H} = {} & \sum_{\mathbf{k},\sigma} (\hbar^2 k^2/2m)c^\dagger_{\mathbf{k}\sigma}c_{\mathbf{k}\sigma} + \sum_{\mathbf{q}\lambda} \hbar\omega_{\mathbf{q}\lambda}(b^\dagger_{\mathbf{q}\lambda}b_{\mathbf{q}\lambda} + \tfrac{1}{2}) \\
& + \sum_{\mathbf{k},\mathbf{q},\sigma} V(\mathbf{q})S(\mathbf{q})c^\dagger_{\mathbf{k}+\mathbf{q},\sigma}c_{\mathbf{k}\sigma} \\
& + \sum_{\mathbf{k},\mathbf{k}',\mathbf{q}} \sum_{\sigma,\sigma'} (4\pi e^2/q^2)c^\dagger_{\mathbf{k}+\mathbf{q},\sigma} c^\dagger_{\mathbf{k}'-\mathbf{q},\sigma'} c_{\mathbf{k}'\sigma'} c_{\mathbf{k}\sigma} \\
& + \sum_{\mathbf{k},\mathbf{k}'} \sum_{\lambda,\sigma} g_{\mathbf{k}\mathbf{k}',\lambda}(b_{\mathbf{k}-\mathbf{k}',\lambda} + b^\dagger_{-\mathbf{k}+\mathbf{k}',\lambda})c^\dagger_{\mathbf{k}'\sigma} c_{\mathbf{k}\sigma}.
\end{aligned}
\tag{3.9}
$$

Here $\omega_{\mathbf{q}\lambda}$ is the bare phonon frequency obtained from $T_n + V_{nn}$. In a metal this does not give a good description of the phonons because of the important role played by the conduction electrons. For example, the bare longitudinal phonon frequency at $q = 0$ has a nonzero value equal to the ionic plasma frequency, $\omega_{0l} = (4\pi NZ^2e^2/M)^{1/2}$. The third term gives rise to the band structure—$V(\mathbf{q})$ is the Fourier transform of the pseudopotential, and $S(\mathbf{q})$ is the structure factor. The fourth term is the bare Coulomb interaction between the electrons. The electron–phonon coupling coefficient in the fifth term is given by the change, per unit displacement, of the total self-consistent crystal potential felt by an electron. In the plane-wave approximation this becomes

$$g_{\mathbf{k}\mathbf{k}',\lambda} = -i \left(\frac{N}{2M\omega_{\mathbf{k}'-\mathbf{k},\lambda}}\right)^{1/2} (\mathbf{k}' - \mathbf{k}) \cdot \hat{\boldsymbol{\epsilon}}_{\mathbf{k}'-\mathbf{k},\lambda} V(\mathbf{k}' - \mathbf{k}). \tag{3.10}$$

When the value of $\mathbf{k} - \mathbf{k}'$ extends beyond the first Brillouin zone it may be transformed back into the first zone by writing $\mathbf{k} - \mathbf{k}' = \mathbf{q} + \mathbf{G}$ where $\mathbf{G}$ is a reciprocal lattice vector. When $\mathbf{G} = 0$ we have only contributions from longitudinal phonons. However, as we shall see, the $\mathbf{G} \neq 0$, or "umklapp" contributions are equally important.

For simple metals, as opposed to transition metals, it is a good approximation to neglect the band terms. This still leaves us with the problem of a phonon field interacting with a self-coupled Fermion field. To solve this problem requires powerful field-theoretical methods which we shall mention only briefly later. The results, however, may be understood intuitively. For example, as a result of the bare Coulomb interaction an electron, or hole, can undergo a variety of scattering events and then return to its original state. We speak of these events as "dressing" the particle, making it a "quasiparticle." The first-order correction to the electron energy is the Hartree–Fock result. If one considers higher order corrections of this same type, one obtains the random phase approximation which has the same form as the Hartree–Fock result except that the bare Coulomb potential is screened by the dielectric "constant." The many-body theorist would also say that the potential should be "vertex corrected" and "renormalized." However, we need not concern ourselves with these. Thus, we think of the quasiparticles as behaving like independent particles interacting through screened interactions. Let us denote these Coulomb corrections to the particle's energy by $\Sigma_c(\mathbf{k}, \omega)$. The average of this correction over the Fermi surface is usually denoted by μ. The values of μ for three different models are shown in Fig. III.1.

The electron–electron interactions also dress the phonon frequencies by the real part of the dielectric constant,

$$\Omega_{q\lambda} = \frac{\omega_{q\lambda}}{\epsilon_1(\mathbf{q}, \omega_{q\lambda})^{1/2}} \simeq \frac{\omega_{q\lambda}}{\epsilon(\mathbf{q}, 0)^{1/2}} \qquad (3.11)$$

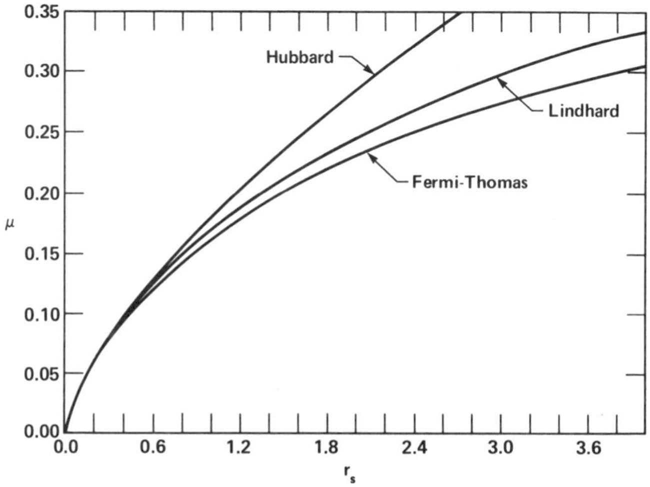

FIG. III.1.   Average Coulomb interaction, μ, as a function of the parameter r_s for three different approximations to the dielectric function (calculations by P. Allen).

This screening of the long range Coulomb force converts the longitudinal phonon into a true Goldstone boson,

$$\Omega_{ql} = \left(\frac{mZ^2}{3M}\right)^{1/2} v_F q. \tag{3.12}$$

The electron–electron interactions also modify the electron–phonon interaction. First of all, the phonon frequency in $g_{kk'\lambda}$ is replaced by its dressed value. Second, the pseudopotential appearing in $g_{kk'\lambda}$ is also screened. Let us denote this dressed electron–phonon coupling constant by $\bar{g}_{kk'\lambda}$.

Let us now consider the phonon contribution to the quasiparticle energy. The energy of a quasiparticle is defined as the energy required to add an extra particle to the system. Let us therefore add an electron to a state k above the Fermi sea and ask how the energy is modified by the virtual emission and reabsorption of a phonon. If this electron is scattered to a state k' and back, the change in energy to second order is $|\bar{g}_{kk'\lambda}|^2$ $(1 - f_{k'})/(\epsilon_k - \epsilon_{k'} - \Omega_{k-k'})$ where the Fermi factor $(1 - f_{k'})$ ensures that the state k' is not occupied. The presence of the extra electron at k, however, now blocks those "vacuum fluctuation" processes in which a particle in a state k' inside the Fermi sea scatters out to k and back again. These contributions must be subtracted from the total energy of the system and therefore contribute to the quasiparticle energy. The field-theoretic arguments alluded to above show that if the quasiparticle energy is denoted by ω the phonon contribution, $\Sigma_p(\omega, k)$, is obtained by replacing ϵ_k in this second-order result by ω itself. Thus,

$$\Sigma_p(\omega, k) = \sum_{k'\lambda} |\bar{g}_{kk'\lambda}|^2 \left(\frac{1 - f_{k'}}{\omega - \epsilon_{k'} - \Omega_{k-k'}} - \frac{f_{k'}}{\epsilon_{k'} - \omega - \Omega_{k'-k}}\right). \tag{3.13}$$

Converting the sum over k' to an integral assuming $\epsilon_{k'} = \hbar^2 k'^2/2m$, and noting that at zero temperature we may replace the numerators within the parentheses by 1 if we replace ϵ_k, by $-\epsilon_k$, in the second term, gives

$$\Sigma_p(\omega, k) = \sum_\lambda \int_0^\infty d\epsilon_{k'} \int \frac{d(\cos\theta_{k'})\, d\varphi_{k'}}{(2\pi)^3} mk' |\bar{g}_{kk'\lambda}|^2$$
$$\left(\frac{1}{\omega + \epsilon_{k'} + \Omega_{k-k'}} - \frac{1}{\epsilon_{k'} - \omega + \Omega_{k-k'}}\right). \tag{3.14}$$

As we shall see, the electron energies of interest will be those less than or of order of the Debye energy $k_B \theta_D$ and $k \sim k_F$. Thus the angular integral becomes an integral over the Fermi surface. The phonon density of states may be represented explicitly by introducing a delta function

$$\sum_p (\omega, \mathbf{k}) = \sum_\lambda \int_0^\infty d\epsilon_{\mathbf{k}'} \int_0^\infty d\Omega \int \frac{dk'^2}{(2\pi)^3 v_F} |\bar{g}_{\mathbf{k}\mathbf{k}'\lambda}|^2 \delta(\Omega - \Omega_{\mathbf{k}-\mathbf{k}'})$$

$$\left(\frac{1}{\omega + \epsilon_{\mathbf{k}'} + \Omega} - \frac{1}{\epsilon_{\mathbf{k}'} - \omega + \Omega} \right). \quad (3.15)$$

Finally, averaging over the Fermi surface gives

$$\sum_p (\omega) = \int_0^\infty d\epsilon_{\mathbf{k}'} \int_0^\infty d\Omega \, \alpha^2(\Omega) F(\Omega) \left(\frac{1}{\epsilon_{\mathbf{k}'} + \Omega + \omega} - \frac{1}{\epsilon_{\mathbf{k}'} + \Omega - \omega} \right)$$

$$(3.16)$$

where $F(\Omega)$ is the phonon density of states

$$F(\Omega) = \sum_\lambda \int [d^3k/(2\pi)^3] \delta(\Omega - \Omega_{\mathbf{k}\lambda}) \quad (3.17)$$

and $\alpha^2(\Omega)$ is the effective electron–phonon coupling constant,

$$\alpha^2(\Omega) F(\Omega) = \frac{\displaystyle\int_{S_F} d^2k \int_{S_F} [d^2k'/(2\pi)^3 v_F] \sum_\lambda |\bar{g}_{\mathbf{k}\mathbf{k}'\lambda}|^2 \delta(\Omega - \Omega_{\mathbf{k}-\mathbf{k}',\lambda})}{\displaystyle\int_{S_F} d^2k}$$

$$(3.18)$$

If this correction is expanded about the Fermi level we find

$$\sum_p (\omega) \simeq -\lambda\omega \quad (3.19)$$

where

$$\lambda \equiv -\frac{\partial \sum_p (\omega)}{\partial \omega} \Bigg|_{\omega=0} = 2 \int_0^\infty \frac{d\Omega \, \alpha^2(\Omega) F(\Omega)}{\Omega} \quad (3.20)$$

is a dimensionless measure of the strength of the electron–phonon interaction. Its value ranges from 0.1 to 1.5 in various metals. It should be clear from the context whether λ refers to the electron–phonon coupling constant or a polarization index. The quasiparticle energy, ω, is obtained by solving the equation $\omega = \epsilon_k + \Sigma(k, \omega)$. If the k-dependence of $\Sigma(k, \omega)$ is negligible, then, using Eq. (3.19), $\omega = \epsilon_k/Z_p$ where $Z_p = 1 + \lambda$. Thus, the mass of an electron on the Fermi surface is renormalized according to

$$m^*/m \equiv \epsilon_k/\omega = 1 + \lambda. \tag{3.21}$$

Note that $\bar{g}_q^2$ is inversely proportional to Ω, so that the first moment of $\alpha^2(\Omega)F(\Omega)$ is independent of the phonon frequencies,

$$\int d\Omega\Omega\alpha^2(\Omega)F(\Omega) = \frac{N(0)\hbar\langle I^2\rangle}{2M} \tag{3.22}$$

where $N(0)$ is the electronic density of states,

$$N(0) = \frac{V}{(2\pi)^3} \int \frac{dS_k}{\nabla_k\epsilon(k)} \tag{3.23}$$

and $\langle I^2\rangle$ is the average of the square of the electron–phonon matrix element,

$$\langle I^2\rangle = \frac{\int_{S_F} d^2k \int_{S_F} d^2k' |\langle\Psi_k|\hat{\epsilon}\cdot\nabla V|\Psi_{k'}\rangle|^2}{\int_{S_F} d^2k \int_{S_F} d^2k'}. \tag{3.24}$$

If we define the average of the square of the phonon frequency by

$$\langle\Omega^2\rangle = \frac{\int d\Omega\Omega\alpha^2(\Omega)F(\Omega)}{\int d\Omega \frac{\alpha^2(\Omega)F(\Omega)}{\Omega}}, \tag{3.25}$$

then λ may be written in the convenient form

$$\boxed{\lambda = \frac{N(0)\langle I^2\rangle}{M\langle\Omega^2\rangle}.} \tag{3.26}$$

We shall frequently refer to this expression for λ in our subsequent discussions.

As we mentioned above, these diagonal corrections to the electron energy do not lead to superconductivity. They have, however, introduced us to the important parameters μ and λ, which are used universally in the literature. Let us now consider the pairing interaction.

2. The BCS Interaction

As we have stressed through our discussion, a proper theory of superconductivity must treat the electron–phonon and Coulomb interactions simultaneously. It is instructive, however, to follow the original BCS argument and then indicate how a more elaborate treatment modifies this result.

The possibility of phonons leading to an attractive interaction between electrons was first noted by Fröhlich.[5] He considered an interacting electron–phonon system without Coulomb interactions:

$$\mathcal{H} = \mathcal{H}_e + \mathcal{H}_{ph} + \mathcal{H}_{e-ph} = \mathcal{H}_0 + \mathcal{H}_{e-ph}. \tag{3.27}$$

As we saw in Chapter I a quadratic Hamiltonian may be diagonalized by a canonical transformation. In the case of a nonlinear Hamiltonian a canonical transformation may enable us to express the Hamiltonian in a more convenient form. This is true of the Fröhlich Hamiltonian, (3.27). Consider the transformation

$$\mathcal{H}' = e^{-S} \mathcal{H} e^{S} = \mathcal{H}_0 + \mathcal{H}_{e-ph} + [\mathcal{H}_0, S] + [\mathcal{H}_{e-ph}, S]$$
$$+ \tfrac{1}{2}[[\mathcal{H}_0, S], S] + \cdots. \tag{3.28}$$

If we require that S be such that the electron–phonon coupling vanishes in lowest order we obtain an effective phonon-mediated interaction

$$\sum_{k,k',q} |\bar{g}_{k,k'}|^2 \frac{\hbar\Omega_q}{(\epsilon_k - \epsilon_{k'})^2 - (\hbar\Omega_q)^2} c^\dagger_{k'+q} c^\dagger_{k-q} c_k c_{k'}. \tag{3.29}$$

Problem III.1

The condition that determines S is

$$\mathcal{H}_{e-ph} + [\mathcal{H}_0, S] = 0.$$

Finding an S that satisfies this relation usually involves a certain amount of trial-and-error.

a. Show that

$$S = \sum_{k,k'} (Ab_{-q}{}^+ + Bb_q)\bar{g}_{kk'} c^\dagger_k c_{k'}$$

can be made to satisfy this condition by finding A and B.

b. Calculate the complete $\mathcal{H}'$ to order $|\bar{g}|^2$.

Notice that when $|\epsilon_k - \epsilon_{k'}| < \hbar\Omega_q$ the interaction (3.29) is attractive. In particular, its average over the Fermi surface is just $-\lambda$. This, incidently, is how Fröhlich obtained the isotope effect from the self-energy. BCS then restored the Coulomb interactions by simply adding μ. Thus the V in Eq. (1.42) is

$$V = \lambda - \mu. \tag{3.30}$$

The sum of these two interactions is illustrated in Fig. III.2a. The BCS approximation to the total interaction is shown in Fig. III.2b.

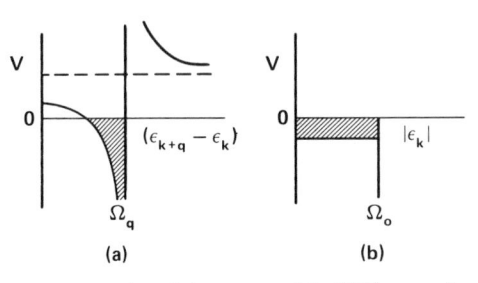

FIG. III.2.   Schematic illustration of the nature of the BCS approximation (b) to the effective electron–electron interaction (a). The effective interaction consists of the phonon interaction given by Eq. (3.29) plus a constant Coulomb interaction. Notice that the attractive region (shaded region) depends upon the relative energies $\epsilon_{k+q} - \epsilon_k$. Ω_0 is the average phonon frequency.

3. Dielectric Formulation

Both the attractive and repulsive interactions may be obtained directly from a dielectric formulation. This is readily seen for the so-called "jellium" model. In this model the ions are assumed to be smeared out into a plasma. The dielectric function associated with the ions will then have a Lindhard form. The screening length will be a Debye–Hückel length which is much longer than the wavelengths of interest. The characteristic frequency is the ionic plasma frequency ω_{ol} introduced above. Thus,

$$\epsilon(q, \omega)_{\text{ions}} \simeq 1 - (\omega_{ol}^2/\omega^2).$$

For the electrons, on the other hand, the screening length is the Fermi–Thomas length, k_s^{-1}, which is very short, while the characteristic frequency is the electronic plasma frequency, ω_p, which is beyond our region of interest. Therefore,

$$\epsilon(q, \omega)_{\text{electrons}} \simeq 1 + (k_s^2/q^2).$$

Adding these and subtracting $\epsilon_\infty = 1$ to insure the proper behavior at high frequencies, the total dielectric function becomes

$$\epsilon(q, \omega) \simeq 1 + (k_s^2/q^2) - (\omega_{ol}^2/\omega^2) \qquad (3.31)$$

In the absence of external charge the Maxwell equation $\epsilon k \cdot E = 0$ is consistent with longitudinal modes, i.e., $k \cdot E \neq 0$, only when ϵ is zero. This occurs at the frequency

$$\Omega_q^2 = \omega_{ol}^2 \, q^2/(q^2 + k_s^2) \qquad (3.32)$$

which, in the long wavelength limit, reduces to Eq. (3.12). The total electron–electron interaction may then be written

$$V(q, \omega) = \frac{4\pi e^2}{q^2 \, \epsilon(q, \omega)} = \frac{4\pi e^2}{q^2 + k_s^2} + \frac{4\pi e^2 \, \Omega_q^2}{(q^2 + k_s^2)(\omega^2 - \Omega_q^2)}. \quad (3.33)$$

When averaged over the Fermi surface these two terms just give μ and λ, respectively. The result for λ may also be obtained from Eq. (3.29). Bardeen and Pines[6] have derived the bare electron–phonon coupling constant:

$$g_{kk'} = -i\frac{4\pi e^2}{q} \left(\frac{\hbar Z^2 N}{2M\omega_{ol}}\right)^{1/2}. \quad (3.34)$$

As we mentioned above, electron–electron interactions are included by replacing ω_{ol} by Ω_q and screening the pseudopotential which means dividing $g_{kk'}$ by $\epsilon(q, \omega)_{\text{electrons}}$. The result is

$$|g_{kk'}|^2 = 4\pi e^2 \, \Omega_q/(q^2 + k_s^2). \quad (3.35)$$

Substituting this into Eq. (3.29) and averaging over the Fermi surface gives the same value for λ obtained above.

4. Retardation

We now know that the Coulomb contribution to the pairing interaction is not simply μ. The point is that the phonon-mediated interaction is "retarded." That is, it takes a finite time, of the order of a phonon frequency, for the interaction to be effective. The Coulomb interaction, on the other hand, is essentially instantaneous. We can see how retardation renormalizes the Coulomb interaction by an argument due to de Gennes.[8] If, in Eq. (3.1), we distinguish between the attractive phonon interaction and the repulsive Coulomb interaction the gap equation becomes

$$\Delta(\omega) = -N(E_F) \int d\omega' \, V_{\text{ph}}\Delta(\omega') \frac{\tanh(\beta\hbar\omega'/2)}{2\omega'}$$

$$- N(E_F) \int d\omega' \, V_c\Delta(\omega') \frac{\tanh(\beta\hbar\omega'/2)}{2\omega'}. \quad (3.36)$$

The integrals over ω' run from 0 to infinity. It is convenient to define the origin at the Fermi level. From Eq. (3.10) we see that V_{ph} will be a function of $\omega - \omega'$. Let us assume that when $|\omega|$, $|\omega'| < \omega_D$ the average gap

[8] P. G. de Gennes, *Superconductivity of Metals and Alloys*, pp. 125–127. Benjamin, New York, 1966.

has the value Δ_1. Then

$$\Delta_1 \simeq - N(0) \, V_{ph} \Delta_1 \ln \left(\frac{2\gamma}{\pi} \frac{\hbar\omega_D}{k_B T_c} \right) + \Delta_2 \tag{3.37}$$

where Δ_2 is the Coulomb contribution to Eq. (3.36). If we take the cutoff frequency for the Coulomb interaction to be ω_p, then

$$\Delta_2 = N(0) \int_{-\omega_F}^{-\omega_D} d\omega' \, V_c\Delta(\omega') \frac{\tanh (\beta\hbar\omega'/2)}{2\omega'}$$

$$+ N(0) \int_{-\omega_D}^{\omega_D} d\omega' \, V_c\Delta(\omega') \frac{\tanh (\beta\hbar\omega'/2)}{2\omega'}$$

$$+ N(0) \int_{\omega_D}^{\omega_F} d\omega' \, V_c\Delta(\omega') \frac{\tanh (\beta\hbar\omega'/2)}{2\omega'}. \tag{3.38}$$

For frequencies beyond ω_D we expect the gap to be approximately given by Δ_2. Therefore, we replace $\Delta(\omega')$ in the first and third integrals by Δ_2 and $\Delta(\omega')$ in the second integral by Δ_1. Thus

$$\Delta_2 = N(0) \, V_c \left[\Delta_2 \ln \left(\frac{\omega_D}{\omega_F} \right) + \Delta_1 \ln \left(\frac{2\gamma}{\pi} \frac{\hbar\omega_D}{k_B T_c} \right) \right]. \tag{3.39}$$

In order that Eqs. (3.37) and (3.39) be consistent, we must have

$$1 = - N(0) \ln \left(\frac{2\gamma}{\pi} \frac{\hbar\omega_D}{k_B T_c} \right) \left[V_{ph} - \frac{V_c}{1 + N(0) \, V_c \ln (\omega_F/\omega_D)} \right]. \tag{3.40}$$

The second term within the brackets is the renormalized Coulomb interaction. This suggests we replace μ in Eq. (3.30) by

$$\mu^* = \frac{\mu}{1 + \mu \ln (\omega_F/\omega_D)} \tag{3.41}$$

as derived by Morel and Anderson.[9]

Once the energy gap is known as a function of temperature various thermodynamic quantities may be determined. The critical-field curve, $H_c(T)$, is an example. Near $T = 0$, $H_c(T)/H_c(0)$ decreases as T^2 while near T_c it decreases linearly with T. At intermediate values the variation may be obtained numerically. It is customary to plot the critical field in terms of its *deviation* from a parabolic relation. This is indicated by the dashed curve in Fig. III.3.[10] Also shown are data for several superconductors. The theory works well for aluminum, but deviates markedly for lead and

[9] P. Morel and P. W. Anderson, *Phys. Rev.* **125,** 1263 (1962).

[10] J. C. Swihart, D. J. Scalapino, and Y. Wada, *Phys. Rev. Lett.* **14,** 106 (1965).

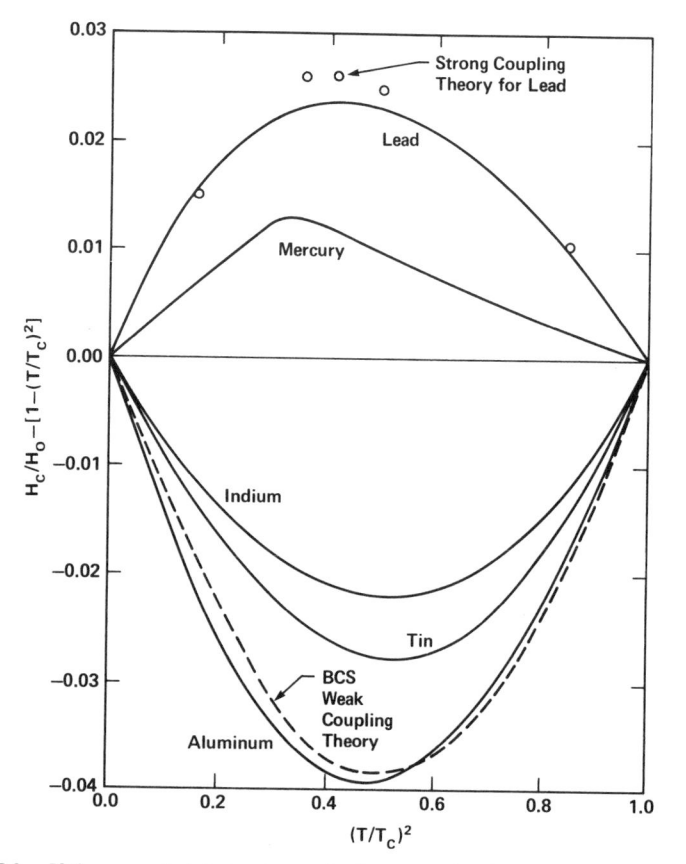

FIG. III.3. If the normal state heat capacity is assumed to be linear in temperature and that of the superconducting state cubic then the critical field $H_c(T)/H_c(0) = 1 - (T/T_c)^2$. It is therefore convenient to plot, as shown above, the *deviation* of the reduced critical field from a parabolic curve as a function of the square of the reduced temperature. The BCS theory represents the "law of corresponding states" in this case. After Swihart *et al.*[10]

mercury. The origin of this discrepancy lies in the fact that electron–phonon interaction in these metals is large. This is known independently from the large phonon enhancement of the effective electron mass at the Fermi surface. When the electron–phonon interaction becomes large the quasiparticle picture on which the BCS weak-coupling theory is based breaks down. In particular, the pairing interaction involves virtually excited quasiparticle states whose level width is of the order of their energy. The extension of the BCS theory to the *strong coupling* regime is a tour de force in many-body physics.

5. Strong-Coupling Theory

Since high T_c superconductors turn out to fall in the strong-coupling regime, we must extend the theory to this regime. The result will be that the BCS gap equation, Eq. (3.2), expands into a set of coupled integral equations, the so-called Eliashberg equations. The reader who has not been exposed to the language of many-body physics may wish to skip to the result, Eq. (3.42). In the next few pages we sketch what is involved in deriving these equations. For details we refer the reader to the paper by Scalapino, Schrieffer, and Wilkins.[11]

Many-body theory utilizes the so-called Green's function, or propagator. We actually made contact with this function in Chapter I when we discussed the superconducting order parameter. The single-particle Green's function is defined as

$$G(\mathbf{r}, t; \mathbf{r}'t') = -\langle T\psi(\mathbf{r}, t)\psi^\dagger(\mathbf{r}', t')\rangle$$

where $\langle \cdots \rangle$ represents a thermal average, $\psi(rt)$ is a Heisenberg field operator, i.e.,

$$\psi(\mathbf{r}t) = e^{i(\mathcal{H}-\mu N)t/\hbar}\psi(\mathbf{r})e^{-i(\mathcal{H}-\mu N)t/\hbar}$$

and T is the time-ordering operator that orders the operators within the braces such that the operators with earlier times are placed to the right. The field could be that of electrons, phonons, or other particles. Since the Green's function is an expectation value of field operators, it is simply a function of the coordinates rt and $\mathbf{r}'t'$ and we might suspect that we have lost detailed information about the ground state. However, a knowledge of the Green's function enables us to calculate the expectation values of any single-particle operators in the ground state, the ground-state energy, and the excitation spectrum—just about anything one would wish to know!

If the Hamiltonian is time independent and the system is homogeneous, then the Green's function will only be a function of the relative coordinates $\mathbf{r} - \mathbf{r}'$ and $t - t'$. In this case it is convenient to work with the Fourier transformed Green's function. It turns out that the Green's function is antiperiodic in the imaginary relative time variable with a period $\beta \equiv \hbar/k_B T$. Therefore, the Fourier transformed Green's function has the form $G(\mathbf{k}, i\omega_n)$ where $\omega_n = (2n + 1)\pi/\hbar\beta$ for electrons.

Another advantage of the Green's function is that it is readily amenable to perturbation techniques that are intuitively appealing. Since $-G$ is the probability amplitude for finding, say, an electron at $r't'$ when it was ini-

[11] D. J. Scalapino, J. R. Schrieffer, and J. W. Wilkins, *Phys. Rev.* **148**, 263 (1966).

tially at **r**t it is represented by a straight line. In the absence of interactions this is a simple straight line:

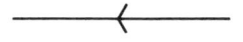

The effect of *electron–electron* interactions is to introduce scattering with other particles. If we represent the interaction between two electrons by a dashed line then the basic interaction has the form

$$\mathbf{k'+g} \xleftarrow{\qquad} \xleftarrow{\qquad} \mathbf{k'}$$
$$\qquad \wedge \; v(q)$$
$$\mathbf{k'-g} \xleftarrow{\qquad} \xleftarrow{\qquad} \mathbf{k}$$

In lowest order there are only two ways of closing two of these lines leaving a "dressed" electron line:

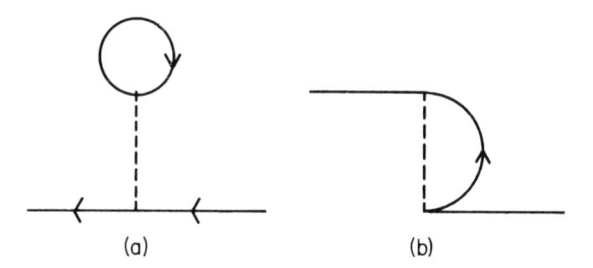

(a)   (b)

In (a) the particle forward scatters against the background particles but leaves them in their original state. In (b) the particle exchange scatters with the background. What makes these diagrams so powerful is that there is a one-to-one correspondence between terms in the perturbation expansion for the Green's function and such diagrams. Thus, one can build the perturbation series for the Green's function by first drawing all the diagrams in which a particle enters from the right, undergoes all topologically distinct interactions, then exits at the left. One then applies certain rules developed by Feynman for evaluating these diagrams. If we represent the interacting Green's function by a double line then the diagramatic series would have the form

The "blob" defines what is called the *self-energy* correction to the Green's function. It is convenient to introduce a *proper* self-energy, Σ^*, which is a self-energy insertion that cannot be separated by cutting a single Green's function line (as could be done, for example, in the last term in the first line above). The "equation" for the Green's function then becomes

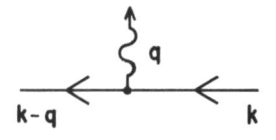

When combined with the Feynman rules this becomes an integral equation, known as Dyson's equation, for the Green's function,

$$G(\mathbf{k},\, i\omega_n) = G_0(\mathbf{k},\, i\omega_n) + G_0(\mathbf{k},\, i\omega_n)\, \Sigma^*\, (\mathbf{k},\, i\omega_n)G(\mathbf{k},\, i\omega_n).$$

Let us now consider the effect of the electron–phonon interaction. For this purpose it is convenient to introduce a phonon field operator.

$$\phi(\mathbf{r}) = \sum_q \left(\frac{\hbar\Omega_q}{2V}\right)^{1/2} (b_q + b^\dagger_{-q})e^{i\mathbf{q}\cdot\mathbf{r}}$$

which enables us to define the phonon Green's function. We shall denote its Fourier transform as $D(\mathbf{q}, i\nu_n)$. It also satisfies a Dyson equation:

$$D(q, i\nu_n) = D_0(q, i\nu_n) + D_0(q, i\nu_n)\, \Pi^*\, (q, i\nu_n)D(q, i\nu_n)$$

where Π^* is the proper phonon self-energy. Let us represent the phonon propagator by a wavy line. Since the electron–phonon interaction involves one phonon amplitude and two electron amplitudes the basic interaction is a *vertex* of the form

with which we associate the coupling constant g_q analogous to the electron–electron interaction $V(\mathbf{q})$. Therefore, the lowest order correction to the electron proper self-energy is

Similarly, the lowest order correction to the phonon proper self-energy is

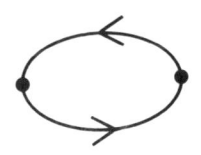

The higher order diagrams have the effect of converting the propagators to their *exact* counterparts and screening one (to avoid double counting) of the vertices. Thus, the proper self-energies become

$$\Sigma^* =$$

$$\Pi^* =$$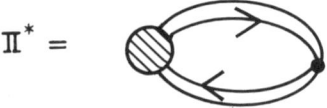

which correspond to a set of nonlinear coupled integral equations. So far, the process appears straightforward: given the interaction, the many-body theorist draws the appropriate diagrams and applies the Feynman rules to convert these to integral equations. He then, hopefully, can solve these equations, and calculate desired physical properties.

The solution of these equations, however, is very difficult. A significant simplification was introduced by Migdal[12] in 1959 when he showed that *phonon* corrections to the vertex are of order $(m/M)^{1/2}$ where m/M is the ratio of the electron mass to ion mass. Consequently, one can use the bare electron–phonon interaction in the equations above. This means that one can obtain accurate expressions for the self-energies in the normal state even when the coupling is strong. From the electron self-energy we obtain the phonon renormalization factor, Z_p, already introduced above Eq. (3.21).

In 1960 both Nambu[13] and Eliashberg[14] extended Migdal's treatment to superconductors. As we saw in Chapter I the order parameter in a super-conductor involves the pair correlation $\langle c_{k\uparrow} c_{-k\downarrow} \rangle$. Nambu, therefore, suggested that we take the electron field operator to be the two-

[12] A. B. Migdal, *Sov. Phys.—JETP (Engl. Transl.)* **7**, 996 (1958).
[13] Y. Nambu, *Phys. Rev.* **117**, 648 (1960).
[14] G. M. Eliashberg, *Sov. Phys.—JETP (Engl. Transl.)* **11**, 696 (1960).

component quantity,

$$\psi_k = \begin{pmatrix} c_{k\uparrow} \\ c^{\dagger}_{-k\downarrow} \end{pmatrix}.$$

When the Hamiltonian is expressed in terms of these field operators it involves the 2×2 Pauli matrices,

$$\sigma_0 = \begin{pmatrix} 1 & 0 \\ 0 & 1 \end{pmatrix}, \quad \sigma_1 = \begin{pmatrix} 0 & 1 \\ 1 & 0 \end{pmatrix}, \quad \sigma_2 = \begin{pmatrix} 0 & -i \\ i & 0 \end{pmatrix}, \quad \sigma_3 = \begin{pmatrix} 1 & 0 \\ 0 & -1 \end{pmatrix}.$$

For example, the one-electron energies become $\Sigma_k \epsilon_k \psi_k^{\dagger} \sigma_3 \psi_k$. The electron–phonon and the electron–electron terms also involve only σ_3. The matrices σ_1, σ_2, and σ_3 transform like the x-, y-, and z-components of a spin. Thus, in this pseudospin representation, the Hamiltonian is invariant with respect to rotations in spin space about the z-axis. The electron Green's function becomes a 2×2 matrix, the diagonal elements G_{11} and G_{22} being the Green's functions introduced above for the up-spin electrons and down-spin holes, while G_{12} and G_{21} are related to the order parameter. The noninteracting Green's function is

$$G_o(\mathbf{k}, i\omega_n) = [i\omega_n \sigma_0 - \epsilon_k \sigma_3]^{-1}.$$

Nambu showed that the many-body theory we outlined above could also be applied to these spinor fields. This leads to proper self-energies that are also 2×2 matrices. The expression for $\Sigma(\mathbf{k}, i\omega_n)$ may be analytically continued with respect to $i\omega_n$ to the real axis by replacing $i\omega_n$ by $\omega + i\delta$. The self-energy then becomes a function of the continuous (real) variable $\mathbf{k}$ and ω. It is convenient to express this self-energy in the general form

$$\Sigma(\mathbf{k}, \omega) = [1 - Z(\mathbf{k}, \omega)]\omega\sigma_0 + Z(\mathbf{k}, \omega)\Delta(\mathbf{k}, \omega)\sigma_1 + \delta\epsilon(\mathbf{k})\sigma_3 \quad (3.42)$$

which defines the renormalization parameter $Z(\mathbf{k}, \omega)$, and the gap parameter $\Delta(\mathbf{k}, \omega)$, both of which involve the electron–phonon coupling constant λ. The phase of the gap parameter has been chosen such as to exclude the term proportional to σ_2. Since σ_1 transforms like σ_x we see that the existence of a gap "breaks the symmetry" of the Hamiltonian. Equation (3.42) represents a set of four coupled integral equations, known as the *Eliashberg equations*. If an average is made over the Fermi surface, the zero temperature equation for the renormalization parameter, for example, is

$$[1 - Z(\omega)]\omega = \int_0^\infty d\omega' \mathrm{Re} \left\{ \frac{\omega'}{[(\omega')^2 - \Delta(\omega')^2]^{1/2}} \right\} K_-(\omega', \omega) \quad (3.42a)$$

where

$$K_{\pm}(\omega', \omega) = \int_0^\infty d\Omega \alpha^2(\Omega) F(\Omega) \left[\frac{1}{\omega' + \omega + \Omega + i\delta} \pm \frac{1}{\omega' - \omega + \Omega + i\delta} \right].$$

Notice that this has the same form as Eq. (3.16), except that $\epsilon_{k'}$ has been written as ω' and a superconducting density of states factor has been introduced to take into account the modification of the ground state. The equation for the gap parameter is

$$\Delta(\omega) = \frac{1}{Z(\omega)} \int_0^\infty d\omega' \mathrm{Re} \left\{ \frac{\Delta(\omega')}{[\omega'^2 - \Delta(\omega')^2]^{1/2}} \right\} K_+(\omega', \omega)$$

$$- \frac{\mu^*}{Z(\omega)} \int_0^\infty d\omega' \mathrm{Re} \left\{ \frac{\Delta(\omega')}{[\omega'^2 - \Delta(\omega')^2]^{1/2}} \right\}. \quad (3.42b)$$

In the strong coupling theory the frequency dependence of $\Delta(\omega)$ is explicitly taken into account, whereas in weak coupling the change in $\Delta(\omega)$ is small compared with $k_B T_c$ so that $\Delta(\omega)$ is taken as $\Delta(\omega = 0) \equiv \Delta_0$.

In Figure III.4 we illustrate how structure in the phonon density of states manifests itself in the gap parameter. The BCS gap would have the

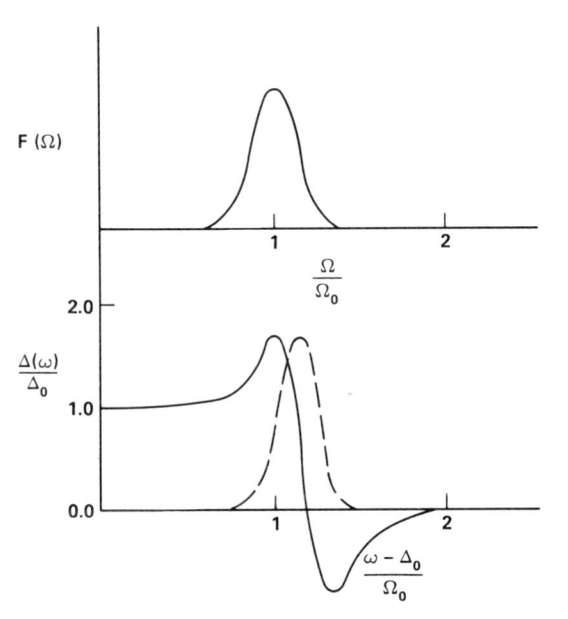

FIG. III.4.   Illustration of how structure in the phonon density of states (top) is reflected in the gap (bottom). The solid line in the bottom plot is the real part of the gap while the dashed line is the imaginary part.

value Δ_0 up to $\omega = \Delta_0 + \Omega_0$ where Ω_0 is a characteristic phonon frequency, and zero above. The difference in the vicinity of $\omega = \Delta_0 + \Omega_0$ is responsible for the anomalous behavior of the critical fields in mercury and lead shown in Fig. III.3. This is illustrated by the calculated points[10] for lead.

The real test of the Eliashberg equations, however, came with the tunneling experiments of Rowell, Anderson, and Thomas.[15] In these experiments they measured the conductance, dI/dV, versus voltage for a Pb–Pb tunnel junction at 1.3 K. The ratio of the superconducting to normal state conductances is equal to the ratio of their densities of states. This ratio is just the coefficient of K in the integrand of Eq. (3.42a).

$$\frac{(dI/dV)_S}{(dI/dV)_N} = \frac{N_T(V)}{N(0)} = \text{Re} \left\{ \frac{|\omega|}{[\omega^2 - \Delta(\omega)^2]^{1/2}} \right\}. \tag{3.43}$$

Here, the subscript T indicates the "tunneling" density of states. Using a simple representation for the phonon spectrum of lead, Schrieffer, Scalapino, and Wilkins[16] calculated $\Delta(\omega)$ and $Z(\omega)$ numerically. Their calculated density of states is compared with the tunneling data of Rowell *et al.* in Fig. III.5.[11] The remarkably good agreement, considering the simplicity of the model, is a tribute to the power of many-body theory. It is interesting to note that the temperature dependence of the gap at the gap edge, $\text{Re}\Delta(\omega = \Delta_0, T)$, follows the BCS curve very closely. That is, the structure near $\omega = \Delta_0 + \Omega_0$, which is very pronounced in tunneling data, does not affect the gap as it is usually defined.

The fact that tunneling data are very sensitive to the phonon density of states plus the fact that the theory is very accurate suggests that one combine these to actually determine the phonon spectrum $\alpha^2 F$ and the Coulomb interaction parameter μ^*. McMillan[17] developed a computer program for doing just this. In Fig. III.6[17,18] we compare the phonon spectrum of Pb obtained from tunneling with that obtained from inelastic neutron scattering. Notice that the transverse phonons are very much in evidence.

Once one has determined $\alpha^2(\Omega)F(\Omega)$ it can be used to calculate the electron–phonon coupling constant λ defined above.

In 1968, McMillan[19] numerically solved the Eliashberg equations for various cases, and constructed an approximate equation relating the su-

[15] J. M. Rowell, P. W. Anderson, and D. E. Thomas, *Phys. Rev. Lett.* **10**, 334 (1963).

[16] J. R. Schrieffer, D. J. Scalapino, and J. W. Wilkins, *Phys. Rev. Lett.* **10**, 336 (1963).

[17] W. L. McMillan and J. M. Rowell, *Phys. Rev. Lett.* **14**, 108 (1965).

[18] R. Stedman, L. Almquist, and G. Nilsson, *Phys. Rev.* **162**, 549 (1967).

[19] W. L. McMillan, *Phys. Rev.* **167**, 331 (1968).

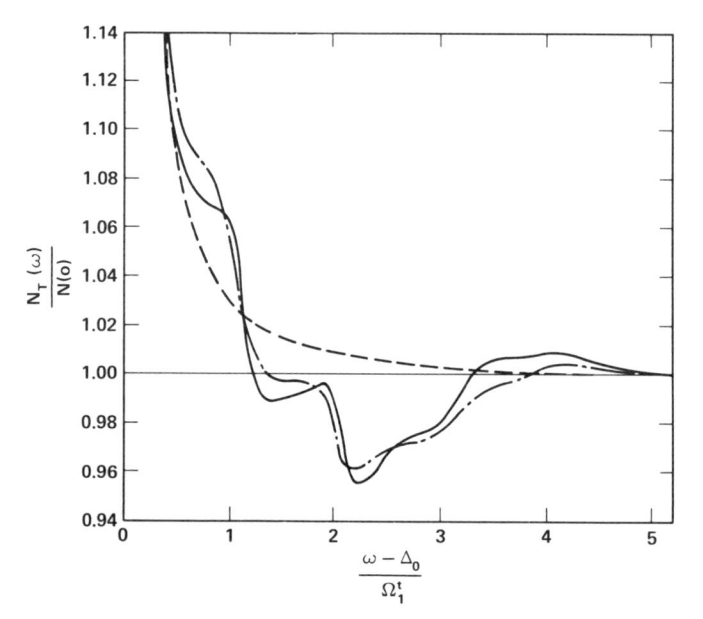

FIG. III.5.   The dot-dashed curve is the experimental ratio of the differential conductance of Pb in the superconducting state to that in the normal state. The solid curve is a theoretical tunneling density of states curve obtained by modeling the phonon density of states by two transverse modes and one longitudinal mode. The dashed curve is just the BCS density of states. After Scalapino *et al.*[11]

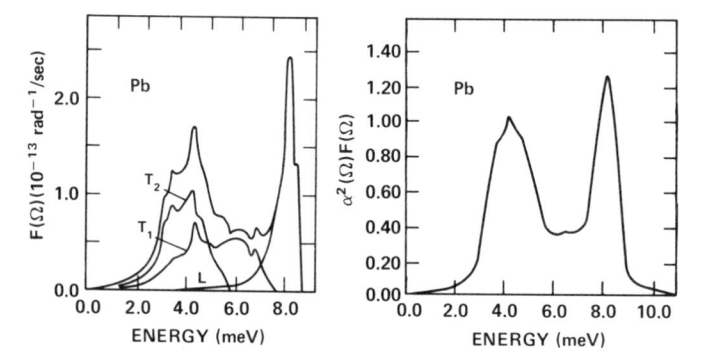

FIG. III.6.   Comparison of the phonon density of states of Pb as obtained from (a) neutron scattering (after Stedman *et al.*[18]) with that obtained from (b) electron tunneling (after McMillan and Rowell[17]).

perconducting transition temperature T_c to the parameters we have already introduced:

$$T_c = \frac{\omega_D}{1.45} \exp\left\{-\frac{1.04(1 + \lambda)}{\lambda - \mu^*(1 + 0.62\lambda)}\right\}. \qquad (3.44)$$

The prefactor, ω_D, is not a good enough measure of the effective phonon frequency which determines T_c. Dynes[20] has shown that the use of $\langle\Omega\rangle/1.2$ in place of $\omega_D/1.45$ improves the agreement with experiment. Here $\langle\Omega\rangle$ is the first moment of the normalized weight function $g(\Omega) = (2/\lambda\Omega)\alpha^2 F$. A still better prefactor involves the logarithmic average denoted $\Omega_{\log}$. This equation clearly reveals the relative roles of the electron–phonon and electron–electron interactions on the superconducting transition temperature.

As we mentioned at the beginning of this chapter the clue to the microscopic origin of superconductivity came from the isotope effect. This dependence enters T_c directly through the factor $\langle\Omega\rangle$ and indirectly through the dependence of μ^* on $\langle\Omega\rangle$. McMillan finds that $T_c \sim M^{-\alpha}$ where the isotope shift coefficient is

$$\alpha = \frac{1}{2}\left[1 - \left(\mu^* \ln\frac{\langle\Omega\rangle}{1.20T_c}\right)^2 \frac{1 + 0.62\lambda}{1 + \lambda}\right]. \qquad (3.45)$$

If we neglect the "strong-coupling" correction $(1 + 0.62\lambda)/(1 + \lambda)$ in this equation, then we can determine the Coulomb parameter from the experimental values of α, T_c, and $\langle\Omega\rangle$. If we then use this value of μ^* in the McMillan equation (3.44) we can determine the electron–phonon parameter λ.

As we saw above, the electron–phonon interaction renormalizes the electron mass averaged over Fermi surface. Since the density of states and, hence, the electronic heat capacity coefficient, γ, is proportional to this mass, it is also renormalized. Therefore, if we know λ we can find the unrenormalized (by phonons) or "band-structure" density of states, $N(0)$, from the experimental value of γ:

$$N(0) = 3\gamma/2\pi^2 k_B^2(1 + \lambda). \qquad (3.46)$$

It is interesting to compute this density of states for a series of alloys with the same crystal structure. This has been done by McMillan for the bcc alloy system Ta-W. Since the band structures of Ta and W are similar, a rigid band model should apply, in which case the effect of alloying is merely to increase the number of electrons from 5 per atom to 6 per atom.

[20] R. C. Dynes, *Solid State Commun.* **10**, 615 (1972).

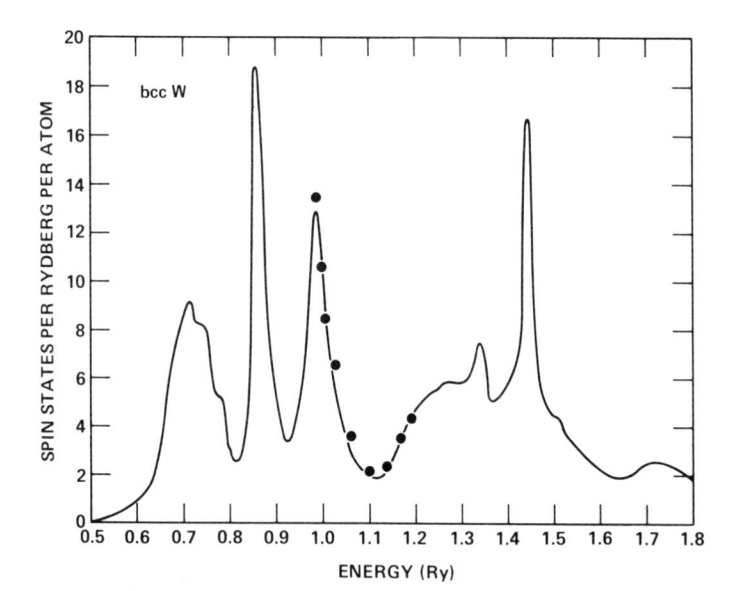

FIG. III.7.   Band density of states for several bcc 5d alloys from tunneling data (dots) superimposed on Mattheiss' theoretical band structure density of states. After McMillan.[19]

In Fig. III.7 we compare the density of states obtained this way with a theoretical calculation for W by Mattheiss[21] using the relativistic APW method. The energy at which to place these points was determined from the theoretical electron/atom ratio. This agreement not only strengthens our confidence in the McMillan equation but in the APW calculation as well.

In view of these successful calculations there is little doubt that the mechanism underlying the superconductors known today involves the electron–phonon interaction.

Problem III.2

In Pb the transition from the superconducting state to the normal state occurs at a temperature $k_BT_c/2\Delta(0) = 1/(4.3)$ which corresponds to a lower value for a given $\Delta(0)$ than that predicted by the weak-coupling BCS result derived below Eq. (3.5). This is due to a broadening of the quasiparticle energies by an amount $\hbar/\tau$, where $\hbar/\tau = 2\lambda\pi k_BT$ is the electron–phonon contribution to the electron scattering time. Since this damping rate is greater at higher temperature, it reduces the transition temperature much more than the energy gap.

a. If the Debye temperature of Pb could somehow be increased relative to the electron–phonon coupling, what would be the "weak coupling" transition temperature?

[21] L. F. Mattheiss, *Phys. Rev.* **139,** A1893 (1965).

b. Suppose the Fermi level lies at (or near) a peak in the density of states of width W. Since τ^{-1} increases strongly with temperature the energy smearing results in $\lambda(T_c)$ being smaller than $\lambda(T = 0)$. Estimate the limit on T_c imposed by this smearing.

c. T_c is also affected by a redistribution of quasiparticles. Explain why T_c increases when the superconductor is exposed to microwave radiation ($h\nu < 2\Delta(0)$).

6. Calculation of λ

From Eq. (3.10) we see that λ is directly related to the Fourier transform of the electron–ion potential. It can be shown that this potential is, in fact, the same potential that determines the Fermi surface. That is, screening and Coulomb renormalization effects enter both potentials in the same way. Therefore, it should be possible to obtain an empirical pseudopotential from Fermi surface data which can then be used to calculate λ. Such a program has been carried out by Allen and Cohen[22] for 16 "simple" metals plus Ca, Sr, and Ba. By "simple" is meant a metal whose conduction electrons may be treated as nearly free. The results of such a pseudopotential calculation are shown in Fig. III.8, where they are plotted as a function of the empirical values obtained from McMillan's formula. The correlation is reasonably good.

Transition metals are characterized by a d-band arising from the strong scattering of electrons having energies in the vicinity of the atomic d-states. In such a case the weak pseudopotential approximation is not valid. Therefore, to calculate λ one must first carry out a full band calculation to determine the Fermi surface. However, since this scattering is strong it means that the electrons spend a relatively long time in the vicinity of the scattering center. Therefore, we might suspect that the electron–phonon interaction is dominated by the local potential. The importance of this local environment in determining λ in transition metals was first emphasized by Hopfield[23] and exploited by Gaspari and Gyorffy.[24]

We begin by writing the electron–phonon matrix element, Eq. (3.24), as

$$\langle I^2 \rangle = \frac{1}{N(0)^2} \int d^3r \int d^3r' \nabla V(r) \cdot \nabla' V(r') \rho(r, r'; E_F) \rho(r', r; E_F) \quad (3.47)$$

where $\rho(r, r'; E_F)$ is the Fermi energy density matrix

$$\rho(r, r'; E_F) = \frac{V_0}{(2\pi)^3} \int d^3k \delta(E_k - E_F) \psi_k^*(r) \psi_k(r') \quad (3.48)$$

[22] P. B. Allen and M. L. Cohen, *Phys. Rev.* **187**, 525 (1969).
[23] J. J. Hopfield, *Phys. Rev.* **186**, 443 (1969).
[24] G. D. Gaspari and B. L. Gyorffy, *Phys. Rev. Lett.* **28**, 801 (1972).

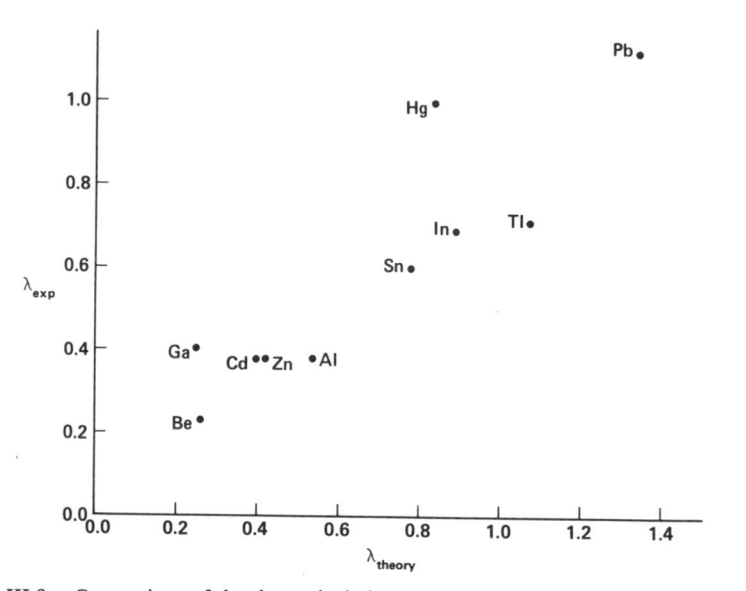

FIG. III.8. Comparison of the theoretical electron–phonon coupling constants obtained from pseudopotentials with those obtained empirically using McMillan's formula.

and V_0 is the volume of a Wigner–Seitz cell. The local aspect is introduced by expanding the Fermi energy density matrix in terms of radial wave functions $R_l(r)$ and cubic harmonics $K_{l\mu}^\Gamma(\hat{r})$, where Γ labels the irreducible representation. The coefficients in this expansion, $T_{ll'}^\Gamma$, also characterizes the decomposition of the density of states, i.e.,

$$N(0) = \int_{V_0} d^3r \rho(r, r; E_F) = \sum_{\Gamma l, l'} T_{ll'}^\Gamma \rho_{ll'}^\Gamma \qquad (3.49)$$

where

$$\rho_{ll'}^\Gamma = \int_{V_0} d^3r R_l(r) R_{l'}(r) \sum_\mu K_{l\mu}^\Gamma(\hat{r}) K_{l'\mu}^\Gamma(\hat{r}). \qquad (3.50)$$

When this expansion is introduced into Eq. (3.47) we obtain radial integrals of the form

$$V_{l,l+1}' = \int r^2 dr R_l(r) R_{l+1}(r) dV/dr. \qquad (3.51)$$

Notice that these lattice-vibration-induced potential fluctuations require $\Delta l = \pm 1$. Gaspari and Gyorffy employ the *rigid muffin-tin approximation* to evaluate these matrix elements. In this approximation one assumes that the total self-consistent change in crystal potential when one nucleus is moved is given by the gradient of the usual band theory potential. That is, each ion carries around its own muffin-tin sphere which scatters the electron. For r greater than the muffin-tin radius the radial wave function may be expressed in terms of phase shifts δ_l according to

$$R_l(r) = \cos \delta_l j_l(r) - \sin \delta_l n_l(r) \qquad (3.52)$$

where j_l and n_l are the spherical Bessel and Neuman function, respectively. Gaspari and Gyorffy show that $V'_{l,l+1}$ then takes the surprisingly simple form

$$V'_{l,l+1} = \sin(\delta_l - \delta_{l+1}). \qquad (3.53)$$

Butler[25] has calculated the coefficients $T^{\Gamma}_{ll'}$ and the phase shifts δ_l and the resulting electron–phonon matrix element $\langle I^2 \rangle$ for eleven 4d transition metals. This involves a fairly standard band calculation. His values are compared with the empirical values in Fig. III.9. The agreement is quite good.

One of the interesting results of Butler's calculation is that the fraction of the total density of states that is f-like at the Fermi energy is larger than one would have expected simply from scattering off a single potential. This is found to be the case for all the elements studied, and is due to the fact that the density of states within a given Wigner–Seitz cell is based on an expansion about the center of that cell. Thus a d orbital "tail" from a neighboring cell will also contribute to the f-like density. This enhanced f-like density of states is important, for, as we noted above, the angular momenta entering the matrix element of the potential gradient must differ by one. In particular, Butler finds that the d ↔ f scattering is more important than the d ↔ p scattering. Thus, the matrix element $\langle I^2 \rangle$ is proportional to the $l = 3$ Fermi energy state density within the Wigner–Seitz cell.

Varma et al.[26] have also employed a localized description to calculate the squared electron–phonon coupling constant, $\langle I^2 \rangle$. They use a tight binding scheme based on a nine orbital s-p-d basis. A two-center approximation is used including the overlap responsible for nonorthogonality. The various parameters entering such a description are determined by fitting the APW band structure of Nb. The results agree surprisingly well with those shown in Fig. III.9. The values for Nb and Mo, for example, are 0.014 and 0.025, respectively. In this approach $\langle I^2 \rangle$ is stronger the more antibonding character of the d states. This is consistent with Butler's result, for an antibonding d-state has f-like character.

[25] W. H. Butler, *Phys. Rev. B* **15,** 5267 (1977).
[26] C. M. Varma, P. Vashishta, W. Weber, and E. I. Blount, to be published.

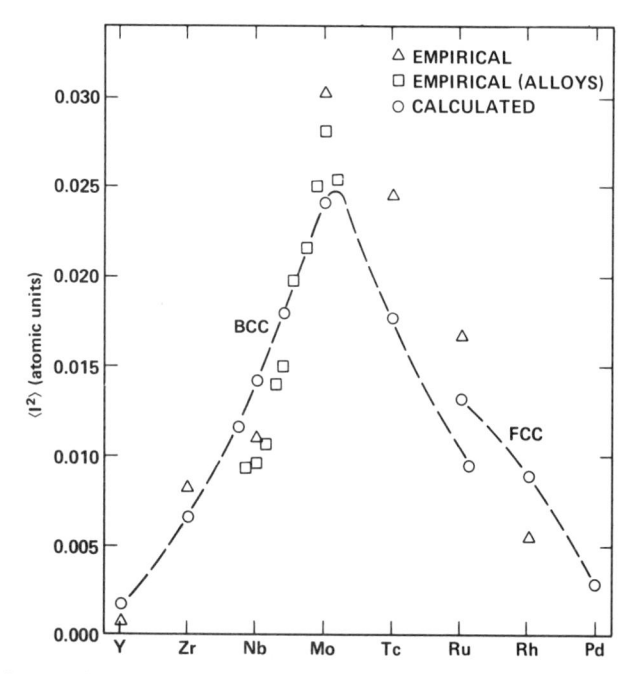

FIG. III.9. Empirical and calculated values of $\langle I^2 \rangle$, the Fermi surface average of the electron–phonon interactions. Triangles are empirical values for elements. Squares are empirical values for ZrNb, NbMo, and MoRe alloys. Circles are calculated values for cubic phases. After Butler.[25]

7. Fröhlich Lattice Wave

The search for high-temperature superconductors has led to a consideration of organic systems. Since these compounds often have a chain-like or one-dimensional character, it is perhaps worthwhile pointing out another mechanism for superconductivity based on the electron–phonon interaction suggested by Fröhlich[27] in 1954, before the BCS theory completely overwhelmed the field. If, in a one-dimensional metal with a lattice period a, the lattice develops an additional periodicity, i.e., a CDW with wavevector $2k_F$, then a gap is introduced in the electron spectrum and we have the Peierls insulator mentioned in Chapter I. In a mean field theory the transition temperature is given[28] by $k_B T_P = 4.55\, E_F \exp(-1/\lambda)$ where the electron–phonon coupling constant, $\lambda = 2|g|^2 a/\pi v_F \omega_{2k_F}$.

Fröhlich noted that since the energy of the CDW does not depend upon its phase there are no restoring forces for a phase fluctuation. In particular, it will move under the action of a dc electric field and carry current.

[27] H. Fröhlich, *Proc. R. Soc.* (*London*) A **223**, 296 (1954).

[28] M. J. Rice and S. Strässler, *Solid State Commun.* **13**, 125 (1973).

Since electron scattering processes are suppressed by the energy gap, superconductivity should result. Lee, Anderson, and Rice[29] point out that this situation may be visualized by considering the dielectric function associated with an oscillator with frequency ω_T:

$$\epsilon = \epsilon_\infty + \frac{Ne^2/m^*}{\omega_T^2 - \omega^2 - i\Gamma\omega}. \tag{3.54}$$

Allowing the phonon frequency $\omega_T \to 0$ as well as the damping Γ,

$$\epsilon \to \epsilon_\infty - (Ne^2/m^*\omega^2). \tag{3.55}$$

This implies that the imaginary part of the conductivity $\sigma_2(\omega) = Ne^2/4\pi m^*\omega$ corresponding to a perfect conductor (see Fig. II.3).

Unfortunately, in real crystals there are various mechanisms that "pin" the CDW. These effectively restore a finite ω_T and destroy the superconductivity. However, at finite temperatures the system can fluctuate into this current-carrying excited state and might give an unusually high conductivity.

A system that displays a Peierls transition is KCP, which stands for $K_2[Pt(CN)_4]Br_{0.3}0.3(H_2O)$. At low temperatures the oscillation of the pinned CDW is observed with an activation energy of 2 meV. Thermal unpinning gradually takes place above 100 K, but there is no evidence of propagating collective modes. Another system, TTF-TCNQ, also shows evidence of a Peierls transition associated with an enhancement in the conductivity.

$NbSe_3$ shows two charge density wave transitions, one at 95 K and one at 145 K. The resistivity increases associated with these transitions can be suppressed by an electric field.[30] This is suggestive of unpinning of the CDW.

8. Maximum T_c

The question of whether there is a natural well-defined temperature limit to the occurrence of superconductivity has intrigued both scientists and science fiction enthusiasts since the unexpected discovery that there can be such a phenomenon as superconductivity:

> . . . the children's wire was a superconductor at room temperature. A thread the size of a cobweb could carry all the current turned out by Niagra without heating up. A heavy-duty dynamo could be replaced by a superconductive dynamo that would almost fit in one's pocket.[31]

[29] P. A. Lee, T. M. Rice, and P. W. Anderson, *Solid State Commun.* **14,** 703 (1974).

[30] P. Monçeau, N. P. Ong, A. M. Portis, A. Meerschant, and J. Rouxel, *Phys. Rev. Lett.* **37,** 602 (1976).

[31] M. Leinster, "Four from Planet 5." Fawcett, Greenwich, Connecticut, 1959.

However, superconductivity, which occurs over almost three decades in temperature, has not yet been found at room temperature. Why not?

Theoretically, it is just as difficult to calculate a maximum T_c as it is to calculate a maximum temperature for any other cooperative phenomena. Nevertheless some interesting predictions can be made from the Eliashberg equations.

Consider first McMillan's equation for T_c, Eq. (3.44). The preexponential and exponential factors of this equation are related since, as we have shown, λ can be written as

$$\lambda = \frac{N(0)\langle I^2\rangle}{M\langle\Omega^2\rangle} \equiv \frac{\eta}{M\langle\Omega^2\rangle}. \tag{3.56}$$

McMillan noted that, experimentally among the transition metals, the product $N(0)\langle I^2\rangle = \eta$ is surprisingly constant. The calculation by Butler, which was discussed above, shows that this is due to a rapid increase in $\langle I^2\rangle$ in going from Nb to Mo which is offset by the decrease in $N(0)$. If $\lambda\langle\Omega^2\rangle$ is held fixed, then, according to McMillan's equation, as λ is increased T_c goes over a broad maximum at $\lambda = 2.6$ where $T_c \sim 0.105(\lambda\langle\omega^2\rangle)^{1/2}$.

Allen and Dynes,[32] however, have reanalyzed the dependence of T_c on λ. McMillan's formula is based on 22 numerical solutions of the Eliashberg equations for $0 \leqslant \mu^* < 0.25$ and $0 < \lambda < 1.5$, using a single shape for $\alpha^2 F$ patterned after the phonon density of states of Nb. Allen and Dynes performed more than 300 numerical solutions of the Eliashberg equations for a number of shapes, and values of λ up to 10^6. Of these they selected, for purposes of fitting, 217 calculations representing the range $0 \leqslant \mu \leqslant 0.20$, $0.3 < \lambda < 10$, and three shapes (Pb, Hg, and Einstein). These more extended solutions of the Eliashberg equation monotonically increase with λ, even if $\lambda\langle\Omega^2\rangle$ is held fixed. The actual maximum is $T_c \sim 0.15(\lambda\langle\Omega^2\rangle)^{1/2}$. This implies that McMillan's "$\lambda = 2.6$ limit" is spurious. A comparison of Allen and Dynes' solution with McMillan's is shown in Fig. III.10. The Allen–Dynes curve was obtained using the experimental value for $^2\alpha(\Omega)F(\Omega)$ for Pb. Note that, despite its spurious large-λ limit, the McMillan equation describes the data well below $\lambda \sim 2$, the range for which it was derived.

From Eq. (3.56) we see that a decrease in $\langle\Omega^2\rangle$ will increase λ. Although a decrease in $\langle\Omega^2\rangle$ will likely be accompanied by a decrease in $\langle\Omega\rangle$, the fact that λ enters T_c exponentially whereas $\langle\Omega\rangle$ only enters linearly [see Eq. (3.44)] makes the variation in λ more important. However, it is now generally believed that variations in η for different classes of materials are more important than variations in $\langle\Omega^2\rangle$ in causing T_c to change. This is illustrated, for example, by Nb and Nb$_3$Sn as shown in Table III.1. The

[32] P. B. Allen and R. C. Dynes, *Phys. Rev. B* **12**, 905 (1975).

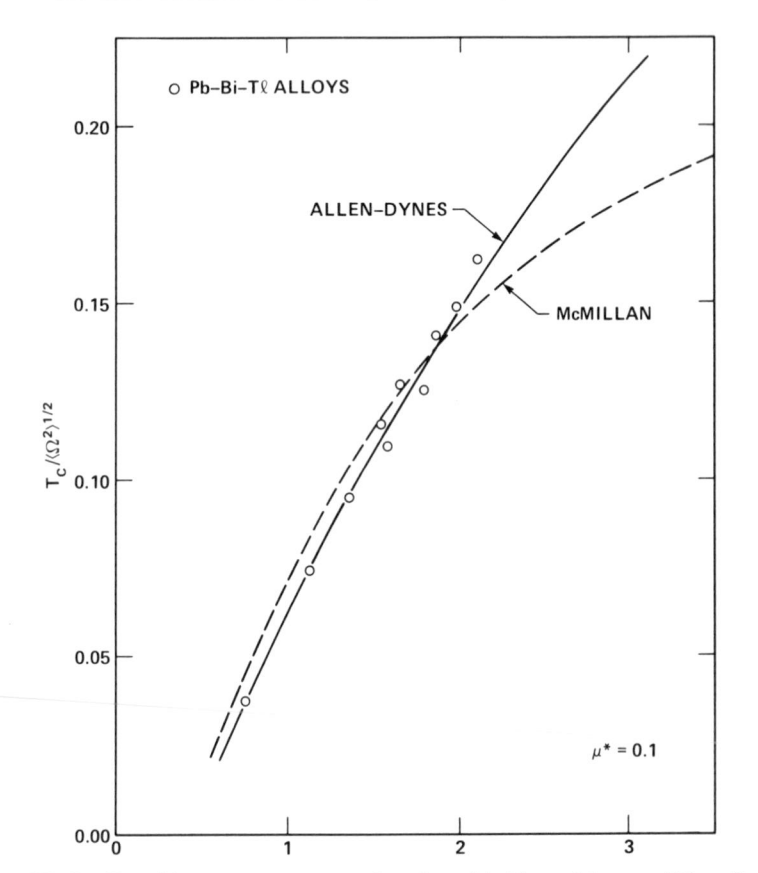

FIG. III.10.   Transition temperature as a function of λ. The solid curve (Allen–Dynes) was calculated using the tunneling density of states of Pb. The dashed curve is the McMillan equation. The open circles are experimental tunneling data. After Allen and Dynes.[32]

reason for the higher T_c's in d-band materials than in s-p band materials such as Pb, is the larger value of $\langle \Omega \rangle$ which enters T_c directly.

Bergmann and Rainer[33] have obtained an expression for the functional derivative $\delta T_c / \delta(\alpha^2 F(\Omega))$ which is proportional to Ω for small Ω. This means that T_c is not sensitive to the low-frequency phonon spectrum.

The failure to find higher T_c's despite their theoretical possibility seems to be connected with metastability. When a trend is found in which the superconducting interactions are increased by simple chemical substitution, inevitably, as the substitution is continued and the superconducting interaction becomes stronger, the derived phase becomes metastable. A new

[33] G. Bergmann and D. Rainer, Z. Physik **263**, 59 (1973).

TABLE III.1

	T_c (K)	$\langle \Omega \rangle$ (K)	η	$\sqrt{\langle \Omega^2 \rangle}$ (K)	λ
Nb	9.2	175	4.7	183	0.85
Nb$_3$Sn	18.1	146	7.9	163	1.67
Pb	7.2	60	2.4	65	1.55

phase with weaker superconducting interactions becomes stable. The Tl-Pb-Bi system is such an example (see Chapter VI). T_c and λ that uniformly increase as Bi is added to the fcc phase suddenly drop when more Bi is added and the ϵ-phase becomes stable. It appears to be the rule that when the electron–phonon interaction becomes too strong in a metal, the electronic energies can be lowered by readjusting (and removing portions of, if not all) the Fermi surface and frequently transforming to a less-symmetrical crystal structure. In the high-temperature A-15 compounds, such phase transformations sometimes occur, but are much gentler. Most of the Fermi surface remains after the transformation. Nb$_3$Ge, for example, can be prepared in a metastable state with an onset of the superconducting transition starting as high as 23 K. The less-favorable-for-superconductivity stable structure contains less Ge and has a $T_c \sim 6$ K.

B. EXCITONIC MECHANISMS

In 1964 Little[34] suggested that the virtual oscillation of electronic charge in the side chains of a long organic molecule could lead to an attractive interaction between electrons moving along the "spine" of the molecule. Such an "excitonic" mechanism was predicted to have a potentially higher superconducting transition temperature because the characteristic energy of the mediating particle, i.e., the exciton, is larger by the square root of the ionic mass to the electronic mass, $\sqrt{M/m}$. Shortly thereafter, Ginzburg[35] suggested that such exciton-induced pairing might also occur at a metal–semiconductor interface.

There are no known superconductors in which the pairing is suspected of being excitonic. In fact, the theoretical basis is still controversial. The situation envisioned by Ginzburg and illustrated in Fig. III.11 has been analyzed by Allender, Bray, and Bardeen[36] (ABB). They consider the interaction between an electron in the metal, whose wave function decays

[34] W. A. Little, *Phys. Rev.* **134**, A1416 (1964).
[35] V. L. Ginzburg, *Phys. Lett.* **13**, 101 (1964).
[36] D. Allender, J. Bray, and J. Bardeen, *Phys. Rev. B* **7**, 1026 (1973).

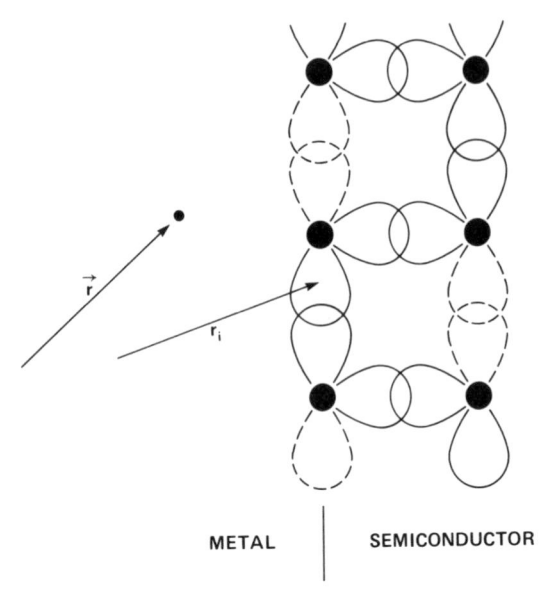

METAL   SEMICONDUCTOR

FIG. III.11.   Interface between a metal and a "resonantly bonded" semiconductor where it is suggested that virtual exciton exchange may contribute to electron pairing.

exponentially into the semiconductor, and the bonding electrons. In general this will have the form

$$\mathcal{H}_{int} = eV(\mathbf{r}) = \sum_i \int d^3r' \, \epsilon^{-1}(\mathbf{r}, \mathbf{r}') \frac{e^2}{|\mathbf{r}' - \mathbf{r}_i|}. \qquad (3.57)$$

The static dielectric function has been used because characteristic electron frequencies, ω_F, are much higher than those in which we are interested.

Introducing Fourier transforms, Eq. (3.57) may be rewritten

$$eV(\mathbf{r}) = \sum_i \sum_{\mathbf{K}} \sum_{\mathbf{K}'} \int d^3q \int d^3q' \, \epsilon^{-1}(\mathbf{q} + \mathbf{K}, \mathbf{q}' + \mathbf{K}')$$

$$\frac{4\pi e^2}{|\mathbf{q}' + \mathbf{K}'|^2} e^{-i(\mathbf{q}'+\mathbf{K}')\cdot(\mathbf{r}-\mathbf{r}_i)} \qquad (3.58)$$

where $\mathbf{K}$ and $\mathbf{K}'$ are reciprocal lattice vectors and $\mathbf{q}$ is restricted to the first Brillouin zone. If the medium is homogeneous so $\epsilon^{-1}(\mathbf{r}, \mathbf{r}') = \epsilon^{-1}(\mathbf{r} - \mathbf{r}')$, then the Fourier transform of the interaction is simply proportional to the product of $\epsilon^{-1}(q)$ and $4\pi e^2/q^2$ and we have no local field effects. That is, the charges being paired and those responsible for the pairing sample the same electric fields. In this case, there is some question as to whether the excitonic mechanism can ever be attractive. Cohen and Louie,[37] for example, have evaluated the Fermi surface average of the interaction using the Lindhard dielectric function. They find that the interaction is always *repulsive*. This suggests that local field corrections may be essential for this mechanism. Nevertheless, ABB write the interaction (3.58) as

$$eV(\mathbf{r}) \to \int d^3q \; S(4\pi e^2/q^2) \, \rho_\mathbf{q} e^{-i\mathbf{q}\cdot\mathbf{r}} \tag{3.59}$$

where S is the average of the inverse of the dielectric function and $\rho_\mathbf{q}$ is the Fourier transform of the electron density in the semiconductor. This interaction scatters an electron from some state $\mathbf{k}$ to a state $\mathbf{k} + \mathbf{q}$ with the excitation of an electron–hole pair or exciton. This leads to an effective electron–electron interaction due to the emission and reabsorption of the exciton. Since this is second order in perturbation theory, the effective interaction is attractive,

$$V_\text{ex} = 2 \sum_N (|M|^2/\omega_{N0}) \tag{3.60}$$

where N labels the excited state. The excitation energy ω_{N0} is approximated by the average gap width, ω_g. The matrix element is evaluated by appealing to certain sum rules which relate the matrix elements of the density operator $\rho_\mathbf{q}$ to the imaginary part of the dielectric function, $\epsilon_2(\omega)$. ABB assume that $\epsilon_2(\omega)$ is sharply peaked at $\omega = \omega_\text{g}$. Such excitations, however, are *transverse* and will therefore not couple to the electron! Coupling will occur if one introduces umklapp scattering associated with local field effects. Assuming this to be the case, ABB obtain for the exciton coupling constant,

$$\lambda_\text{ex} = \alpha \, S(\omega_\text{p}^2/\omega_\text{g}^2) \, \mu \tag{3.61}$$

where α is a constant related to the fraction of time that an electron is in the semiconductor, and ω_p is the electron plasma frequency. Notice that since this mechanism is basically Coulombic it is proportional to μ.

Without a more careful analysis of local field effects, it is difficult to determine just how large λ_ex might become. ABB suggest that the effect should be largest in materials such as PbTe which have large lattice and ionic polarizabilities. However, a study[38] of the superconductivity of ultrathin layers of Pb, In, and Tl on PbTe showed no excitonic enhancement of their T_c's.

[37] M. L. Cohen and S. G. Louie, *in* "Superconductivity in d- and f-band Metals" (D. H. Douglas, ed.), p. 7. Plenum, New York, 1976.

[38] D. L. Miller, M. Strongin, O. F. Kammerer, and B. G. Streeman, *Phys. Rev. B* **13**, 4834 (1976).

IV. The Curie and Néel Temperatures

The quantitative aspects of paramagnetism and ferromagnetism were established by Pierre Curie[1] in Paris at the turn of the century. In particular, using a Faraday balance, he measured the temperature dependence of the magnetic susceptibilities of a wide variety of substances. He discovered that the susceptibility of many of these substances followed a $1/T$ law which now bears his name. A few years later, in 1905, Langevin[2] developed the classical theory of paramagnetism which explained this behavior.

Curie had also found that the susceptibility of ferromagnets above their ordering temperatures varied at $1/(T - T_c)$. Weiss[3] explained this behavior in 1907 by introducing the concept of the "molecular field." In particular, Weiss assumed that the effective field acting on the magnetic moments in a ferromagnet is $H + \lambda M$ where H is the applied field, M the magnetization, and λ a proportionality factor. By combining this concept with Langevin's theory Weiss was able to describe all the salient experimental features of ferromagnetism. In particular, the Curie temperature is given by

$$T_c = N\mu^2\lambda/3k_B \qquad (4.1)$$

where μ is the magnetic moment, $\mu = g\mu_B J$. The origin of the proportionality factor λ, however, remained a mystery. The only interaction between magnetic moments known at that time was the dipole–dipole interaction. The value of λ associated with this mechanism is of the order of 4π while the actual value of λ for iron is 5×10^3. It was not until the birth of quantum theory that the origin of ferromagnetism was clarified.

A. Direct Exchange

In 1926, Heisenberg[4] and Dirac[5] independently recognized that the state of a system composed of n identical particles is potentially $(n!)$-fold

[1] P. Curie, *Ann. Chim. Phys.* [2] **5**, 289 (1895).

[2] P. Langevin, *J. Phys.* (*Paris*) **4**, 678 (1905).

[3] P. Weiss, *J. Phys.* (*Paris*) **6**, 667 (1907).

[4] W. Heisenberg, *Z. Phys.* **38**, 411 (1926).

[5] P. A. M. Dirac, *Proc. Soc. London, Ser. A* **112**, 661 (1926).

degenerate. This "exchange degeneracy," as it is called, arises from the fact that this system is indistinguishable from that in which two of the identical particles have been interchanged. The state of the system will therefore, in general, be a linear combination of these $n!$ different permutations. Of the $n!$ possible linearly independent combinations, only one is completely symmetric and one is completely antisymmetric. If the identical particles are electrons, then the Pauli exclusion principle requires that the completely antisymmetric state must be chosen. Slater recognized that this unique combination could be written as a determinant in which the columns refer to particles and the rows to states.

In 1927 Heitler and London,[6] who were interested in the quantum mechanics of the hydrogen molecule, laid the foundation for the valence bond method of calculating the electronic properties of molecules. To illustrate how this approach relates to the problem of ferromagnetism, consider two electrons moving in a potential arising from two overlapping core potentials. The Hamiltonian is

$$\mathcal{H} = \sum_{i=1}^{2} \left[\frac{p_i^2}{2m} + \sum_n V(\mathbf{r}_i - \mathbf{R}_n) \right] + \frac{e^2}{|\mathbf{r}_1 - \mathbf{r}_2|}. \tag{4.2}$$

Let us assume that $\phi(\mathbf{r} - \mathbf{R}_n)\chi_\alpha(\sigma)$ is the eigenfunction associated with the lowest eigenvalue ϵ of the Hamiltonian $\mathcal{H}_0 = p^2/2m + V(\mathbf{r} - \mathbf{R}_n)$. Here $\chi_\alpha(\sigma)$ is a spin function. Each electron has four states available to it. Thus there would be 16 possible two-electron states. However the antisymmetrization requirement restricts these to only four:

$$\psi(\mathbf{r}_1, \sigma_1; \mathbf{r}_2, \sigma_2)$$

$$= \begin{cases} \frac{1}{\sqrt{2}} [\phi(\mathbf{r}_1 - \mathbf{R}_1)\phi(\mathbf{r}_2 - \mathbf{R}_2) \\ \qquad - \phi(\mathbf{r}_2 - \mathbf{R}_1)\phi(\mathbf{r}_1 - \mathbf{R}_2)] \left\{ \frac{1}{\sqrt{2}} \begin{bmatrix} \alpha(1)\alpha(2) \\ [\alpha(1)\beta(2) + \beta(1)\alpha(2)] \\ \beta(1)\beta(2) \end{bmatrix} \right\} \\ \\ \frac{1}{\sqrt{2}} [\phi(\mathbf{r}_1 - R_1)\phi(\mathbf{r}_2 - \mathbf{R}_2) \\ \qquad + \phi(\mathbf{r}_2 - \mathbf{R}_1)\phi(\mathbf{r}_1 - \mathbf{R}_2)] \left\{ \frac{1}{\sqrt{2}} [\alpha(1)\beta(2) - \beta(1)\alpha(2)] \right\} \end{cases} \tag{4.3}$$

where $\alpha(\sigma) \equiv \chi_{+1/2}(\sigma)$ and $\beta(\sigma) = \chi_{-1/2}(\sigma)$. These are the familiar triplet ($S = 1$) and singlet ($S = 0$) states. The energies associated with these

---

[6] W. Heitler and F. London, Z. Phys. **44**, 455 (1927).

states are

$$\frac{\langle S, M_s | \mathcal{H} | S, M_s \rangle}{\langle S, M_s | S, M_s \rangle} = E_{0,1} = 2\epsilon + \frac{2v + U \pm (2Sw + J)}{1 \pm S^2} \tag{4.4}$$

where the plus signs are associated with the singlet and where

$$v = \int d^3r \phi^*(\mathbf{r} - \mathbf{R}_1) V(\mathbf{r} - \mathbf{R}_2) \phi(\mathbf{r} - \mathbf{R}_1)$$

$$w = \int d^3r \phi^*(\mathbf{r} - \mathbf{R}_1) V(\mathbf{r} - \mathbf{R}_1) \phi(\mathbf{r} - \mathbf{R}_2)$$

$$U = \int d^3r_1 d^3r_2 \phi^*(\mathbf{r}_1 - \mathbf{R}_1) \phi^*(\mathbf{r}_2 - \mathbf{R}_2) \frac{e^2}{|\mathbf{r}_1 - \mathbf{r}_2|} \phi(\mathbf{r}_1 - \mathbf{R}_1) \phi(\mathbf{r}_2 - \mathbf{R}_2)$$

$$\tag{4.5}$$

$$J = \int d^3r_1 d^3r_2 \phi^*(\mathbf{r}_1 - \mathbf{R}_1) \phi^*(\mathbf{r}_2 - \mathbf{R}_2) \frac{e^2}{|\mathbf{r}_1 - \mathbf{r}_2|} \phi(\mathbf{r}_2 - \mathbf{R}_1) \phi(\mathbf{r}_1 - \mathbf{R}_2)$$

and

$$S = \int d^3r \phi^*(\mathbf{r} - \mathbf{R}_1) \phi(\mathbf{r} - \mathbf{R}_2).$$

Even this simple case illustrates the complexities introduced by the anti-symmetrization demanded by quantum mechanics. Let us first consider what happens if we assume the two functions $\phi(\mathbf{r} - \mathbf{R}_1)$ and $\phi(\mathbf{r} - \mathbf{R}_2)$ are *orthogonal*. The overlap integral S is then zero and the energy becomes $2\epsilon + 2v + U \pm J$. The quantity U is referred to as the Hartree term or direct Coulomb integral, while J is the *exchange* integral. Thus the exchange is the difference between the singlet and triplet energies. This exchange integral is a *positive* quantity because it is the self-energy of the charge density $e\phi^*(\mathbf{r} - \mathbf{R}_1)\phi(\mathbf{r} - \mathbf{R}_2)$. Thus, the energy is lowest when the two electrons have parallel spin, i.e., are ferromagnetic.

An example of such orthogonal functions are atomic orbitals on the same site. In particular, consider two p-electrons with orbitals $\phi_{m_l}(\mathbf{r})$. These give rise to a multiplet structure consisting of a ${}^3P$, ${}^1D$, and ${}^1S$. Thus there are 15 states each represented by some combination of Slater determinants. These are worked out, for example, in Tinkham.[7] The ${}^1D$ state with $M_L = 1$ is

$$|{}^1D, M_L = 1\rangle = \frac{1}{\sqrt{2}} \begin{vmatrix} \phi_1(1)\alpha(1) & \phi_0(1)\beta(1) \\ \phi_1(2)\alpha(2) & \phi_0(2)\beta(2) \end{vmatrix} - \frac{1}{\sqrt{2}} \begin{vmatrix} \phi_1(1)\beta(1) & \phi_0(1)\alpha(1) \\ \phi_1(2)\beta(2) & \phi_0(2)\alpha(2) \end{vmatrix}.$$

[7] M. Tinkham, "Group Theory and Quantum Mechanics," Chapter 6. McGraw-Hill, New York, 1964.

The average Coulomb energy will consist of a "direct" term and an exchange term represented schematically as follows:

$$\left\langle \psi \left| \frac{e^2}{r_{12}} \right| \psi \right\rangle = \sum_{a>b} \left[\left(ab \left| \frac{e^2}{r_{12}} \right| ab \right) - \left(ab \left| \frac{e^2}{r_{12}} \right| ba \right) \right].$$

Thus the Coulomb energies of the three multiplets are

$$E(^1D) = U(1^+1^-)$$

$$E(^1S) = U(1^+ -1^-) + U(1^- -1^+) + U(0^+0^-) - U(1^+1^-)$$
$$\qquad - U(1^+0^+) + J(1^+0^+)$$

$$E(^3P) = U(1^+0^+) - J(1^+0^+)$$

where, for example,

$$U(1^+1^-) = \sum_{\sigma_1=\pm} \sum_{\sigma_2=\pm} \iint d^3r_1 d^3r_2 \phi_1(1)^* \phi_1(2)^* \frac{e^2}{r_{12}} \phi_1(1)\phi_1(2)$$

$$\alpha^+(\sigma_1)\beta^+(\sigma_2)\alpha(\sigma_1)\beta(\sigma_2).$$

By expanding $1/r_{12}$ in spherical harmonics these integrals may be reduced to a small number of radial integrals denoted by F_k. Thus,

$$U(1^+1^-) = F_0 + \tfrac{1}{25}F_2$$

$$J(1^+0^+) = \tfrac{3}{25}F_2 \qquad \text{etc.}$$

Notice that the only exchange integral entering these energies is $J(1^+0^+)$ and that it does not involve F_0. The only exchange integrals that do involve F_0 are $J(\pm 1^\pm \pm 1^\pm)$ and $J(0^\pm 0^\pm)$. But these are canceled by the direct integrals.

The orbital degeneracy in this problem is dealt with by averaging over the various orbital states. Slater[8] has shown that the *difference* between the average up- and down-spin energies is

$$\left[-\frac{1}{2l} \sum_{k\neq 0} C_k(l0; l0) F_k(nl; nl) \right] (n_\uparrow - n_\downarrow) \qquad (4.6)$$

where the C_k's are the angular integrals resulting from the expansion of $1/r_{12}$. $n_\uparrow - n_\downarrow$ is essentially the magnetization, m, so Eq. (4.6) may be written as Im. Notice that I is *not* the same as the average of the exchange integrals. The latter would include, for example, F_0 whereas I does not. This distinction is dramatically illustrated by two s-electrons on the same

[8] J. C. Slater, *Phys. Rev.* **165**, 658 (1968).

site. The exchange integral J is well-defined. However, the energy required to turn one of these spins over is not defined.

Allowing the up- and down-spin densities to be unequal constitutes what is called the *unrestricted* Hartree–Fock approach. This approach is essential in understanding the magnetic hyperfine field. Mn^{2+}, for example, is found to have a large negative spin density at its nucleus. This ion has a half-filled 3d shell. If one made a conventional Hartree–Fock calculation one would find that the spatial variations of the charge densities of the inner s-electrons were the same. However, with a spin-dependent calculation one finds that the d-electrons polarize these core s-electrons which in turn produces the large contact hyperfine field.

When we have a two-center situation the two functions are, in general, *not* orthogonal. Furthermore, if the potentials are attractive, then w in Eq. (4.4) is negative and the total energy may, in fact, be lower for *antiparallel* spins. This corresponds to the *covalent bond,* which is generally the situation that prevails. The oxygen molecule is one of the few diatomic molecules with a spin $S = 1$ in its ground state. In this case the relevant orbitals are a p_x orbital on one oxygen and a p_y orbital on the other where z is the direction between their centers.

The paper by Heitler and London[6] stimulated Heisenberg to suggest that the exchange interaction was the long-awaited explanation of ferromagnetism, and, in 1928 he published[9] the first microscopic theory of this phenomenon. This interaction, originating in the Coulomb interaction, is certainly strong enough to explain ferromagnetism—the Coulomb interaction between two electrons separated by 1 Å is approximately 10^5 K. Furthermore, Dirac[10] made the spin dependence explicit. The group theoretical approach to exchange had already been initiated by Heitler[11] who showed that the spin dependent contribution to the energy could be represented as a linear function of permutation operators, which have the effect of interchanging particles among their orbital basis, with coefficients given by integrals of the form U and J. The average energy is then related to the average of these permutation operators. However, the Pauli exclusion principle couples the permutations of orbital variables with those of spin. Dirac, assuming an orthogonal orbital basis, showed that the permutation operator that interchanges particles 1 and 2 is equivalent to $-(1 + \sigma_1 \cdot \sigma_2)/2$ where the σ_1 are the Pauli spin operators. Since there are only two independent states of spin for each electron, this is a tremendous simplification.

[9] W. Heisenberg, *Z. Phys.* **49**, 619 (1928).
[10] P. A. M. Dirac, *Proc. R. Soc. London, Ser. A* **123**, 714 (1929).
[11] W. Heitler, *Z. Phys.* **46**, 47 (1927).

For n electrons there are $n!$ permutations. However, if the electronic wavefunctions are reasonably localized the dominant terms are those involving the interchange of a single pair of electrons. The effective spin Hamiltonian then becomes

$$\mathcal{H} = -2 \sum_{i>j} J_{ij} \mathbf{s}_i \cdot \mathbf{s}_j. \tag{4.7}$$

A detailed discussion of the validity of this form is given by Herring.[12]

Problem IV.1

Consider a one-dimensional Ising system with near-neighbor exchange,

$$\mathcal{H} = -2J \sum_i s_i^z s_{i+1}^z.$$

A useful "trick" in dealing with this system is to make the change of variables

$$2s_i^z s_{i+1}^z \rightarrow S_i^z + \tfrac{1}{2}.$$

a. Calculate the partition function and the heat capacity.
b. Calculate the correlation function $\langle s_i^z s_{i+n}^z \rangle$ and determine the correlation length for this model.

If we have several electrons localized to a given ion their ferromagnetic intraionic exchange will give a ground state characterized by a total spin S. Since this intraionic exchange is generally much stronger than the interionic exchange, it is desirable to express the exchange interaction in terms of these ionic spins. Thus, if S_A is the total spin of ion A and S_B that of ion B one often sees the exchange interaction written $J\mathbf{S}_A \cdot \mathbf{S}_B$. This, however, is only true if the electronic states of ions A and B are orbital singlets. When this is not the case the problem is very complex and one can obtain additional contributions of the form $\mathbf{S}_A \cdot \mathbf{D} \cdot \mathbf{S}_B$ (anisotropic exchange) or $\mathbf{D} \cdot (\mathbf{S}_A \times \mathbf{S}_B)$ (antisymmetric exchange). For a discussion of exchange between orbitally degenerate electrons we refer the reader to the article by Levy.[13]

Let us now consider how this interaction determines the critical temperature. Suppose the spins interact through the isotropic exchange,

$$\mathcal{H}_{ex} = -2 \sum_A \sum_{B>A} J_{AB} \mathbf{S}_A \cdot \mathbf{S}_B. \tag{4.8}$$

[12] C. Herring, *in* "Magnetism" (G. Rado and H. Suhl, eds.), Vol. 2B, p. 1. Academic Press, New York, 1966.
[13] P. Levy, *in* "Magnetic Oxides" (D. J. Craik, ed.), Part 1, p. 181. Wiley, New York, 1975.

It is convenient to introduce Fourier transforms according to

$$S_R = \frac{1}{N} \sum_k e^{ik \cdot R} S_k . \qquad (4.9)$$

Then, since

$$\sum_A \exp[i(k + k') \cdot R_A] = N\Delta(k + k') \qquad (4.10)$$

where $\Delta(k)$ is the Kronecker delta function (which is one if k is some multiple of a reciprocal lattice vector and zero otherwise),

$$\mathcal{H}_{ex} = -2 \sum_k J(k) S_k \cdot S_{-k} \qquad (4.11)$$

where

$$J(k) = \frac{1}{N} \sum_{B>A} J_{AB} e^{-ik \cdot (R_B - R_A)}. \qquad (4.12)$$

The Zeeman interaction may also be expressed in Fourier variables:

$$\mathcal{H}_Z = g\mu_B \sum_A S_A \cdot H_A = g\mu_B \sum_k H_k \cdot S_{-k}. \qquad (4.13)$$

Notice that we are allowing the applied field to vary from site to site. In practice one can only apply a uniform ($k = 0$) field. However, Eq. (4.13) enables us to calculate the response of the system at any wave vector. Notice that neither the exchange interaction (4.11) nor the Zeeman interaction (4.13) couples *different* Fourier components. This is a consequence of the assumption of an infinitely periodic lattice which enters Eq. (4.10).

By comparing Eq. (4.11) with Eq. (4.13) we see that the exchange acts as an effective field $-2J(k)S_k/g\mu_B$. In the absence of interactions between the spins their response to an applied field is characterized by the Brillouin function[14] which at high temperature reduces to the Curie law, $\chi_0(k) = C/T$ where $C = (N/V)\mu^2/3k_B$ and $\mu^2 = g^2\mu_B^2 S(S + 1)$. The mean field approximation replaces S_k in Eq. (4.11) by its average value $\langle S_k \rangle$. The magnetization is given by

$$M(k) = -g\mu_B \langle S_k \rangle = \chi_0(k) H_k^{eff} \qquad (4.14)$$

[14] See, for example, C. Kittel, "Introduction to Solid State Physics," 5th ed., ch. 14. Wiley, New York, 1976.

where $H_k^{\text{eff}} = H_k - 2J(k)\langle S_k\rangle/g\mu_B$. Solving for $\langle S_k\rangle$ gives the generalized susceptibility for the interacting system, Eq. (1.15). This contains a divergence at the temperature $T_C = J(k)CV/g^2\mu_B^2$. Suppose J_{AB} is *positive* and nonzero only for z nearest neighbors. Then as the temperature decreases the divergence first occurs for $k = 0$ (ferromagnetism) at the temperature

$$T_C = \frac{2zS(S + 1)J}{3k_B}. \qquad (4.15)$$

If J_{AB} is *negative* the instability occurs when $k = G/2$, G being a reciprocal lattice vector. This corresponds to antiferromagnetism. The Néel temperature has the same form as (4.15) with J replaced by $|J|$.

Notice that the quantity entering the exchange interaction is the electron *spin*. In ions with strong spin–orbit coupling the spin is not a good quantum number, but rather the total angular momentum, $J = L + S$. Using this definition of J plus the definition of the Landé g-factor, $g_J J = L + 2S$, means $S = (g_J - 1)J$. Therefore in such systems the transition temperature, or more appropriately, the Curie–Weiss constant, should be proportional to $(g_J - 1)^2 J(J + 1)$. Fig. IV.1[15] compares the Curie–Weiss constant of the rare-earth metals with this factor.

It would appear that to calculate the Curie temperature, one need only evaluate the exchange integral J. Unfortunately, most magnetic materials cannot be understood in terms of this Heisenberg direct exchange. For example, if one assumes that the d-electrons in the transition metals are localized enough to make a Heitler–London description valid one can calculate the exchange from the singlet–triplet splitting as defined by Eq. (4.4). Freeman and Watson[16] have carried out such a calculation for two 3d electrons situated on two Co^{2+} ions 2.37 Å apart. They considered the ions sitting in an octahedral environment in which the eigenfunctions become linear combinations of the five components of the $l = 2$ orbital. The exchange integral between two d_{xy} orbitals (see Fig. IV.6), for example, is -602 K. The important point is that these exchange integrals are *negative*, i.e., antiferromagnetic. Although one is faced with the problem of identifying an orbitally averaged exchange, as we mentioned above, any reasonable combination still gives an antiferromagnetic coupling. Attempts to improve these calculations by incorporating the presence of the other electrons tends to reduce these values. But they never become large and positive. Therefore, Freeman and Watson concluded that direct ex-

[15] R. J. Elliott, *in* "Magnetism" (G. Rado and H. Suhl, eds.), Vol. 2A, p. 396. Academic Press, New York, 1965.

[16] A. J. Freeman and R. E. Watson, *Phys. Rev.* **124**, 1439 (1961).

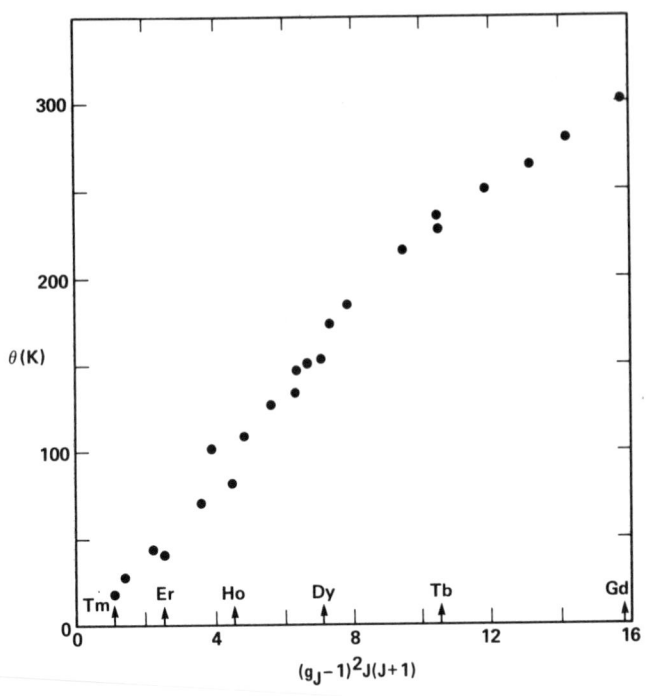

FIG. IV.1.   Dependence of the Curie–Weiss constant on the factor $(g_J - 1)^2 J(J + 1)$ for various heavy rare earths and alloys with yttrium. After Elliott.[15]

change, at least in the Heitler–London description, is *not* the dominant source of ferromagnetism in the transition metals. We shall return to this question in Section D.

Despite the difficulties in calculating the exchange constant, the exchange Hamiltonian considered as a phenomenological tool provides a great deal of information about the dynamics of magnetism. Bloch,[17] for example, showed that the elementary excitations of a Heisenberg system are the spin waves whose existence we deduced hydrodynamically in Chapter II. The frequency of these spinwaves in a cubic lattice with an interspin spacing a is given by $\hbar\omega_q = Dq^2$ where $D = 2JSa^2$. Bloch showed that many of the low-temperature properties of ferromagnets were governed by these excitations.

At the time of Heisenberg's theory the only known ferromagnets were metallic. Of course, the most ancient magnetic material is magnetite, Fe_3O_4, which is actually a ferrite, or ferromagnetic "insulator." However, its conductivity is still relatively high, of the order of $10^2(\Omega \cdot cm)^{-1}$

---

[17] F. Bloch, *Z. Phys.* **61**, 206 (1930).

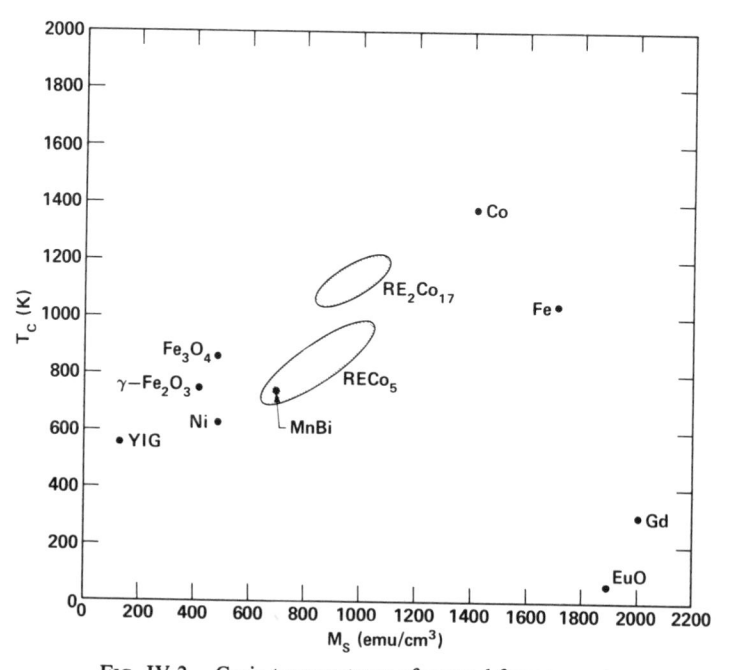

FIG. IV.2. Curie temperatures of several ferromagnets.

at room temperature (that of Cu is $6 \times 10^5 (\Omega \cdot cm)^{-1}$). Therefore, it is not surprising that the following year Bloch[18] applied Heisenberg's exchange concept to the case in which the electron wave functions are plane waves. In particular, he approximated the state of the system by a determinant of plane waves, an approximation which neglects "correlation." Bloch found that such an itinerant system will not be ferromagnetic unless the interatomic spacing exceeded about 5 Å. This is consistent with the non-magnetic natures of K and Na, for example, but, of course, does not explain Fe or Ni. Several years later Wigner[19] showed that when correlation effects are taken into account, even the low density system will not be ferromagnetic. This problem of itinerant ferromagnetism was considered in greater detail later by Slater and Stoner and we shall return to this.

B. SUPEREXCHANGE

As we have mentioned above, the Heisenberg Hamiltonian immediately became the basis for numerous "mathematical" investigations. One

[18] F. Bloch, *Z. Phys.* **57**, 545 (1929).
[19] E. P. Wigner, *Trans. Faraday Soc.* **34**, 678 (1938).

of these was the nature of the ground state of a Heisenberg system with a *negative* exchange constant. Bethe[20] solved this problem exactly for a one-dimension chain of spins ½ with nearest neighbor coupling. The following year Néel[21] generalized the Weiss molecular field concept to a system with two sublattices and derived what we now recognize as the susceptibility of an antiferromagnet. At that time the susceptibility, in conjunction with the specific heat, was the only way that one had to identify an antiferromagnet. Since the development of nuclear reactors, however, neutron scattering has revealed that antiferromagnets are very abundant. In Fig. IV.3 we indicate the appearance of antiferromagnetism among the transition metal oxides and fluorides. The fact that TiO and VO are paramagnetic metals is possibly due to the overlap of their d-bands

STRUCTURE								
ROCKSALT	TiO para	VO para		MnO 118 AF	FeO 198 AF	CoO 293 AF	NiO 520 AF	CuO (MONOCLINIC) 230 AF
SPINEL				Mn₃O₄ 41 Ferri	Fe₃O₄ 850 Ferri 119 (Verwey)	Co₃O₄ 40 AF		
CORUNDUM	Ti₂O₃ 500 para	V₂O₃ 150 AF	α-Cr₂O₃ 308 AF	α-Mn₂O₃ 80 AF	α-Fe₂O₃ 963 Canted AF			
RUTILE	TiO₂ dia	VO₂ 340 para	CrO₂ 393 F	β-MnO₂ 84 AF				

RUTILE	VF₂ 27-42 AF	CrF₂ 53 AF	MnF₂ 67 AF	FeF₂ 78 AF	CoF₂ 37 AF	NiF₂ 73 AF	CuF₂ 69 AF

2-DIMENSIONAL			K₂MnF₄ 42 Rb₂MnF₄ 38 AF	Rb₂FeF₄ 56 AF	K₂CoF₄ 107 Rb₂CoF₄ 101 AF	K₂NiF₄ 97 Rb₂NiF₄ 90 AF	K₂CuF₄ 6 F

☐ INSULATING (OR SEMICONDUCTING)   ▨ METALLIC

FIG. IV.3. Summary of the magnetic states of various transition metal oxides and fluorides. The numbers refer to the magnetic transition temperatures (compiled by J. W. Allen).

[20] H. A. Bethe, *Z. Phys.* **71**, 205 (1931).
[21] L. Néel, *Ann. Phys.* (*Paris*) [10] **18**, 5 (1932).

with their s-bands as suggested by the band calculations shown in Fig. IV.4.[22] This tendency of the cation d-states to move down away from the s-states as one moves across the periodic table is found in other systems as well.

One of the interesting features of many antiferromagnets is the fact that they are ionic with anions, such as O^{2-} and F^-, separating the magnetic cations. The magnetic direct Heisenberg exchange is too small at these separations to account for the Néel temperatures. In 1934 Kramers[23] suggested a *superexchange* mechanism to explain this indirect antiferromagnetic coupling. This mechanism was later developed quantitatively by Anderson.[24]

To illustrate how Kramers' mechanism works let us consider two cations separated by an anion. For simplicity let us focus on four electrons whose possible ground state configurations are shown in Fig. IV.5a. We have assumed that the two anion electrons occupy one of the three p-orbitals, while the cation electrons are d-like. Kramers introduces a perturbing Hamiltonian that consists of a direct Heisenberg exchange interaction between anion and cation electrons as well as an electron

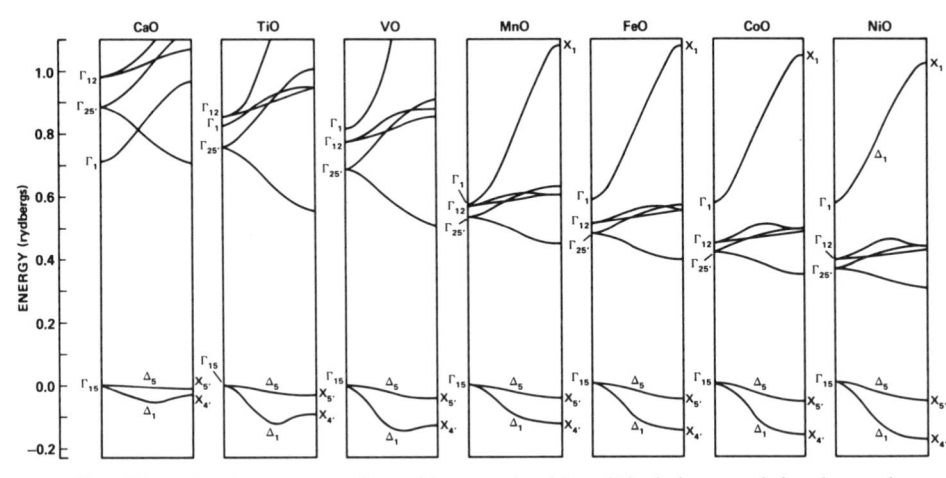

FIG. IV.4. Band structures of transition metal oxides. Calculations used the observed lattice constants, except for VO where the lattice constant of the hypothetical insulating state was used (after Mattheiss[22]). Starting on the right with NiO and moving to the left we see the d-bands (labeled Γ_{12}, Γ_{25}) moving up into the s-band (Γ_1).

[22] L. F. Mattheiss, *Phys. Rev.* **5**, 290 (1972).
[23] H. Kramers, *Physica (Utrecht)* **1**, 182 (1934).
[24] P. W. Anderson, *Phys. Rev.* **79**, 350 (1950).

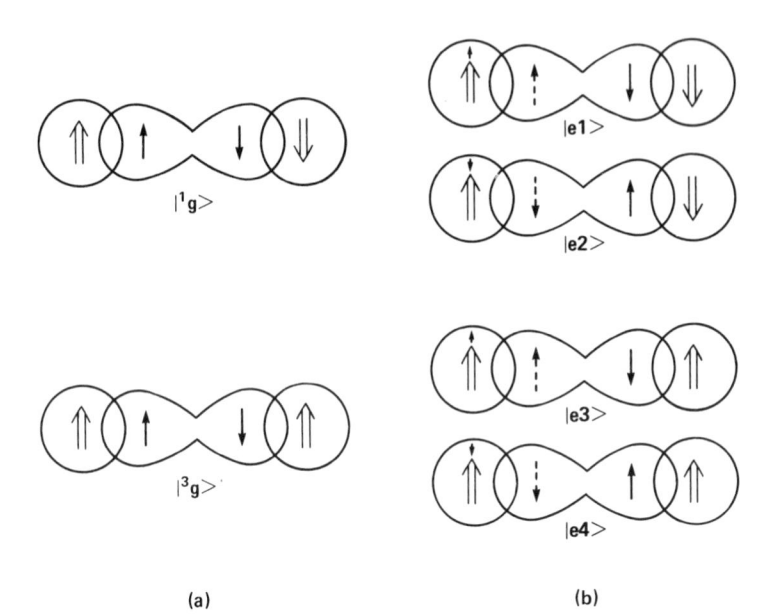

FIG. IV.5.   Illustration of the (a) two ground and (b) four excited states entering the calculation of Kramers cation–anion–cation superexchange.

transfer between anion and cation,

$$\mathcal{H} = t \sum_{\mu} \sum_{S} |e_{\mu}\rangle\langle^{2S+1}g| - 2\sum_{pd} J_{pd}\mathbf{s}_p \cdot \mathbf{s}_d \qquad (4.16)$$

where t is the matrix element for the transfer of an electron from the state $|^{2S+1}g\rangle$ to the state $|e_{\mu}\rangle$. J_{pd} is the exchange integral. Since the orbitals shown in Fig. IV.5a do not have the same centers they may not necessarily be orthogonal. Thus J_{pd} may be positive or negative. This Hamiltonian has no first-order matrix elements between the two possible ground states. It does, however, have matrix elements to and among, the four excited states shown in Fig. IV.5b.

We now employ perturbation theory to calculate the corrections to the two ground states associated with the Kramers Hamiltonian. In first order they are zero. In second order these corrections are the same for both states. In third order, however, the difference in energies between the triplet and singlet ground states is

$$\Delta E(^1g) - \Delta E(^3g) \equiv \Delta E = \frac{t^2 J_{pd}}{4}\left(\frac{1}{E_{e2}^2} - \frac{1}{E_{e1}^2}\right) \qquad (4.17)$$

where E_{e1} and E_{e2} are the energies of the excited cation in which the transferred spin is parallel or antiparallel to the total cation moment, respectively. The effective exchange interaction between the two cation spins may, therefore, be written as $-2J_{dd'}\mathbf{s}_d \cdot \mathbf{s}_{d'}$ where $J_{dd'} = \Delta E$.

Notice that if the parallel excited configurations $|e1\rangle$ and $|e3\rangle$ are excluded by virtue of the fact that the d-shell is already half-filled or more than half-filled, then $E_{e1} = \infty$ and the sign of the d–d exchange is the same as that of the p–d exchange. Since there is a large anion–cation overlap in transition metal oxides, sulfides, and fluorides we expect, on the basis of the arguments presented in the last section, that J_{pd} will be negative (antiferromagnetic). Indeed, most of the oxides, sulfides, and fluorides of Mn^{2+}, Fe^{2+}, Co^{2+}, and Ni^{2+} are antiferromagnets.

For cations with less than half-filled d-shells the Hund's rule coupling will exclude the antiparallel transfer configurations, i.e., $E_{e2} = \infty$, and we might expect ferromagnetism in this part of the periodic table. However, as we see from Fig. IV.3 this is not the case. Thus, antiferromagnetism is more prevalent than this theory would suggest.

Both Goodenough[25] and Kanamori[26] pointed out that it was important to consider the symmetry of the cation and anion orbitals. In particular, rather than using the spherical atomic cation d-orbitals as we did above, we must use the proper crystal field orbitals. If, for example, the cation sits in a site which has cubic symmetry then the fivefold degenerate atomic d-state is split into a triplet, denoted t_2, and a doublet, denoted e. In Fig. IV.6 we indicate the symmetry of the d_{xy} component of t_2. Also shown is an anion p-orbital directed along the axis between the two ions, a so-called p_σ orbital. From the signs of the phases it is easy to see that these two orbitals, even though they have different centers, are orthog-

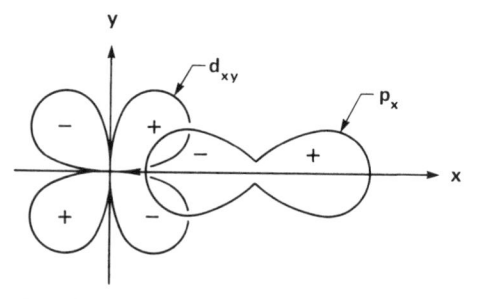

FIG. IV.6. Illustration of an anion p-orbital overlapping a cation crystal field d_{xy} orbital. The plus and minus signs indicate the relative phases of the wave functions. As a result of these phases $\langle d_{xy}|p_x\rangle = 0$.

[25] J. B. Goodenough, *Phys. Rev.* **100**, 564 (1955).
[26] J. Kanamori, *Phys. Chem. Solids* **10**, 87 (1959).

onal. Therefore the exchange is ferromagnetic. Thus, if we have Cr^{3+} ($3d^3$) cations in which the t_2 states lie lower than the e states, and the "Hund's rule" intraatomic exchange is strong, then the Kramers' mechanism will, in fact, give antiferromagnetic superexchange.

By such considerations Goodenough and Kanamori were able to establish a set of empirical rules for determining the sign of the exchange in a wide variety of materials.

The fact that superexchange is very sensitive to the nature of the cation and anion wave functions, plus the fact that the exchange effect appears in high order of perturbation theory led Anderson[27] to reformulate the problem from a different point of view. He suggested that we first determine the one-electron state for the crystal as a whole, excluding the effects of exchange. These will be Bloch functions from which we can construct Wannier functions. Each Wannier function is associated with a particular unit cell and represents the solution to the ligand field problem involving the magnetic cation with its surrounding anion complex. Kaplan and co-workers[28] have questioned Anderson's Hartree–Fock prescription for obtaining these ligand field functions. Rather than solve for the Bloch functions directly they suggest that one write the Hamiltonian in a Wannier representation and treat the matrix elements as variational parameters. The Hamiltonian is

$$\mathcal{H} = \sum_{i,j} \sum_{\sigma} b_{ij} c_{i\sigma}^{\dagger} c_{j\sigma} + U \sum_{i} n_{i\uparrow} n_{i\downarrow}. \qquad (4.18)$$

Here $c_{i\sigma}^{\dagger}$ is an operator which creates an electron in a Wannier function w_i with spin σ, $n_{i\sigma} = c_{i\sigma}^{\dagger} c_{i\sigma}$, and b_{ij} and U are the variational parameters. Physically, the coefficient U represents an average Coulomb energy. This includes *both* direct and exchange contributions. In fact, for a d-orbital one often sees this coefficient written as $(U + 4J)/5$. This Hamiltonian contains most of the ingredients required for a magnetic state. Anderson himself discussed this Hamiltonian in his book, *Concepts in Solids*, in 1962. The nature of the eigenstates of this Hamiltonian, however, was established by Hubbard in 1964 and it is, therefore, now referred to as the Hubbard Hamiltonian. Notice that since this model neglects orbital degeneracy it explicitly does *not* include the intraatomic exchange interaction which, as we saw above, only involves *parallel* spins. If we were to include such degeneracy we would have additional terms of the form

$$J \sum_{\sigma} \sum_{m} \sum_{m' \neq m} n_{im\sigma} n_{im'\sigma}. \qquad (4.19)$$

[27] P. W. Anderson, *Phys. Rev.* **115**, 2 (1959).
[28] See, for example, N. P. Silva and T. A. Kaplan, *AIP Conf. Proc.* **18**, 656 (1973).

Such terms will become important in less localized systems and we shall return to this later.

The magnetic properties associated with the Hubbard Hamiltonian are intimately related to the hopping term. Let us begin by assuming that $b_{ij} = 0$. The lowest state of a system with just one electron per atom will then consist of one electron at each site. Since the spins are independent this state is 2^N-fold degenerate. One of these configurations, for example, is the Néel state $|\uparrow \downarrow \uparrow \downarrow \cdots\rangle$. Another is the ferromagnetic state $|\uparrow \uparrow \uparrow \uparrow \cdots\rangle$. The next excited state consists of that in which one electron–hole pair has been produced. This has an energy U. We shall assume that when b_{ij} becomes nonzero it is still much smaller than U so that its effect may be treated by perturbation theory. Since b_{ij} has no matrix elements *within* the ground state manifold we must go to second order. There are two types of second-order terms—diagonal terms in which an electron hops over to a neighbor and then returns and off-diagonal terms in which the "visiting" electron hops over but the "visited" electron returns. The problem of actually finding the eigenstates to second order in b_{ij} even for this manifold is formidable. However, we can recast the problem into a spin representation that shows that the eigenvalue problem is that of an antiferromagnetically coupled set of spins.

To show this equivalence we project out only those degrees of freedom in which we are interested. In this case it is the two sites involved in the hopping. Let $|m\rangle$ and $|n\rangle$ represent two of the 2^N states in the ground state manifold excluding the pair under consideration, and $|\mu\rangle$ an appropriate electron–hole state. Then the second-order matrix elements are

$$\sum_\mu \frac{\langle n| \sum_{kl\sigma'} b_{kl} c_{l\sigma'}^\dagger c_{k\sigma'} |\mu\rangle \langle \mu| \sum_{i,j,\sigma} b_{ij} c_{j\sigma}^\dagger c_{i\sigma} |m\rangle}{-U}. \tag{4.20}$$

Clearly $k = j$ and $l = i$, but σ' may or may not equal σ. Thus,

$$\mathcal{H}_1^{\text{eff}} = -\sum_{ij} \frac{|b_{ij}|^2}{U} [c_{i\sigma}^\dagger c_{j\sigma} c_{j\sigma}^\dagger c_{i\sigma} + c_{i-\sigma}^\dagger c_{j-\sigma} c_{j\sigma}^\dagger c_{i\sigma}]. \tag{4.21}$$

The first term is $c_{i\uparrow}^\dagger c_{j\uparrow} c_{j\uparrow}^\dagger c_{i\uparrow} + c_{i\downarrow}^\dagger c_{j\downarrow} c_{j\downarrow}^\dagger c_{i\downarrow}$. Using commutation relations this becomes

$$-[n_{i\uparrow}(n_{j\uparrow} - 1) + n_{i\downarrow}(n_{j\downarrow} - 1)]$$
$$= -[\tfrac{1}{2}(\sigma_i^z \sigma_j^z - 1) + \tfrac{1}{2}(1 - n_i)(1 - n_j)] \tag{4.22}$$

where $n_i \equiv n_{i\uparrow} + n_{i\downarrow}$. The second term is

$$c_{i\downarrow}^\dagger c_{j\downarrow} c_{j\uparrow}^\dagger c_{i\uparrow} + c_{i\uparrow}^\dagger c_{j\uparrow} c_{j\downarrow}^\dagger c_{i\downarrow} = -\tfrac{1}{4}[\sigma_i^- \sigma_j^+ + \sigma_i^+ \sigma_j^-]. \tag{4.23}$$

Therefore

$$\mathcal{H}_1^{\text{eff}} = -2 \sum_{i,j} J_{ij} \left(\frac{1}{4} - \mathbf{s}_i \cdot \mathbf{s}_j \right) \tag{4.24}$$

where

$$J_{ij} = -\frac{2|b_{ij}|^2}{U} \tag{4.25}$$

which is always antiferromagnetic. Because of its origin in the hopping term Anderson called this *kinetic exchange*. Since this results from a virtual process, any other electrons in the vicinity do not have time to screen the doubly occupied site. Therefore, the U that enters this expansion is the unscreened Coulomb energy, of the order of 20 eV. In addition to this antiferromagnetic contribution to the superexchange there is also a ferromagnetic contribution arising from terms not included in the Hubbard Hamiltonian of the form

$$\sum_{i,j} \sum_{\sigma,\sigma'} U_{ij} c_{i\sigma}^\dagger c_{j\sigma'}^\dagger c_{i\sigma} c_{j\sigma'}. \tag{4.26}$$

This contribution Anderson called *potential exchange*.

Chapellmann[29] has employed the Nambu formalism we introduced in the previous chapter to discuss the Hubbard Hamiltonian. In addition to obtaining the Anderson exchange, he also obtains a "gap equation" for the expectation value of the localized moment on a site. In the weak hopping case this has the form

$$|\mathbf{S}_i| = -\frac{\hbar}{2} + \frac{\hbar}{2} \sum_j \frac{|b_{ij}|^2}{U^2} \left(1 - \frac{\mathbf{S}_i \cdot \mathbf{S}_j}{|\mathbf{S}_i||\mathbf{S}_j|} \right). \tag{4.27}$$

This illustrates an important feature of the Hubbard model. Namely, that the condition for local moment formation, given by a relation such as Eq. (4.27), is not the same as that for long range order, which is related to an exchange parameter such as Eq. (4.25). This distinction between local moment formation and long range order is often not appreciated. We return to this later in the discussion of itinerant magnetism.

Neither Anderson's nor Kaplan's prescription for evaluating the exchange parameters is particularly easy. In practice one assumes that the Wannier functions underlying this theory may be approximated by a linear combination of atomic orbitals. Nai Li Huang Liu and Orbach,[30] for

[29] H. Chapellmann, *J. Phys. F* **4**, 1112 (1974).
[30] Nai Li Huang Liu and R. Orbach, *AIP Conf. Proc.* **10**, 1238 (1972).

example, have carried out explicit evaluations of the numerous matrix elements entering such an approach. The results of such calculations are surprisingly good. They obtain a net exchange coupling of 9 K for the interaction between V^{2+} near-neighbor pairs in MgO in comparison with an experimental value of 5 K.

Since the superexchange interaction depends upon cation–anion overlap, it should be sensitive to the interionic spacing d and to the cation–anion–cation angle. This manifests itself in a variety of ways such as the pressure dependence of the ordering temperature and the magnetization as well as effects of thermal expansion on the magnetization. The variation of the near-neighbor (J_1) and next near-neighbor (J_2) exchange in the Eu chalcogenides as determined from spin wave spectra is shown in Fig. IV.7.[31] Quantitatively this dependence of the exchange may be char-

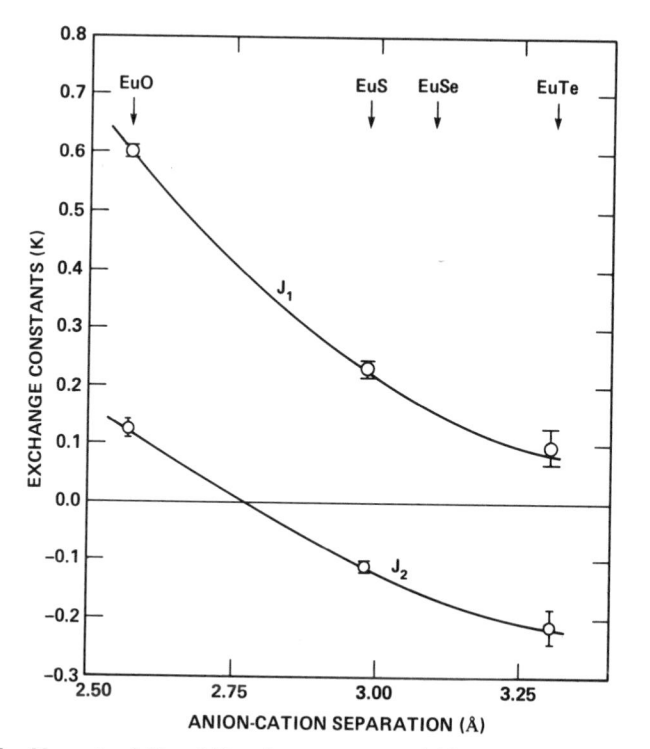

FIG. IV.7. Nearest neighbor (J_1) and next nearest neighbor (J_2) exchange constants in the Eu chalcogenides as a function of anion–cation separation (after L. Passell et al.[31]). The ferromagnetic transitions fall from EuO (T_c = 70 K) as J_2 becomes negative. EuTe is in fact antiferromagnetic (T_N = 9 K).

[31] L. Passell, O. W. Dietrich, and J. Als-Nielsen, Phys. Rev, B **14**, 4897 (1976).

acterized by a "magnetic Gruneisen constant,"

$$\gamma_m = - \partial \ln J / \partial \ln V \qquad (4.28)$$

where V is the volume. From an analysis of the magnetic properties of certain transition metal oxides D. Bloch[32] suggested that $\gamma_m \approx 10/3$, which corresponds to an interionic spacing dependence of d^{-10}. Johnson and Sievers,[33] for example, have measured the shift of the far-infrared antiferromagnetic resonance as a function of applied hydrostatic pressure in MnO and MnF_2. They find that the volume dependence of the next nearest neighbor exchange in MnF_2 which has the value $J_2 = 1.2$ K is described by $\gamma_m(J_2) = 3.4$. In MnO the next nearest neighbor has the volume dependence $\gamma_m(J_2) = 3.6$. If more than one exchange parameter is involved in determining the transition temperature the pressure dependence of T_C can be either positive or negative. This is illustrated in Fig. IV.8 where we indicate how T_C changes with pressure. In the chromium

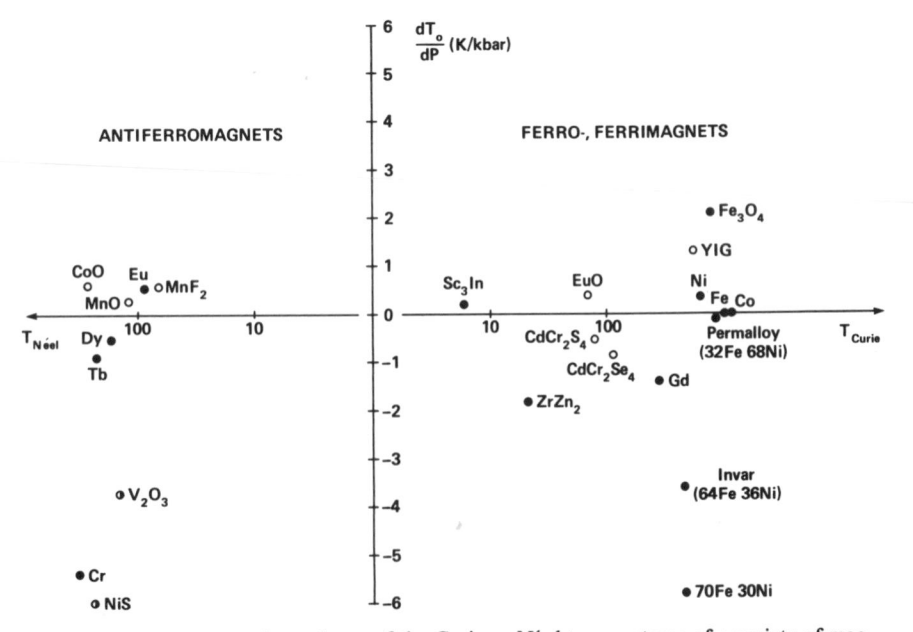

FIG. IV.8. Pressure dependence of the Curie or Néel temperatures of a variety of magnetic materials. The open circles indicate insulators, the black circles metals and the half-and-half circles show materials with metal–insulator transitions. Data were taken from Amer. Inst. of Phys. Handbook, Third Ed., McGraw-Hill, New York (1972), Table 5f-17.

[32] D. Bloch, *J. Phys. Chem. Solids* **27**, 881 (1966).
[33] K. D. Johnson and A. J. Sievers, *Phys. Rev. B* **10**, 1027 (1974).

spinels, for example, the nearest neighbor Cr^{3+} ions couple antiferromagnetically through a direct exchange and ferromagnetically through a 90° Cr–X–Cr superexchange. The antiferromagnetic exchange is very sensitive to cation–cation overlap and increases rapidly with decreasing separation. This is why dT_c/dP is negative for $CdCr_2S_4$ and $CdCr_2Se_4$.

In itinerant systems we must also consider the possibility that the moment itself changes with pressure. This may be described by writing the "equation of state" $\Delta V_M = kM^2$, where ΔV_M is the volume change associated with the presence of the magnetization M and k, the coefficient, is taken to be independent of temperature, pressure, and magnetic field. It then follows from thermodynamics that

$$dT_c/dP = -2k\chi_0 T_c(P = 0)$$

where χ_0 is the high-field susceptibility at $T = 0$. Notice that Fe–Ni alloys with 30–40% Ni exhibit a dramatic pressure dependence. In the case of Invar the positive volume change associated with the establishment of a moment almost entirely cancels the normal thermal contraction near room temperature. This is the reason why Invar is used in situations where dimensional stability with variations in the ambient temperature is important.

C. VIRTUAL PHONON EXCHANGE

If phonons are so effective in producing superconducting pairs might they also produce spin–spin coupling? The answer is affirmative, but the interaction is generally weak. Let us consider a paramagnetic ion with a d-electron configuration surrounded by an octahedron of charges. Of all the different vibrational modes of this cluster only those which transform according to the irreducible representations Γ_{3g} (or E_g) and Γ_{5g} (or T_{2g}) will couple to such a d-configuration. The electron–phonon, or more appropriately, orbit–lattice interaction has the same origin as in Eq. (3.1). For the jth cluster this is written

$$\mathcal{H}_{OL} = \sum_{l=2,4,6} \sum_{m=\theta,\epsilon} V(\Gamma_{3q}l)C(\Gamma_{3q}l, m)_j Q(\Gamma_{3g}, m)_j$$

$$(4.29)$$

$$+ \sum_{l=2,4,6} \sum_{m=0,\pm1} (-1)^m V(\Gamma_{5q}l)C(\Gamma_{51}l, n)_j Q(\Gamma_{5q}, -m)_j.$$

The C's are linear combinations of the spherical harmonics for the n

d-electrons. For example

$$C(\Gamma_{3q}, 2, \theta) = \sum_{i=1}^{n} \left(\frac{4\pi}{5}\right)^{1/2} Y_2^0(\theta_i). \tag{4.30}$$

The Q's are the normal modes of vibration which are eventually projected onto plane waves. The V's are then the coupling constants.

Since the Q's are linear combinations of the anion displacements they give rise to one-phonon terms [see, Eq. (3.4)]. If $\mathcal{H}_{0L}$ is treated to second order in pertrubation theory it leads to an effective ion–ion interaction. It can be shown that only emission and absorption of identical phonons leads to a nonzero interaction. As a result, all reference to phonon occupation numbers cancels leaving only zero-point vibrations to provide the coupling. The details are worked out by Orbach and Tachiki.[34] There are a number of interesting features about the results. First of all, if the excited states all have the same total quantum number S (or $J = L + S$) as the ground state then the sum over the spherical harmonics in Eq. (4.29) may be replaced by a combination of total spin operators chosen such as to give the same matrix elements among these states as the original expression. For example,

$$\sum_{i=1}^{n} Y_2^0(\theta_i) \rightarrow \frac{\alpha}{4} \sqrt{\frac{5}{\pi}} [3S_z^2 - S(S + 1)] \tag{4.31}$$

where α is a constant which depends upon the configuration. Therefore, the virtual phonon exchange may involve a coupling between total spins S_A and S_B of the form $(S_A \cdot S_B)^2$. This is refered to as *biquadratic exchange*. Such exchange has been invoked by Allen[35] to account for the interactions between the uranium ions in UO_2. In Fig. IV.9[35–37] for example, we show the spin wave spectrum of UO_2 without (dashed curves) and with (solid curves) biquadratic exchange.

D. Coulomb Interactions among Itinerant Electrons

Itinerant electrons are electrons having a Fermi surface. Fermi surface studies of Fe, Co, and Ni indicate that both the 3d- and 4s-electrons in these materials possess Fermi surfaces. We are, therefore, faced with the

[34] R. Orbach and M. Tachiki, *Phys. Rev.* **158**, 524 (1967).

[35] J. Allen, *Phys. Rev.* **166**, 530 (1968).

[36] R. A. Cowley and D. G. Dolling, *Phys. Rev.* **167**, 464 (1968).

[37] R. J. Birgeneau, M. T. Hutchings, J. M. Baker, and J. D. Riley, *J. Appl. Phys.* **40**, 1070 (1969).

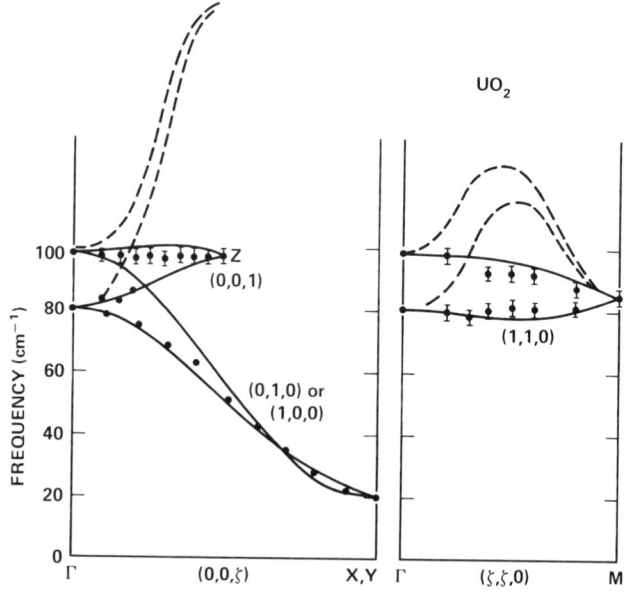

FIG. IV.9. Spin wave spectra of UO_2. The points are experimental values as measured by Allen[35] and Cowley and Dolling.[36] The dashed lines are the theoretical dispersion curves calculated assuming only bilinear exchange and single-ion anisotropy, whereas the solid lines are those calculated by Allen including the quadrupole–quadrupole interaction. After Birgeneau *et al.*[37]

problem of understanding the nature of the magnetic ordering in such materials.

Let us begin by considering the simplest case, that of free electrons. In the absence of any interactions (with ions or electrons) the ground state consists of two identical Fermi surfaces for spin-up and spin-down. Now let us add the Coulomb interaction between the electrons. If the electron density is *high* the kinetic energy will be large and we might expect to be able to treat the Coulomb interactions as a perturbation. We therefore assume a ground state wave function in the form of a Slater determinant constructed from the spin-up and spin-down plane waves. We shall, however, allow the relative sizes of the spin-up and spin-down Fermi spheres to be a variable as illustrated in Fig. IV.10. This corresponds to the Hartree–Fock approximation. If we define $m = (N_\uparrow - N_\downarrow)/N$ then the average energy in atomic units (two Rydberg units) is given by

$$\frac{E_{HF}}{N} = \frac{0.553}{r_s^2}[(1 + m)^{5/3} + (1 - m)^{5/3}] - \frac{0.227}{r_s}[(1 + m)^{4/3} + (1 - m)^{4/3}]$$

$$(4.32)$$

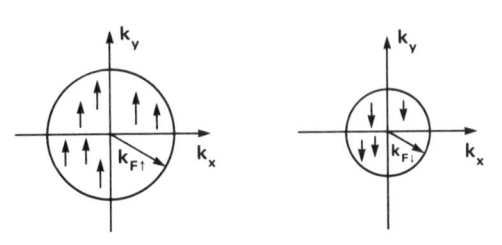

FIG. IV.10.　Up- and down-spin Fermi spheres of a polarized free-electron system. Compare with the Fermi surfaces of Ni in Fig. IV.26.

where

$$r_s = r_0/a_0, \qquad a_0 = \hbar^2/me^2, \qquad r_0 = (3V/4\pi N)^{1/3}. \qquad (4.33)$$

The first term is the kinetic energy contribution while the last term is the exchange. The direct Coulomb energy has been canceled out by the uniform background positive charge. The difference between this energy and that for $m = 0$ is shown in Fig. IV.11.[38] We see that for $r_s > 5.45$ the ferromagnetic state has the lower energy. This is the result found by Bloch that we mentioned above.

The problem with the Hartree–Fock ground state is that it forces the electrons into one particular determinantal form. If we had included all possible determinants, then the system could adopt a lower energy configuration. This difference between the Hartree–Fock energy and the true ground state is referred to as the *correlation energy*. Since the exclusion principle already keeps parallel spins apart the effect of correlation is to lower the energy of antiparallel spins. This obviously weakens the condition for ferromagnetism. Many-body treatments indicate that at metallic densities ($r_s \approx 4$) the correlation corrections are comparable to the exchange itself. In particular, as we mentioned above, correlation considerations indicate that an electron system interacting via the Coulomb interaction is never ferromagnetic. Conversely, incidentally, the fact that exchange is comparable to correlation in this metallic regime indicates that the Hubbard Hamiltonian we discussed above, which included *only* correlation in the sense that parallel spins never "see" one another, would *not* be an appropriate model for this free electron case.

In the 1930's Stoner[39] applied the Weiss mean field concept to a degenerate electron system thereby deriving a condition for itinerant ferromag-

[38] C. Herring, in "Magnetism" (G. Rado and H. Suhl, eds.), Vol. 4, p. 17. Academic Press, New York, 1966.

[39] E. C. Stoner, *Philos. Mag.* [7] **15**, 1018 (1933); *Proc. R. Soc. London, Ser. A* **154**, 656 (1936); **165**, 372 (1938); **169**, 339 (1938).

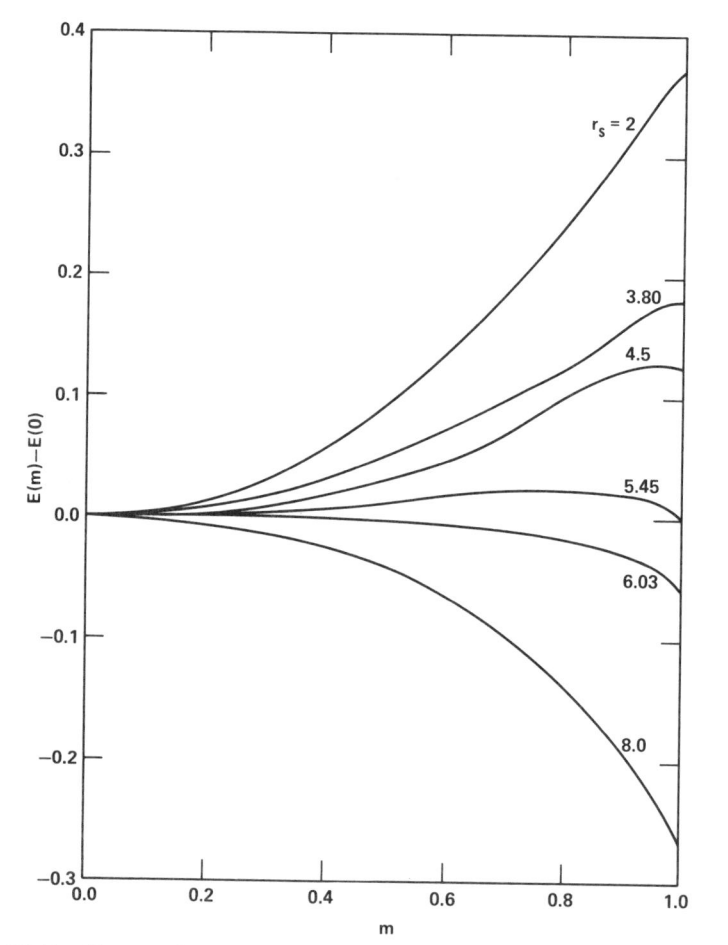

FIG. IV.11.   Energy of a magnetic state relative to the unmagnetized state versus the functional magnetization. The energy is in units of the kinetic energy at $m = 0$, i.e., $1.106/r_s^2$. Typical metallic electron densities correspond to $1.8 < r_s < 5.5$. After Herring.[38]

netism. Basically, the Stoner model assumes that the spin-up and spin-down energy bands are split in proportion to the magnetization $m = n_\uparrow - n_\downarrow$. The coefficient of proportionality is the difference between the up- and down-spin exchange energies, and is usually written I. Stoner's criterion emphasizes the necessity of having a high density of states at the Fermi level, as we shall see below.

In 1936 Slater[40] pointed out how sensitive the properties of an itinerant

[40] J. C. Slater, *Phys. Rev.* **49**, 537 (1936).

model are to the band structure. Since then, a variety of approaches [orthogonalized plane waves (OPW), augmented plane waves (APW), linear combinations of atomic orbitals (LCAO), Green's functions (KKR), etc.] and large computers have provided us with sophisticated band calculations.[41]

If Φ refers to a determinantal state constructed from Bloch functions, as opposed to the plane waves used above, the Hartree–Fock energy may be written

$$E(\Phi) = \langle \mathcal{H}_0 \rangle_\Phi + V_{\text{dir}}(\Phi) + V_x(\Phi). \qquad (4.34)$$

Here $\mathcal{H}_0$ is the one-electron Hamiltonian including the kinetic energy as well as the interaction with the nuclei. V_{dir} and V_x are direct and exchange energies, respectively. These latter contributions as well as the effects of correlation must be calculated accurately if one is to understand the variety of long range order in itinerant systems.

In 1951 Slater[42] suggested that we approximate the effect of exchange by the potential

$$V_x(\mathbf{r}) = -6[(3/8\pi)\rho(\mathbf{r})]^{1/3}. \qquad (4.35)$$

This approximate form has been used extensively both in atomic[43] and solid state calculations. Physically, this density to the one-third power arises from the fact that in the Hartree–Fock approximation parallel spins are kept farther apart than antiparallel spins. Therefore, there is an "exchange hole" around any particular spin associated with a deficiency of similar spins. The radius of this hole must be such that $(\frac{4}{3})\pi r^3 \rho = 1$. Since the potential associated with this deficiency is proportional to $1/r$, we obtain an exchange potential proportional to $\rho^{1/3}$.

If the number of up- and down-spins are unequal, then the *difference* in their exchange potentials is proportional to

$$\rho(\mathbf{r})_\uparrow^{1/3} - \rho(\mathbf{r})_\downarrow^{1/3} \simeq \frac{2^{2/3}}{3} \rho(\mathbf{r})^{1/3} \left(\frac{\rho(\mathbf{r})_\uparrow - \rho(\mathbf{r})_\downarrow}{\rho(\mathbf{r})} \right). \qquad (4.36)$$

This has the same form as the atomic result, Eq. (4.6), i.e., $V_{x\uparrow} - V_{x\downarrow} = Im$. A self-consistent band calculation employing such an exchange potential may result in spin-split bands. Such a splitting does not necessarily imply long range order. In the Stoner model for itinerant magnetism, the

[41] See, for example, W. A. Harrison, "Solid State Theory." McGraw-Hill, New York, 1970.

[42] J. C. Slater, *Phys. Rev.* **81**, 385 (1951).

[43] See, for example, F. Herman and S. Skillman, "Atomic Structure Calculations." Prentice-Hall, Englewood Cliffs, New Jersey, 1963.

order parameter is assumed to be the *local* magnetization or, equivalently, the exchange splitting. Therefore, in this model, an exchange splitting automatically implies long range order. Cr and NiS are among the few metallic materials where local moment formation does coincide with the onset of long range order. In EuS, however, the spin-splitting of the f-electrons, which determines local moment formation, is 10^5 K while the Curie temperature is only 16.5 K. Thus, a full theory of itinerant magnetism should allow for local moment formation that is separate from the appearance of long range order. This distinction is beautifully illustrated by Fig. IV.12. Here the exchange splitting in nickel, as measured[44] by angle-resolved photoemission, is plotted as a function of temperature. We see that this splitting has only decreased by 40% at the Curie point.

In this same context it is interesting to note that the BCS theory imposes long range order when pairing occurs by requiring that the pairs all have the same phase ϕ [see discussion above Eq. (1.35)]. We are not aware of any situation where "incoherent" pairing occurs.

In 1965 Kohn and Sham[45] rederived the exchange potential by a dif-

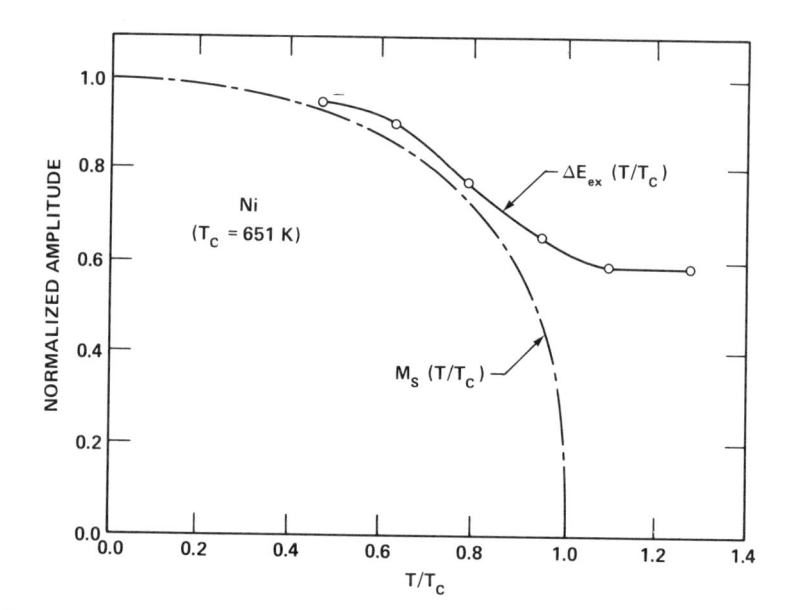

FIG. IV.12.   Experimental temperature–dependent d-band exchange splitting in nickel compared with the measured saturation magnetization. After Eastman *et al.*[44]

[44] D. E. Eastman, F. J. Himpsel, and J. A. Knapp, *Phys. Rev. Lett.* **40**, 1514 (1978).
[45] W. Kohn and L. J. Sham, *Phys. Rev.* **140**, A1133 (1965).

ferent method and obtained a value two thirds that of Slater. This led workers to multiply Eq. (4.35) by an adjustable constant, α, which can be determined for each atom by requiring that the, so-called, $X\alpha$ energy be equal to the Hartree–Fock energy for that atom. An interesting application of the $X\alpha$ method has been made by Hattox.[46] He has calculated the magnetic moment of bcc vanadium as a function of lattice spacing. The result is shown in Fig. IV.13. We see that the moment falls suddenly to zero at a spacing 20% larger than the actual observed spacing. This decrease is due to the broadening of the 3d band as the lattice spacing decreases. This is consistent with the fact that bcc vanadium, where the vanadium–vanadium distance is $\sqrt{3}\ a_0/2 = 2.49$ Å, is observed to be nonmagnetic. In Au_4V, however, the vanadium–vanadium distance has increased to 3.78 Å and the vanadium has a moment near one Bohr magneton.

If one uses the "approved method," as Slater calls it, to find α for the magnetic transition metals one finds that the resulting magnetic moments are not in agreement with those observed. This is not surprising when we consider that we have replaced a nonlocal potential (the Hartree–Fock

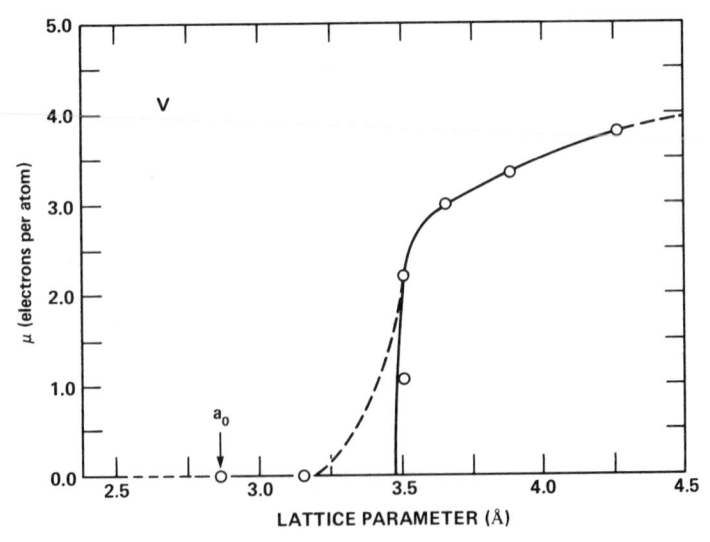

FIG. IV.13. Calculated magnetic moment of vanadium metal as a function of lattice parameter. The two points for $a = 3.5$ Å correspond to two distinct self-consistent solutions associated with different starting potentials. These two solutions are the result of a double minimum in the total energy versus magnetization curve. At 4.25 Å and 3.15 Å the calculations converged to unique values. After Hattox et al.[46]

[46] T. M. Hattox, J. B. Conklin, Jr., J. C. Slater, and S. B. Trickey, J. Phys. Chem. Solids 34, 1627 (1973).

potential) by a local potential (the Slater $\rho^{1/3}$). Furthermore, the fact that these are different means that the Slater potential must, by definition, include some correlation. However, we do not know if it is in the right direction. The density of states of Ni calculated by the $X\alpha$ method with α determined to give the observed moment is shown in Fig. IV.14.[47]

The arbitrariness inherent in the $X\alpha$ method may be avoided by using what is called the local spin density (LSD) approximation. This essentially enables one to utilize the results of many-body calculations for the homogeneous electron gas in determining the exchange and correlation potentials in transition metals. This approach is based on a theorem by Hohenberg and Kohn[48] which states that the ground state energy of an inhomogeneous electron gas is a functional of the electron density $\rho(r)$ and

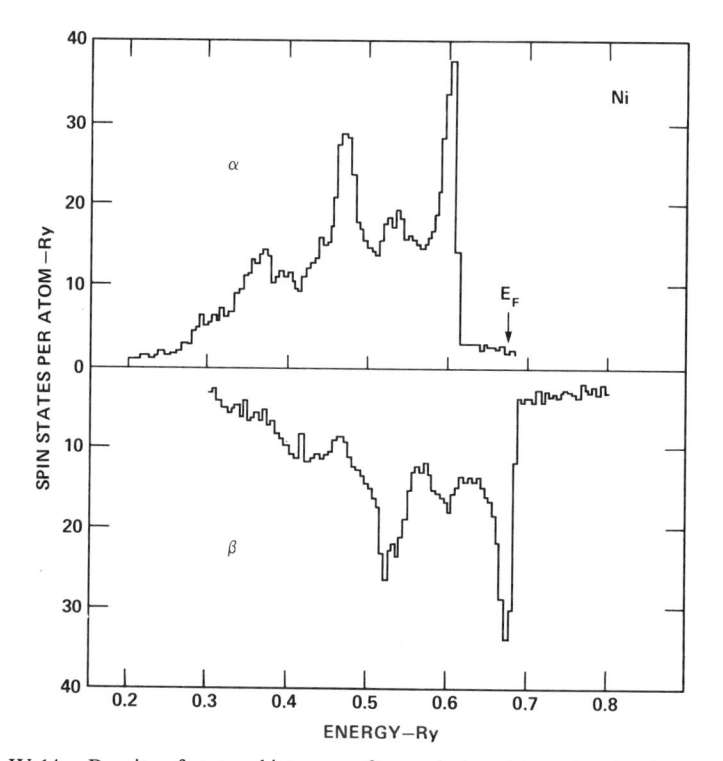

FIG. IV.14. Density-of-states histogram for majority (α) and minority (β) spin electrons in nickel. After Wakoh and Yamashita.[47]

[47] S. Wakoh and J. Yamashita, *J. Phys. Soc. Jpn.* **21**, 1712 (1966).
[48] P. Hohenberg and W. Kohn, *Phys. Rev.* **136**, B864 (1964).

spin density $m(r)$. The effective potential, for example, is given by

$$V_{\text{eff}} = v(\mathbf{r}) + e^2 \int \frac{\rho(\mathbf{r}')d^3r}{|\mathbf{r} - \mathbf{r}'|} + \frac{\delta}{\delta\rho(\mathbf{r})} [\rho(\mathbf{r})\epsilon_{xc}(\rho(\mathbf{r}))] \qquad (4.37)$$

where $v(\mathbf{r})$ is the one-particle potential.

This potential is then used to compute the one-electron eigenfunction $\phi_{k\sigma}$. The density is then evaluated self-consistently from

$$\rho(\mathbf{r}) = \sum_{k,\sigma} \phi_{k\sigma}^*(\mathbf{r})\phi_{k\sigma}(\mathbf{r}).$$

The unknown exchange-correlation functional, $\epsilon_{xc}(\rho(\mathbf{r}))$, appearing in this problem is approximated by the many-body exchange-correlation function for the *homogeneous* electron gas. Von Barth and Hedin,[49] for example, have obtained an analytic expression for this potential that corresponds to a shifted and rescaled spin dependent Slater potential. What the density function approach does, essentially, is to provide one with a density dependent α. Gunnarsson,[50] for example, finds that to simulate correlation effects α must be *less* than $\frac{2}{3}$ and decrease with r_s. The results of this approach are very impressive, at least for zero-temperature properties. In Table IV.1 we list the bulk modulus, $B = -dP/d(\ln V)$, the spin moment and their pressure derivatives. These were obtained by calculating the spin polarization and the electron pressure at varying atomic volumes. The *only input was the atomic numbers and the crystal structures!* Callaway and Wang[51] have calculated the energy bands in ferromagnetic iron using both the $X\alpha$ method and the von Barth–Hedin exchange-correlation potential. They find that the results are not particularly sensitive

TABLE IV.1

THEORETICAL AND EXPERIMENTAL VALUES OF COHESIVE AND
MAGNETIC PROPERTIES OF IRON AND NICKEL

	bcc Iron		fcc Nickel	
	Theory	Exp.	Theory	Exp.
B(Mbar)	2.60	1.68	2.09	1.86
dB/dP	4.5	3.7	4.7	2.9
$m(\mu_B/\text{atom})$	2.17	2.21	0.68	0.61
$d\ln m/dP$(Mbar^{-1})	-0.24	-0.25	-0.17	-0.2

[49] U. von Barth and L. Hedin, *J. Phys. C* **5**, 1629 (1972).
[50] O. Gunnarsson, *J. Phys. F* **6**, 587 (1976).
[51] J. Callaway and C. S. Wang, *Phys. Rev. B* **16**, 2095 (1977).

to the form of the exchange potential used. They find, for the majority
spins, a width of 5.1 eV for both potentials. The minority spin bands are
about 1 eV wider for both potentials. They also find that the exchange
splitting varies substantially over the d-band. Thus, the common approximation of a constant splitting is not a good one for iron. The band structure associated with the von Barth–Hedin potential is shown in Fig.
IV.15. The density of states was constructed from this band structure.
The experimental value of $N(E_F)$ turns out to be larger than that calculated
by a factor of 1.87. The origin of this large difference is not understood.

In an itinerant system the magnetic susceptibility as a function of wave
vector is

$$\chi_0(\mathbf{q})/2\mu_B^2 = \sum_{\mathbf{k}} \frac{f_{\mathbf{k}} - f_{\mathbf{k+q}}}{\epsilon_{\mathbf{k+q}} - \epsilon_{\mathbf{k}}} \qquad (4.38)$$

where $f_{\mathbf{k}}$ is the Fermi function and $\epsilon_{\mathbf{k}}$ the one-electron band energies. For
free electrons $\chi_0(\mathbf{q} = 0)$ is just the Pauli susceptibility, $2\mu_B^2 N(E_F)$. If

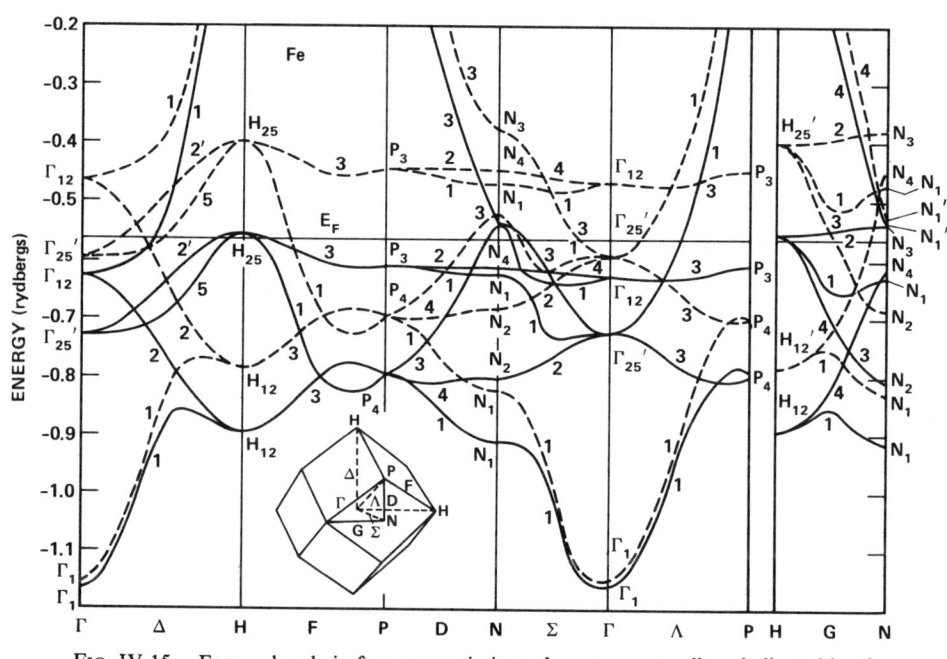

FIG. IV.15.   Energy bands in ferromagnetic iron along symmetry lines indicated by the
Brillouin zone insert. Exchange and correlation were included by using the von Barth–
Hedin potential. Solid lines are majority spin states, dashed are minority spin states. After
Callaway and Wang.[51]

electron–electron interactions are now introduced in the form of the band-splitting parameter I, the interacting susceptibility takes the "exchanged enhanced" form

$$\chi(\mathbf{q}) = \frac{\chi_0(\mathbf{q})}{1 - I\chi_0(\mathbf{q})}. \qquad (4.39)$$

Therefore, if $\chi_0(\mathbf{q})$ exceeds $1/I$ for a particular value of $\mathbf{q}$ the system is potentially unstable with respect to the appearance of a spontaneous magnetization with wave vector $\mathbf{q}$. If we take $\mathbf{q} = 0$ and use the $T = 0$ Pauli expression for $\chi_0(0)$ the condition for ferromagnetism becomes

$$IN(E_F) > 1. \qquad (4.40)$$

In Fig. IV.16 we show the results of an LSD calculation of I for various elements. We see that the Stoner criterion correctly predicts the oc-

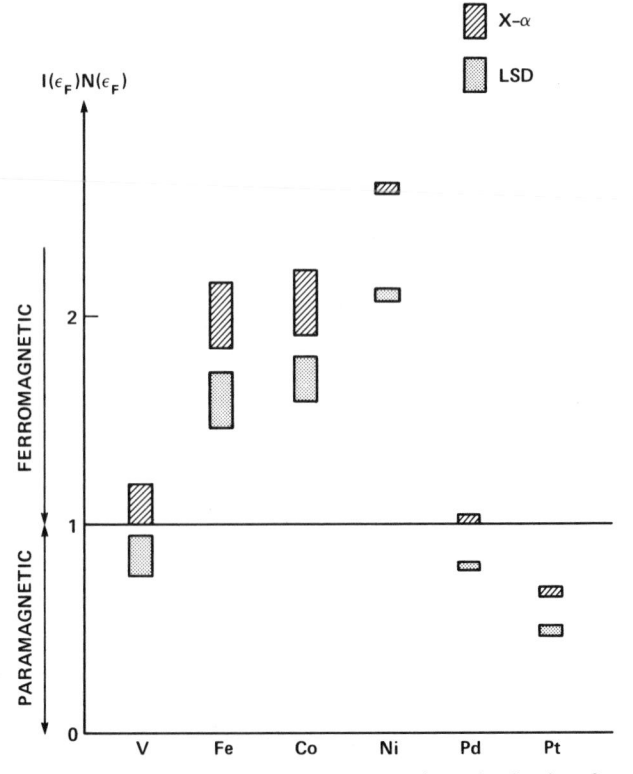

FIG. IV.16.   Calculations of the Stoner parameter, I, times the density of states for several transition metals (courtesy of O. Gunnarsson).

curence of ferromagnetism at $T = 0$. Since Eq. (4.39) was obtained by associating the order parameter with the exchange splitting its divergence indicates local moment formation as opposed to long range order. At $T = 0$, of course, long range order will coincide with the appearance of a moment. If Eq. (4.39) is used to determine T_c, however, the resulting T_c's are higher than those observed. The reason for this is that the Stoner theory neglects the effect of spin fluctuations on the equilibrium states of the system, i.e., they are only treated as fluctuations about the Stoner state. Moriya and Kawabata[52] have investigated the effect of such spin fluctuations on the generalized susceptibility and find

$$\chi(\mathbf{q}) = \frac{\chi_0(\mathbf{q})}{1 - I\chi_0(\mathbf{q}) + \lambda} \tag{4.41}$$

where

$$\lambda = \chi_0 \frac{\partial^2 \Delta F}{\partial^2 M} \tag{4.42}$$

and ΔF is the correction to the Hartree–Fock free energy. The presence of the term λ leads to a lower transition temperature and also a much stronger Curie–Weiss behavior as shown in Fig. IV.17. This figure also shows that strong Curie–Weiss behavior can occur in weak ferromagnets such as $ZrZn_2$ and nearly ferromagnetic metals such as $HfZn_2$.

Returing now to the ground state, the susceptibility $\chi_0(\mathbf{q})$ provides information on whether the long range order will be ferromagnetic or anti-

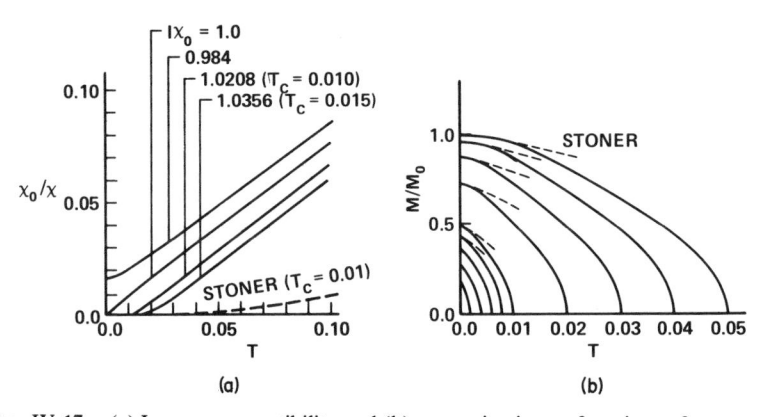

FIG. IV.17.   (a) Inverse susceptibility and (b) magnetization as functions of temperature for various values of the Stoner parameter. After Moriya and Kawabata.[52]

[52] T. Moriya and A. Kawabata, *Proc. Int. Conf. Magn.*, ICM-73 Vol. IV, p. 5, Moscow, 1974.

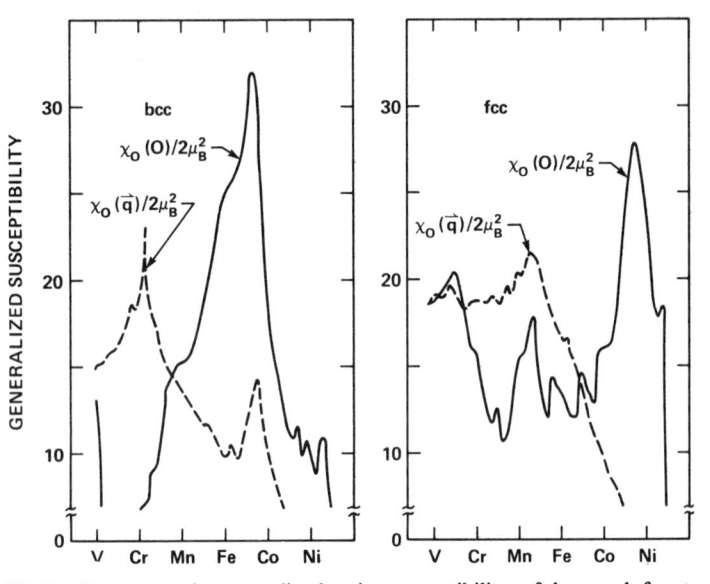

FIG. IV.18.   Paramagnetic generalized spin susceptibility of bcc and fcc transition metals. After Asano and Yamashita.[53]

ferromagnetic. Figure IV.18 shows Asano and Yamashita's[53] calculation of $\chi_0(\mathbf{q})$ for $\mathbf{q} = 0$ and $\mathbf{q} = \mathbf{Q} = (2\pi/a)(001)$. We see that fcc Fe is expected to be antiferromagnetic. This appears to be the case as shown in Fig. VI.1. Similarly, Fig. IV.18 shows that Cr should be antiferromagentic. This antiferromagnetic instability corresponds to a commensurate spin density wave (CSDW). Pure chromium actually exhibits a slightly incommensurate spin density wave which becomes commensurate upon alloying, for example, with the addition of a few percent of Al.

Overhauser and Arrott[54] were the first to suggest that the ground state of chromium was a spin density wave. The detailed origin of the large value of $\chi(\mathbf{Q})$ was pointed out by Lomer.[55] In Fig. IV.19 we show the electron and hole Fermi surfaces of chromium. Lomer noted that the pocket of holes at H will "nest" into a corner of the electron surface at Γ. As a result there will be a large number of states where $\epsilon_{\mathbf{k}} = \epsilon_{\mathbf{k}+\mathbf{q}}$ in Eq. (4.38). The properties of such a two-band model have been studied by Fedders and Martin.[56] Nesting is incorporated by introducing an order parameter which is off-diagonal in these band indices as well as the spin:

[53] S. Asano and J. Yamashita, *Prog. Theor. Phys.* **49,** 373 (1973).
[54] A. W. Overhauser and A. Arrott, *Phys. Rev. Lett.* **4,** 226 (1960).
[55] W. M. Lomer, *Proc. Phys. Soc. London* **80,** 489 (1962).
[56] P. A. Fedders and P. C. Martin, *Phys. Rev.* **143,** 245 (1966).

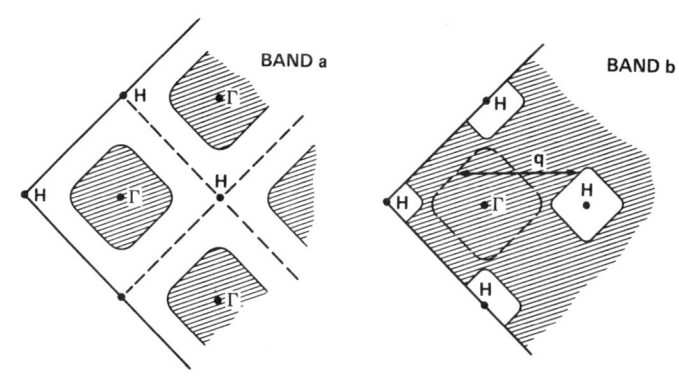

FIG. IV.19. Two-band model for chromium. Band a consists of pockets of electrons around Γ while band b consists of pockets of holes around H. If band b is shifted by the wave vector **q** the hole pockets will "nest" into the electron pockets. After Lomer.[55]

$\Sigma_{\mathbf{k}}\langle c_{\mathbf{k}+\mathbf{Q},1\uparrow}\, c_{\mathbf{k},2\downarrow}\rangle$. The mathematics is analogous to the BCS theory. In particular, the Néel temperature is related to the energy gap by $3.5\, k_B T_N = 2\Delta(0)$. Below T_N the amplitude of the SDW is proportional to $\Delta(T)$.

The reader may recognize the right-hand side of Eq. (4.38) as the same expression that enters the *electric* susceptibility, i.e., the Lindhard function. A divergence in the electric susceptibility would imply a charge density wave instability. Figure IV.20[57] shows the generalized susceptibility

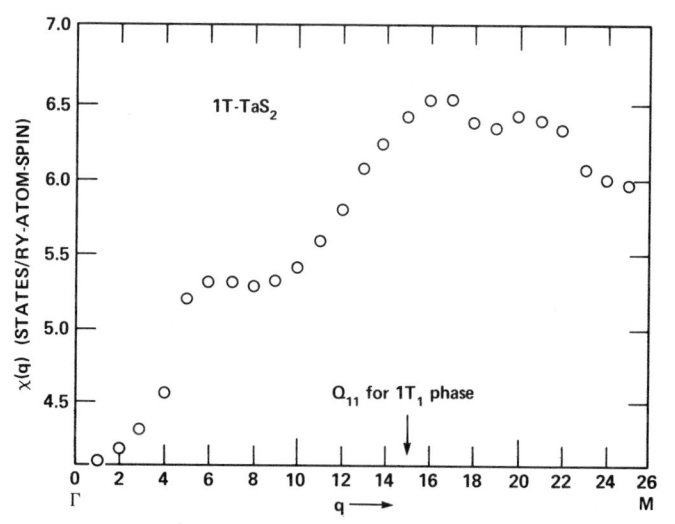

FIG. IV.20. Generalized electrical susceptibility of 1T-TaS$_2$ along the ΓM direction. After Myron et al.[57]

[57] H. W. Myron, J. Rath, and A. J. Freeman, *Phys. Rev. B* **15**, 885 (1977).

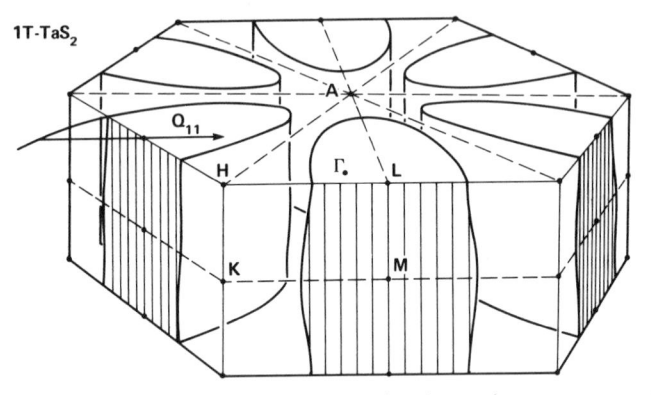

FIG. IV.21.   Fermi surface of 1T-TaS$_2$ showing the nesting wave vector Q_{11}.

for 1T $-$ TaS$_2$. We see that $\chi_0(\mathbf{q})$ does show a maximum in the vicinity of the wave vector that characterizes the ICDW. This peak arises from the "nesting" of those faces of the Fermi surface "spanned" by the vector Q_{11} illustrated in Fig. IV.21. The fact that the Fermi surface consists of vertical portions is a consequence of the two-dimensional nature of the crystal structure. This makes such materials susceptible to electronic instabilities.

What determines whether we obtain spin density waves or charge density waves? The answer is that it depends critically upon how we treat the direct and exchange terms in Eq. (4.34). Unfortunately, this is still controversial. Overhauser,[58] for example, argues that correlation effects favor CDW's over SDW's. Chan and Heine,[59] on the other hand, argue that correlation is unfavorable for CDW's. In fact, they claim that a CDW cannot occur unless it is accompanied by a soft phonon. In Fig. IV.22 we contrast

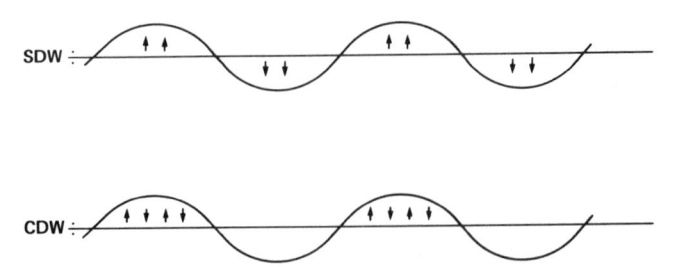

FIG. IV.22.   Comparison of a charge density wave and a spin density wave in an eight-electron system.

[58] A. W. Overhauser, *Phys. Rev.* **167**, 691 (1968).
[59] S.-K. Chan and V. Heine, *J. Phys. F* **31**, 795 (1973).

the SDW with the CDW. It is clear that Coulomb matrix elements between up and down spins will be larger in the CDW case because of the larger overlap. In Overhauser's treatment these matrix elements enter as their square in a second-order contribution to the energy, while in Chan and Heine they enter linearly through a screened interaction.

E. INDIRECT EXCHANGE

So far we have considered exchange between Bloch electrons and between localized electrons. In many systems both types of electrons coexist and are coupled together by what is generally referred to as s–d or s–f exchange. This leads to an indirect coupling between the localized moments themselves, known as the Ruderman–Kittel–Kasuya–Yosida (RKKY) interaction after those who proposed it and explored its consequences.

The form of the RKKY interaction is easily obtained within the framework of the generalized susceptibility. Let us assume that the interaction between a localized spin S_A located at $R_A = 0$ and the conduction spins s_i has the form of a contact interaction

$$-J \sum_i \mathbf{S}_A \cdot \mathbf{s}_i \delta(\mathbf{r}_i). \tag{4.42}$$

Each conduction spin therefore experiences an effective field given by

$$\mathbf{H}_{\text{eff}}(\mathbf{r}) = -(J/g\mu_B)\mathbf{S}_A\delta(\mathbf{r}). \tag{4.43}$$

The response of an electron gas to such a field is determined by its susceptibility $\chi(\mathbf{q})$. Since the Fourier transform of this field is

$$\mathbf{H}_{\text{eff}}(\mathbf{q}) = -(J/g\mu_B)\mathbf{S}_A \tag{4.44}$$

the spin density at r is

$$\mathbf{s}(\mathbf{r}) = (J/g^2\mu_B^2 V) \sum_{\mathbf{q}} \chi(\mathbf{q})e^{i\mathbf{q}\cdot\mathbf{r}}\mathbf{S}_A. \tag{4.45}$$

The spin susceptibility of a free-electron gas is[60]

$$\begin{aligned}
\chi(\mathbf{q}) &= \frac{3(N/V)g^2\mu_B^2}{8\epsilon_F} F\left(\frac{q}{2k_F}\right) \\
&= \frac{3(N/V)g^2\mu_B^2}{8\epsilon_F} \left[\frac{1}{2} + \frac{k_F}{2q}\left(1 - \frac{q^2}{4k_F^2}\right) \log\left|\frac{2k_F + q}{2k_F - q}\right|\right].
\end{aligned} \tag{4.46}$$

[60] See, for example, R. M. White, "Quantum Theory of Magnetism," p. 80. McGraw-Hill, New York, 1970.

The sum over $\mathbf{q}$ in Eq. (4.45) is evaluated by converting it into an integral. The result is

$$\frac{1}{V} \sum_{\mathbf{q}} \chi(\mathbf{q}) e^{i\mathbf{q}\cdot\mathbf{r}} = \frac{3g^2\mu_B^2(N/V)}{8\epsilon_F} \frac{1}{2\pi^2 r} \int dq\, qF\left(\frac{q}{2k_F}\right) \sin qr$$

$$= \frac{3g^2\mu_B^2(N/V)}{8\epsilon_F} \frac{k_F^3}{16\pi} \left\{ \frac{\sin 2k_F r - 2k_F r \cos 2k_F r}{(k_F r)^4} \right\}. \quad (4.47)$$

Thus, when a localized moment is introduced into a metal, the conduction spins develop an oscillating polarization in the vicinity of the moment. The total induced moment, obtained by integrating Eq. (4.45) over all space is $JN(E_F)\mu_B S$.

If there is another localized spin S_B at $\mathbf{r} = \mathbf{R}_B - \mathbf{R}_A$, it interacts with this induced spin density, leading to an effective interaction between the localized spins of the form

$$\boxed{\mathcal{H}_{RKKY} = -\frac{J^2}{g^2\mu_B^2 V} \sum_{\mathbf{q}} \chi(\mathbf{q}) e^{i\mathbf{q}\cdot\mathbf{r}} \mathbf{S}_A \cdot \mathbf{S}_B.} \quad (4.48)$$

Mary Beth Stearns[61] has suggested that this coupling is basically responsible for the ferromagnetism in Fe, Co, and Ni. In particular, she argues that the band structures such as that of iron in Fig. IV.15, show that the d-electrons are primarily localized with only about 5% having itinerant character. Because the density of these itinerant d-electrons is small their k_F is also small which places the first node in the RKKY oscillations beyond the nearest neighbors giving ferromagnetic coupling. The s-electrons, on the other hand, have a higher density which brings the node in giving antiferromagnetic coupling. However, since the d–d atomic exchange is presumably larger than the s–d atomic exchange, the ferromagnetic coupling wins. Stearns also argues that the amount of itinerant d-electrons increases as one moves to the left in the periodic table. This increases k_F weakening the coupling and eventually producing antiferromagnetism in manganese.

In Chapter I we indicated that this indirect exchange is responsible for spin–glass formation and in Chapter VI we shall see how it governs the complex spin structure in the rare earth metals. This coupling can also occur in a semiconductor by the virtual excitation of electrons into the conduction band. In this case the oscillatory behavior is replaced by the exponential $\exp(-k_g|\mathbf{R}_A - \mathbf{R}_B|)$ where $k_g^2 = 2(m_c + m_v)E_g/\hbar^2$, and E_g

[61] M. B. Stearns, *Phys. Rev.* **129**, 1136 (1963); **147**, 439 (1966); *Phys. Rev. B* **4**, 4069 and 4081 (1971); **8**, 4383 (1973).

is the energy gap. Notice that if either of the bands are flat, i.e., have infinite masses, the interaction vanishes.[62]

Problem IV.2
Discuss how you would expect the RKKY interaction to be modified by:
a. a large exchange enhancement in the "host";
b. a finite electron mean free path.

F. ANISOTROPY

Nickel crystallizes in a face centered cubic lattice. Below the Curie point the spontaneous magnetization develops along a cube diagonal. This tendency of the magnetization to align in a preferred crystallographic direction is referred to as *magnetocrystalline anisotropy*.

The rare earth garnets, insulating oxides with complex unit cells, are also cubic. However, it is observed that these materials have anisotropies *lower* than cubic that depend upon the direction of crystal growth. This is referred to as *growth-induced anisotropy*. A more dramatic example of growth-induced anisotropy occurs in thin films of amorphous ferromagnets. Evaporated films of Gd, Tb, and Ho alloyed with Fe and Co, for example, have strong anisotropy relative to the plane of the films.

In the case of magnetocrystalline anisotropy the free energy may be written as a function of the direction of the magnetization with respect to the crystallographic axes. Thus, if $\mathbf{M}/|\mathbf{M}| = \boldsymbol{\alpha} = (\alpha_1, \alpha_2, \text{ and } \alpha_3)$, the energy $E(\boldsymbol{\alpha})$ may be expanded in powers of α_1, α_2, and α_3. If the lattice has cubic symmetry the lowest order terms are

$$E(\boldsymbol{\alpha}) = E_0 + K_1(\alpha_1^2\alpha_2^2 + \alpha_2^2\alpha_3^2 + \alpha_3^2\alpha_1^2) + K_2\alpha_1^2\alpha_2^2\alpha_3^2. \quad (4.49)$$

The coefficients K_1 and K_2 are called the first- and second-order anisotropy constants. The equilibrium direction of the magnetization is obtained by minimizing the energy (4.49). The result will depend upon the relative values of K_1 and K_2. These results are summarized in Fig. IV.23. The anisotropy constants have a wide range in value. The values of K_1 for some typical materials are shown in Fig. IV.24.

Another feature of the anisotropy "constants" is that they are often strongly temperature dependent at least relative to the magnetization. This is illustrated in Fig. IV.25.[63] Notice that the value for Co changes sign. This results in a reorientation of the magnetization at this temperature.

[62] R. Sokel and W. A. Harrison, *Phys. Rev. Lett.* **36**, 61 (1976).
[63] C. Zener, *Phys. Rev.* **96**, 1335 (1954).

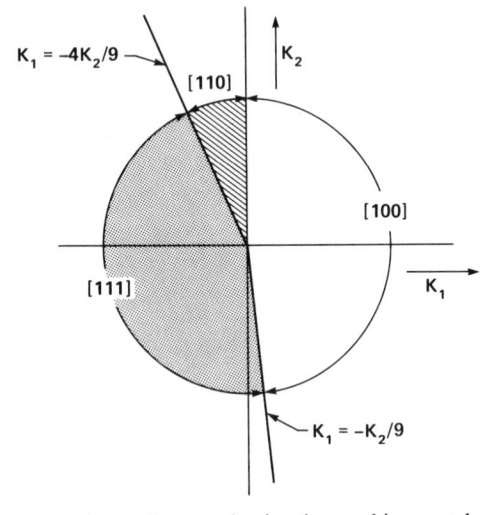

FIG. IV.23.   Easy directions of magnetization in a cubic crystal as a function of K_1 and K_2.

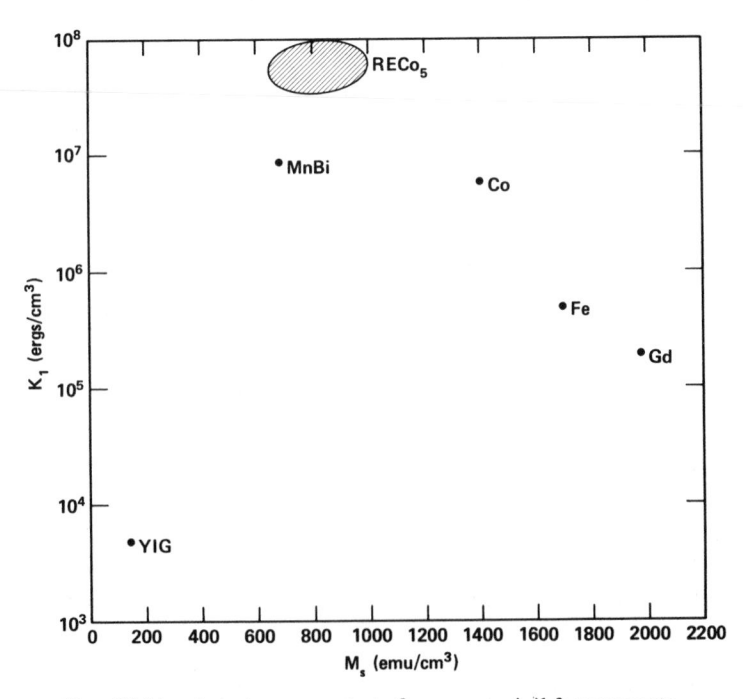

FIG. IV.24.   Anisotropy constants for some typical ferromagnets.

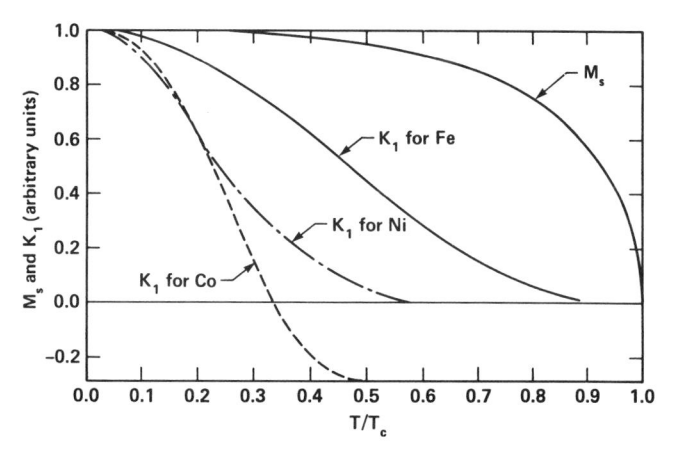

FIG. IV.25. Temperature dependence of the anisotropy constants of Fe, Ni, and Co compared with the normalized magnetization. After Zener.[63]

The interaction generally responsible for magnetocrystalline anisotropy is the spin–orbit interaction,

$$\mathcal{H}_{s-o} = \frac{\hbar}{4m^2c^2}\, \boldsymbol{\sigma} \cdot \nabla V \times \mathbf{p} \tag{4.50}$$

where $\mathbf{p}$ is the electron momentum, $\boldsymbol{\sigma} = 2\hbar\mathbf{S}$ and V is a sum of atomic potentials

$$V = \sum_i v(\mathbf{r} - \mathbf{R}_i). \tag{4.51}$$

Assuming the $v(\mathbf{r} - \mathbf{R}_i) = v(|\mathbf{r} - \mathbf{R}_i|)$ the interaction may be written in terms of the orbital angular momentum defined with respect to a point $\mathbf{R}_i$ as

$$\mathcal{H}_{s-o} = \sum_i \xi(|_4 - \mathbf{R}_i|)\mathbf{S} \cdot \mathbf{L}_i \tag{4.52}$$

where

$$\zeta(r) = (\hbar^2/2m^2c^2)(1/r)(\partial v/\partial r).$$

In itinerant systems an accurate calculation of the anisotropy constants requires a detailed knowledge of the band structure. During the 1960's a number of band calculations were carried out for the ferromagnetic transition metals, some of which we mentioned earlier. These calculations, together with various experimental Fermi surface studies, have established the general nature of the electronic states in these metals. Due to the presence of the itinerant d-electrons the Fermi surfaces consist of several "sheets" as illustrated for Ni in Fig. IV.26.[64]

There have been a number of calculations of the magnetic anistropy of Ni, the most recent being that of Kondorskii and Straube.[65] A similar calculation was made by Furey.[66] Furey begins by considering a basis consisting of a linear combination of atomic orbitals (LCAO), $\Phi_{\mu k}(\mathbf{r})$, to describe the d-electrons plus orthogonalized (to these d-states) plane waves (OPW), $\Phi_{k K}(\mathbf{r})$, for the s-electrons:

$$\psi_{nk}(\mathbf{r}) = \sum_{\mu=1}^{5} a_{n\mu}\Phi_{\mu k}(\mathbf{r}) + \sum_{K} a_{nK}\Phi_{kK}(\mathbf{r}). \qquad (4.53)$$

Four reciprocal lattice vectors are used to characterize the OPW's. This basis (5 LCAO's and 4 OPW's for each spin) is used to construct the 18×18 matrix of the one-electron Hamiltonian. This matrix contains

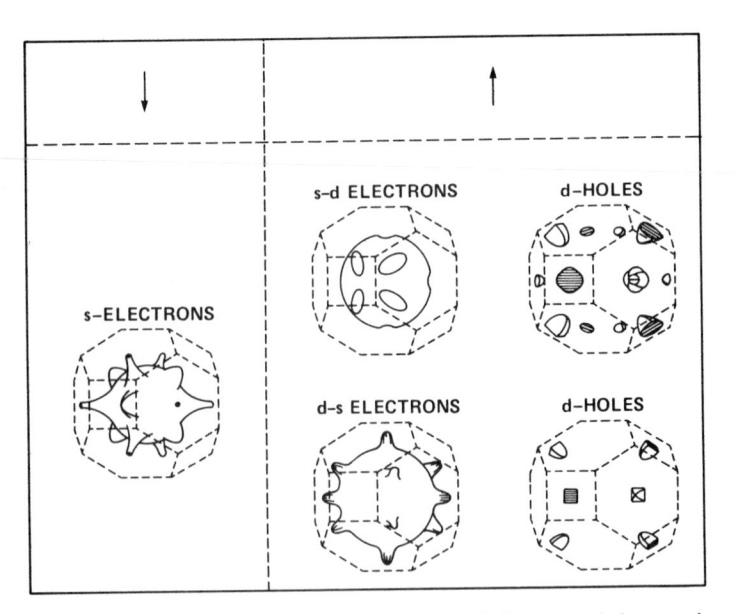

FIG. IV.26. Fermi surfaces for ferromagnetic nickel for up and down spins. After Ehrenreich et al.[64]

[64] H. Ehrenreich, H. R. Philipp, and D. J. Olechna, *Phys. Rev.* **131**, 2469 (1963).

[65] F. I. Kondorskii and E. Straube, *Sov. Phys.—JETP (Engl. Transl.)* **36**, 188 (1973).

[66] W. N. Furey, Ph.D. Thesis, Harvard University, Cambridge, Massachusetts (1967).

adjustable parameters which are "determined" by fitting the resulting band structure to a first-principles calculation[67] carried out only at certain symmetry points. Correlation is introduced by adding adjustable parameters along the diagonal. For example

$$\langle \mathbf{k}\mu\sigma | \mathcal{H}_{corr}^{ferro} - \mathcal{H}_{corr}^{para} | \mathbf{k}\mu\sigma \rangle = U_{eff}^{d-d} [\langle n_{\mu,-\sigma}\rangle^{ferro} - \langle n_{\mu,-\sigma}\rangle^{para}]. \quad (4.54a)$$

and

$$\langle \mathbf{k}\mathbf{K}\sigma | \mathcal{H}_{corr}^{ferro} - \mathcal{H}_{corr}^{para} | \mathbf{k}\mathbf{K}\sigma \rangle = -J_{eff}^{s-d} \sum_{\mu} [\langle n_{\mu\sigma}\rangle^{ferro} - \langle n_{\mu\sigma}\rangle^{para}]. \quad (4.54b)$$

Notice that since these depend upon the band index μ, and since the t_{2g} and e_g composition in the wave function is $\mathbf{k}$-dependent, the resulting band splittings will be $\mathbf{k}$-dependent. The eigenvalues $E_{n\mathbf{k}}$ are obtained by diagonalizing the 18×18 matrix.

The anisotropy is calculated by considering the shift in the eigenvalues due to the spin–orbit interaction. Since ζ is small one can use perturbation theory. For cubic symmetry the lowest order correction to the energy appears in the fourth order, $E_{n\mathbf{k}}^{(4)}$, Furey shows that K_1 is proportional to

$$\sum_{\substack{n,\mathbf{k} \\ (E_{n\mathbf{k}}>E_F)}} \left[\Delta E_{n\mathbf{k}}^{(4)}\Big|_{\alpha=(1/\sqrt{2},1/\sqrt{2},0)} - \Delta E_{n\mathbf{k}}^{(4)}\Big|_{\alpha=(1,0,0)} \right]. \quad (4.55)$$

Numerical evaluation shows that the main contribution to K_1 comes from the region near the X-point where there is a degeneracy of the bands. Furthermore, the contributions from this region appear with different signs. As the temperature increases these contributions tend to cancel one another which would explain the observed decrease.

A similar calculation[68] for the incommensurate spin density wave state of chromium, but using a simplified band model, shows that the temperature dependence of the anisotropy is that of the square of the antiferromagnetic gap. In particular, it does not go through zero at 120 K where the magnetization rotates from one direction to another. A more realistic band structure is probably necessary in order to understand this spin reorientation.

In ionic systems the spin–orbit interaction also plays an important role, although magnetic dipole or electric quadrupole interactions often contribute. For a review of these mechanisms, see Kanamori.[69]

[67] L. Hodges, H. Ehrenreich, and N. D. Lang, *Phys. Rev.* **152**, 505 (1966).

[68] J. W. Allen and C. Y. Young, *Phys. Rev. B* **16**, 1103 (1977).

[69] J. Kanamori, *in* "Magnetism (G. Rado and H. Suhl, eds.), Vol. I, p. 127. Academic Press, New York, 1963.

V. Experimental Considerations

The cumulative effects of long range order can give rise to gross macroscopic properties that are not even detectable by the most sophisticated microscopic techniques. For example, the distortion of the unit cell of ferromagnetic single crystal iron can, via the effect of magnetostriction, be measured with a ruler, but is too small to be detected by X-ray diffraction. Magnetic domains in films of cubic garnets tend to orient themselves in the direction perpendicular to the plane presumably due to a growth-induced anistropy. This result is essential for the formation of magnetic bubbles, but could never have been anticipated from microscopic measurements. If the typical energy gaps of superconductivity and/or superfluidity were two orders of magnitude larger, these phenomena, like magnetism, would have been familiar to the ancients—once the temperature is comparable to the energy gap, one can scarcely avoid being aware of their existence.

In this chapter we turn to a consideration of various experimental techniques for probing long range order. The perspective and intellectual insight gained by viewing ordering phenomena in a unified way do not preclude using a wide variety of experimental methods based on different physical properties. In fact, solid state phenomena are often so complex that a variety of measurements are required before a satisfactory picture can be assembled. While references and texts are available for many of those methods, it is of value to collect them and illustrate their complementary nature. We shall do this, for the most part, by considering situations which emphasize the strength of the particular method being described. Other experimental techniques are also discussed in later chapters in the context of specific problems. Two classes of experimental techniques most frequently used measure either i) the macroscopic system average response, or ii) the microscopic response of individual atoms or nuclei to an applied field under the control of the experimenter. Examples of the former are thermodynamic and transport properties; examples of the latter are electron spin and nuclear magnetic resonance. Another class of useful techniques involves the inelastic excitation of the system by an external probe such as a beam of neutrons. We discuss these examples in the sections below.

166

A. CALORIMETRY

The most general understanding of a phase transition is gained by studying the heat capacity. The excited states of the system continuously manifest their presence as the temperature excites the system out of its ground state by the controlled input of heat. Unfortunately, however, there is generally no unique way to analyze and apportion the heat capacity, and other derived thermodynamic quantities, among different degrees of freedom which may exist and thus to identify the microscopic mechanisms. At low temperatures, where the phonon contributions can be more easily subtracted, heat capacity measurements are particularly valuable. This is true in the investigation of new superconductors where other features conspire to make calorimetry particularly helpful. The diverging heat capacity near the λ-point of ^{4}He has been followed all the way to $\epsilon = 10^{-7}$, which is so close to the transition that one has to worry about the spread in T_c due to a pressurehead of < 1 mm ^{4}He. Contributions of short range order to the heat capacity are also visible above the ordering temperature as a high temperature tail which connects smoothly to the ordering peak. Frequently, as in Fig. I.1, it is obvious that the heat capacity associated with the high temperature ordering cannot easily be separated from the underlying lattice heat capacity although, as a first approximation we would like to be able to separate the magnetic and lattice degrees of freedom. Sometimes the separation can be accomplished experimentally by subtracting the measured heat capacity of a sample which might be expected to have the same lattice properties, but is nonmagnetic.

Low-temperature calorimetry was turned into a quantitative tool by Nernst at the turn of the century and continued its development under Giauque, Simon, and others.[1] A full description of the isothermal calorimeter is given by Giauque in his collected works.[2] Today more sensitive thermal detectors and circuits continue to add new capability to this venerable field. It is possible to measure samples of only a few milligrams in a number of convenient ways.[3]

1. Example: Identification of Superconductivity

Provided adequate low-temperature facilities are available, superconducting transitions are easy to observe because of the dramatic and sud-

---

[1] See, for example, W. F. Giauque, "Some Consequences of Low-Temperature Research in Chemical Thermodynamics," p. 91. Les Prix de Nobel, 1949.

[2] "Scientific Papers of W. F. Giauque," Vol. I, Paper 38. Dover, New York, 1969.

[3] C. N. King, R. B. Zubeck, and R. L. Greene, *Proc. Int. Conf. Low Temp. Phys., 13th, 1972* Vol. 4, p. 626 (1974).

den disappearance of electrical resistance, or because of the appearance of a strong diamagnetic moment in an ac magnetic field. These techniques may also give a large and spurious ac or dc response from a small amount of superconducting phase present perhaps even with a concentration insufficient to be detected by X-ray diffraction, or other metallurgical techniques. If the minority phase happens to exist in connected loops throughout the sample (as it would if it existed in the grain boundaries of a polycrystalline sample, for example) it could give zero resistance and a complete (volumewise) diamagnetic signal corresponding to screening all the applied flux. Unless care is taken, the superconductivity would be attributed to the perhaps-not-even-superconducting majority phase. Heat capacity is a more reliable measurement because it is an extensive (bulk) property, and not so sensitive to trace phases or impurities. In particular, the magnitude of the heat capacity anomaly at the superconducting transition is directly proportional to the amount of the superconducting phase present (except in the unusual circumstance where "gapless" superconductivity is found).

As an example consider elemental metallic uranium which has three polymorphs: the cubic γ-phase stable at high temperatures, a tetragonal β-phase stable between 661 and 769 °C, and the low temperature α orthorhombic phase. ac susceptibility measurements of high purity samples of polycrystalline α-U give a response which indicates the whole volume the sample is excluding flux, i.e., a full diamagnetic (superconducting) signal. The resistive superconducting transition occurs from 0.7 to 1 K. Until the heat capacity investigations were undertaken, it was therefore assumed that α-U was superconducting with a $T_c \simeq 0.7$ to 1 K. The data of Fig. V.1[4], however, suggest that the transitions observed by magnetic signals in α-U are probably due to trace amounts of β-phase present in too small a concentration to be detected other than by superconductivity. It was necessary to stabilize the β-phase in bulk so that its properties could be investigated. This was done with either Pt or Cr impurities. By studying the concentration dependence in both cases and extrapolating to zero concentration the transition of pure meta-stable, β-U was established as near 0.8 K. The heat capacity discontinuities and the magnetic and resistive transitions of the stabilized β-phases are in agreement, although, as usual, the latter are sharper.

The $\gamma \rightarrow \beta \rightarrow \alpha$ transitions in uranium occur necessarily with volume changes and with the build up of some stress as the solid is cooled. If impurities are rejected to grain boundaries they can stabilize the β-phase

[4] B. T. Matthias, T. H. Geballe, E. Corenzwit, K. Andres, G. W. Hull, J. C. Ho, N. E. Phillips, and D. K. Wohlleben, *Science* **151**, 958 (1966).

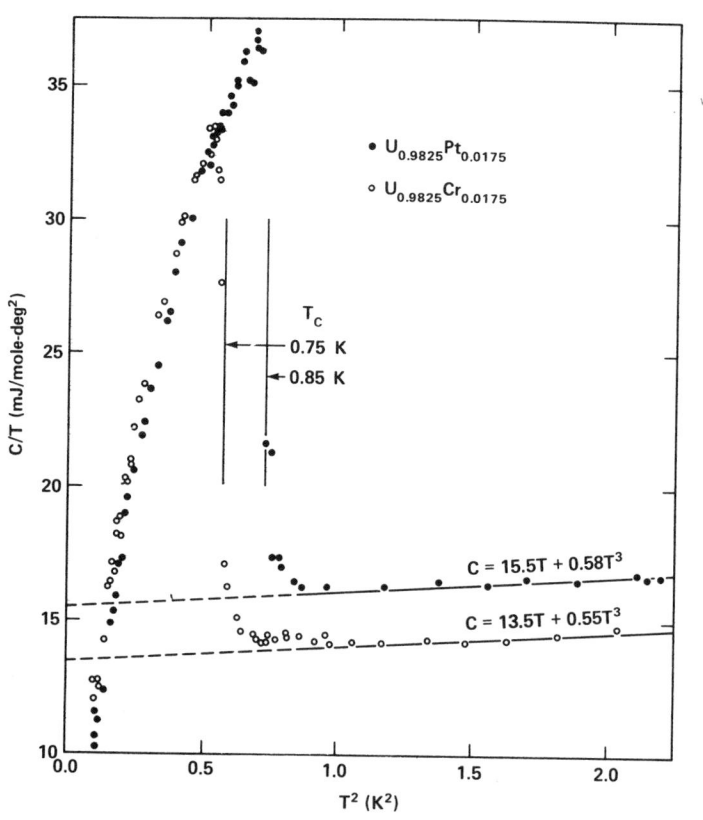

FIG. V.1. The heat capacities of $U_{0.9825} Pt_{0.0175}$ and $U_{0.9825} Cr_{0.0175}$ between 0.3 and 1.5 K showing the superconducting transitions. The normal-state intercept gives the values of γ, the electronic heat capacity coefficient. The equations for the normal-state heat capacities were fitted to data taken between 1 and 6 K. After Matthias et al.[4]

in a grain boundary just as the entire β-phase can be stabilized by a percent or so of impurity in the bulk. In a resistive measurement it is only necessary for one "filament" of superconducting phase to connect the voltage leads in order to observe a "complete" superconducting transition, i.e., to give 0-resistance. With ac magnetic susceptibility measurements or measurements of the magnetic moment in an applied field, the complete volume also gives a superconducting signal indicating the filaments are connected in grain boundaries.

In some cases samples can be prepared which are homogeneous and which do not trap magnetic flux. Magnetization curves are then reversible, and the *explusion* of magnetic flux is observed when the sample becomes superconducting as it is cooled in a magnetic field (the Meissner effect to be discussed in the following section). The magnetic measurements then give the bulk thermodynamic quantities just as the heat capacity does. The condition of homogeneity, however, seldom applies unless the superconductors can be prepared as annealed single crystals. Inhomogeneities trap flux and generally cannot be distinguished from filaments of a superconducting second phase. Nevertheless, the magnetic measurements are so simple and useful that they are commonly employed in the search for new superconducting phases. Fairly crude methods, such as fracturing or "powdering" samples of intermetallic compounds which are usually brittle can be effective in breaking up loops of super conducting filaments. Therefore, it has become a common and fairly effective way to distinguish "bulk" from "filamentary" superconductivity simply by observing the effective volume of superconducting signals as the sample is powdered into finer and finer fragments. If the powdering destroys the superconductivity it is quite possible that the majority phase is not superconducting.

B. Magnetic Susceptibility

The magnetic susceptibility has been a valuable diagnostic property for understanding materials ever since Pierre Curie's landmark investigations. There are a number of convenient methods for measuring the susceptibility in use today.

The *Faraday balance* has continued to improve, benefiting from the development of microbalances with large dynamic range. Faraday balances are particularly useful for small samples. In the typical setup one measures the force on a sample placed in a field gradient.

$$\mathbf{F}(\text{dynes}) = \nabla(\mathbf{M}(\text{emu}) \cdot \mathbf{H}(\text{Oe})). \qquad (5.1)$$

If the field is in the x direction, for example, with a gradient in the z direction, the force in the z direction normalized to 1 gram of sample is

$$F_g = (M_s + \chi_g H_x)(dH_x/dz) \qquad (5.2)$$

where M_s may be an easily saturated ferromagnetic component (M_s independent of H) and $\chi_g H_x$ is the induced magnetization.

Problem V.1

Calculate the error caused by the accidental inclusion of 10^{-6} grams of Fe in a 1 gram sample of Cu. (This 1 ppm of Fe might have come from using steel tweezers or just as a speck of dust.) Use typical values for an electromagnet with inhomogeneous pole pieces:

$$H_x = 10^4 \ Oer,$$

$$\chi_g(Cu) \simeq 0.1 \times 10^{-6} \ emu/g$$

$$dH_x/dz = 10^3 \ Oe/cm.$$

b. Describe a simple procedure that can eliminate the unwanted effect of the Fe (by changing H_x or dH_x/dz).

Vibrating sample magnetometers are as sensitive as a balance and also rugged. They operate by detecting the dipole field of an oscillating sample in a uniform field. Integration of the signal can be used to measure saturation moments and anisotropies in high fields.[5]

There are a few special situations where greater sensitivity is needed than that provided by the two techniques already mentioned. Examples might be the detection of superconducting fluctuations above T_c or the measurement of nuclear susceptibility. In such cases one can use a so-called SQUID (Superconducting QUantum Interference Device) magnetometer which exploits the sensitivity of Josephson tunnel junctions to magnetic fields. The principle of a SQUID is described at the end of this chapter.

1. Long Range Magnetic Order

As we mentioned in Chapter I long range magnetic order takes various forms—ferromagnetism, antiferromagnetism, ferrimagnetism, etc. Measurement of the magnetization of such systems obviously constitutes an important piece of information. In systems with simple configurations the magnetization responds in a characteristic way to the applied field, H_a. In a paramagnet, for example, M is a linear function of H_a. In a ferromagnetic, with the field applied along the easy axis, the magnetization first increases with a slope determined by the demagnetizing factor, $N_\parallel$, and then saturates. In dealing with ferromagnetic systems one must keep in mind the differences between the applied field, H_a, and the internal field, $H_i = H_a - 4\pi NM$. In systems with complex ordering the magnetization may display unusual behavior. Figure V.2,[6] for example, shows the magnetization of holmium as a function of field. These abrupt changes are associated with the transition from a helical spin configuration to a ferromagnetic one.

The susceptibility is the differential increase in magnetization with

[5] See, for example, S. Foner and E. J. McNiff, *Rev. Sci. Instrum.* **39**, 171 (1968).
[6] D. L. Strandburg, B. Legvold, and F. H. Spedding, *Phys. Rev.* **127**, 2046 (1962).

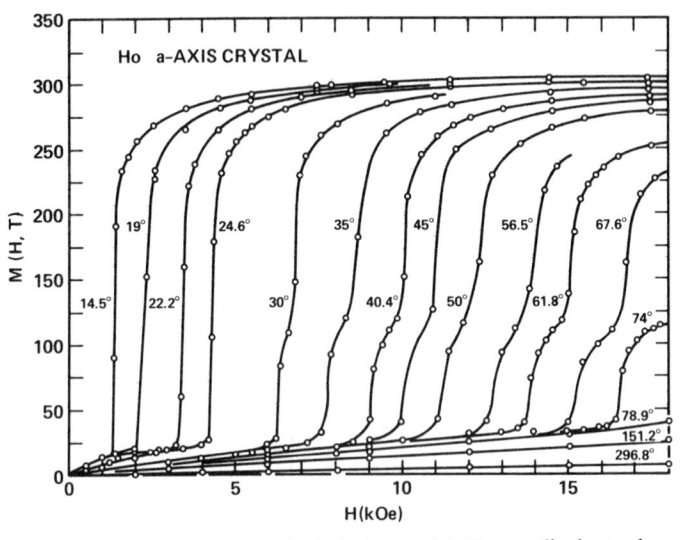

FIG. V.2. Magnetization isotherms for holminum with H perpedicular to the c-axis and along the hard planar direction. After Strandburg et al.[6]

field, and for linear systems can be obtained simply by dividing the magnetization by the field. Far above their ordering temperatures the inverse susceptibility of all magnetic systems is linearly proportional to temperature as illustrated in Fig. V.3. From the slope of such high-temperature data one determines the *effective moment*, p_{eff}, or the effective number of Bohr magnetons, $n_{eff} = p_{eff}/\mu_B$, using the Curie–Weiss relation

$$\chi^{-1} = \frac{3k_B(T - \theta)}{Nn^2_{eff}\mu_B^2}. \tag{5.3}$$

If N is Avogardro's number, χ is the molar susceptibility, i.e., emu/mole, χ_m. (Experimentally, χ is often measured per gram of material; hence, N is the number of atoms per gram and the susceptibility is denoted χ_g. Susceptibilities per cm^3, χ_v, are frequently used in analysis; and, of course, in that case, N is the number of atoms per cm^3.)

The measured specific susceptibility of silver, for example, at room temperature is $\chi_g = -0.19 \times 10^{-6}$ emu/g. When such measured susceptibilities are used to determine properties of valence electrons such as density of states, etc., it is necessary to subtract the diamagnetism of the core electrons. The susceptibilities of various ionic cores have been determined by measuring the susceptibilities of ionic compounds and assuming

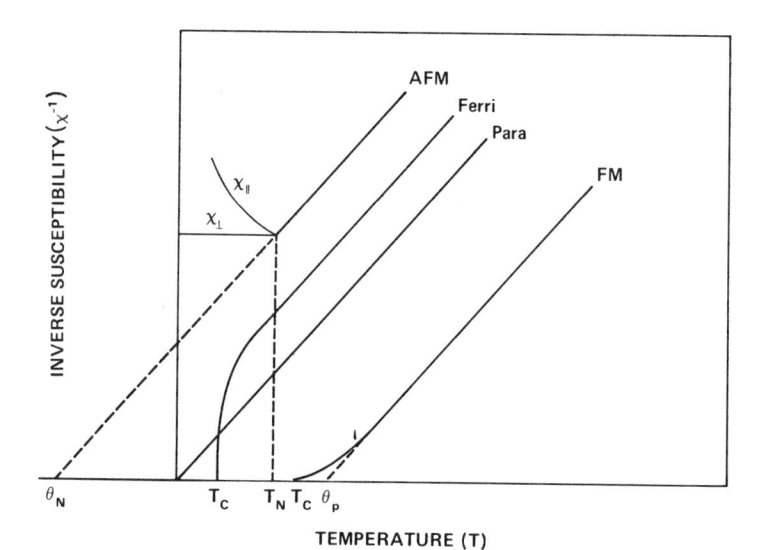

FIG. V.3. Characteristic inverse susceptibilities associated with magnetic ordering.

additivity. A table of such values is given by Selwood.[7] This table gives the molar susceptibility of Ag^+ as -24×10^{-6} emu/mole. This corresponds to -0.22×10^{-6} emu/g. This implies that the valence electron susceptibility of silver is $+0.03 \times 10^{-6}$ emu/g.

The actual *saturation moment, n_A*, is obtained either from the value of the magnetization at low temperatures or by plotting M vs. $1/H$ and extrapolating the data to infinite field, i.e., the zero intercept of M.

Frequently there are deviations from the Curie law near the ordering temperature due to short range order as illustrated by the curvature in χ^{-1} for the ferromagnetic example in Fig. V.3. The intercept of the linear extrapolation of χ^{-1} from high temperatures defines the *paramagnetic, θ_p*. To measure the actual ordering temperatures one should use a calorimetric method. However, T_c can be reasonably accurately determined from magnetization data by plotting the data in a manner suggested independently by Belov and Goryaga[8] and Arrott.[9] In Chapter II we saw that at T_c the order parameter, ψ, and its conjugate field, η, were related by $\psi^2 = (1/C)\eta/\psi$. In magnetic terms this means that if M^2 is plotted as a

[7] P. W. Selwood, "Magnetochemistry," p. 78. Interscience, New York, 1956.
[8] K. P. Belov and A. N. Goryaga, *Fiz. Met. Metalloved.* **2**, 3 (1956).
[9] A. Arrott, *Phys. Rev.* **108**, 1394 (1957).

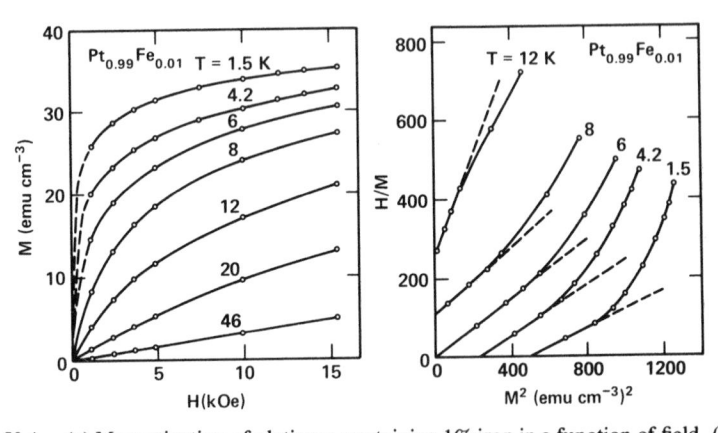

FIG. V.4.   (a) Magnetization of platinum containing 1% iron in a function of field. (b) Arrott plot of the same data. After Fawcett and Sherwood.[10]

function of H/M the data will pass through the origin at $T = T_c$. In Fig. V.4a we show the magnetization data for platinum containing 1% Fe. Although platinum itself is not magnetic, 1% Fe is sufficient to induce ferromagnetism. It would be difficult to say what T_c is from these data. Figure V.4b[10] shows the corresponding "Arrott plot" which clearly indicates $T_c = 6$ K. We might mention that palladium is even more unstable with respect to such impurity-induced ferromagnetism—in this case 1% Fe gives a Curie temperature of 40 K. Further, if N in Eq. (5.3) is taken as the number of Fe atoms per mole of alloy, then n_{eff} is experimentally found to be 12. The so-called "giant-moment" of Fe is due to the induced moments on the Pd atoms being added to the Fe moment.

Data such as n_{eff}, θ_p, etc. are conveniently tabulated in the series of reference volumes entitled *Landolt–Bornstein* published by Springer.

2. Meissner Effect

The variety of ways of studying superconductivity by its magnetic response are discussed in the classic monograph of Schoenberg.[11] The description of the magnetic properties can be formulated in two ways, both of which are equivalent in recognizing the observable physics that $B_i = 0$ inside the superconductor. For detecting superconductivity as we have already discussed, and in studying thermodynamic properties, it is convenient to treat the superconductor as having a uniform magnetization

[10] E. Fawcett and R. C. Sherwood, *Phys. Rev. B* **1**, 4361 (1970).

[11] D. Shoenberg "Superconductivity," 2nd ed. Cambridge Univ. Press, London and New York, 1960 (reprint).

$M = -H_{\text{applied}}/4\pi$ (assuming a needle-shaped sample to avoid introducing unnecessary demagnetizing effects here). The alternative (and microscopically correct) approach recognizes the flow of screening currents explicitly. This description is necessary for samples with dimensions less than λ, or situations where the order parameter varies spatially (such as the intermediate and mixed states). For bulk samples the equivalent equations are summarized below.

Region	Uniform diamagnetic description	Shielding current description
Outside superconductor	$B = H_a$	$B = H_a$
Inside superconductor	$B_i = 0$ $H_i = -4\pi M$ $\chi = -\frac{1}{4}\pi$	$B_i = 0$ $H_i = 0$ $\sigma_{\text{surface}} = (c/4\pi)H_a \times n$ (n is normal to surface)

Well-annealed homogeneous samples of both types I and II superconductors undergo reversible transitions, both as a function of field and temperature, into the superconducting state. In zero field, the transitions are, of course, second order with the discontinuous behavior already discussed. In the presence of a magnetic field, they can be thermodynamically reversible and described by the phase diagrams shown in Fig. V.5.

From the fact that the normal and superconducting free energies differ

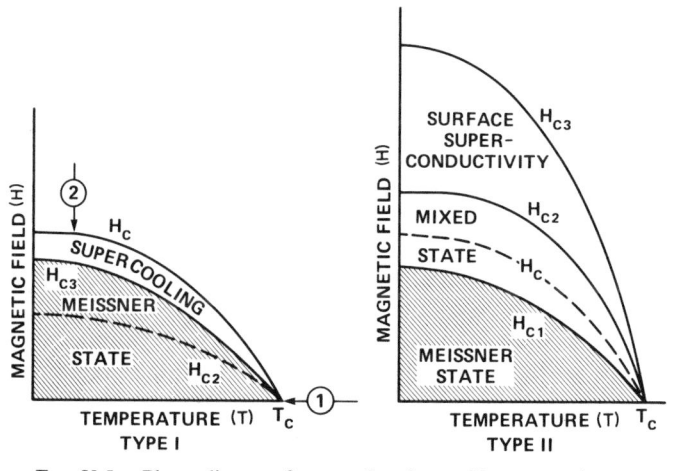

FIG. V.5.   Phase diagram for type I and type II superconductors.

by the condensation energy $H_c^2/8\pi$ ergs/cm³, we find that the entropy difference between the normal and Meissner superconducting states is

$$\frac{\Delta S}{V} = -\frac{H_c}{4\pi}\left(\frac{dH_c}{dT}\right).$$ (5.4)

At $T = 0$, $\Delta S = 0$ means $(dH_c/dT) = 0$. Differentiating the expression for ΔS gives

$$\frac{\Delta C}{V} = -\frac{T}{4\pi}\left(H_c\frac{d^2H_c}{dT^2} + \left(\frac{dH_c}{dT}\right)^2\right).$$ (5.5)

At T_c, $H_c = 0$ and Eq. (5.5) reduces to

$$\frac{\Delta C}{V}\bigg|_{T_c} = -\frac{T_c}{4\pi}\left(\frac{dH_c}{dT}\right)^2_{T_c}.$$ (5.6)

ΔC can be measured magnetically from the slope of the critical field curve *providing the transition is reversible.* In fact, the extent of the agreement between the magnetic and thermal ΔC's is a measure of the reversibility. For highly irreversible type II transitions such as La, or β-uranium (Fig. V.1), the measured dH_c/dT can be factors of 10 to 100 too large even though the transition itself is quite sharp in zero field.

In finite fields and temperature, the expression for ΔS is finite, and the transition becomes first order. Experimentally, one observes latent heats in magnetic fields and, in fact, by reducing the field along path 2 of Fig. V.5, type I superconductors can be kept in the normal state well below H_c; that is, they can be "supercooled." The supercooling occurs because of the positive surface energy it costs to nucleate a superconducting region which must, by definition, be a volume at least of the order of a coherence length. For Al, where $\xi \sim 16 \times 10^{-4}$ cm is about thirty times greater than λ, supercooling fields as low as $\sim 0.03\ H_c$ were found by Faber.[11a]

It was hypothesized by Silsbee long ago, and shown to be correct, that for reversible type I cylindrical samples, a critical current exists whose magnitude is just that necessary to generate the field H_c at the surface of the superconductor. The shielding currents and transport currents necessary to suppress the transition have the same magnitude and the same spatial distribution.

If the sample is not a long thin needle parallel to the applied field so that the field at the surface is not of uniform magnitude due to demagnetizing effects, the sample will enter into an *intermediate state* with coexisting

[11a] T. E. Faber, *Proc. Roy. Soc. (London)* A **241**, 531 (1957).

normal and superconducting regions as soon as the field is high enough to become critical on at least part of the surface. The intermediate state bears some relationship to a ferromagnetic multidomained structure, as we shall see in Chapter VIII.

3. Superconducting Fluctuations

The large correlation length in superconductors makes fluctuation effects above the critical temperature very small. According to the fluctuation–dissipation theorem the amplitude of fluctuations of the order parameter is related to the response of the order parameter to its conjugate field, the "Δ susceptibility," χ_Δ. Unfortunately, because the order parameter is "off-diagonal," as we have pointed out in Chapter I, there is no conjugate thermodynamic field available to the experimentalist. Thus, one cannot easily measure χ_Δ and study fluctuations. Ferrell[12] and Scalapino[13] have suggested ways of measuring the Δ susceptibility by coupling a superconductor just above its critical temperature to another well below its T_c through a Josephson junction. The current–voltage characteristic then turns out to be a measure of the Fourier transform of χ_Δ. However, in practice this is difficult. It is possible to observe fluctuations above T_c by their contribution to both the electrical conductivity,[14] and the magnetic susceptibility.[15] Since these effects are so small it has only been through the close collaboration of theory and experiment that they have been established.

Superconducting fluctuations above T_c can be seen experimentally,[15] as a diamagnetic contribution to the static susceptibility. The detection of the fluctuations is helped by the fact that the normal state (Pauli) susceptibility is usually small and temperature independent. A favorable (decreasing) fluctuation in entropy of the order of k_B over a small region with dimensions characteristic of the order parameter will cause a fluctuation into the superconducting state with free energy G_S if

$$k_B T = G_S - G_n.$$

Such fluctuations will give a diamagnetic contribution to χ and will increase as T_c is approached from above and $(G_s - G_n) \to 0$. The divergence of χ at T_c will go as $\epsilon^{1/2}$ for small regions with a volume ξ^3 (i.e.,

[12] R. A. Ferrell, *Proc. Conf. Sci. Supercond., 1969* p. 265 (1971).
[13] D. J. Scalapino, *Phys. Rev. Lett.* **24**, 1052 (1970).
[14] R. E. Glover, *Phys. Lett. A* **25**, 542 (1967).
[15] J. P. Gollub M. R. Beasley, R. W. Newbower, and M. Tinkham, *Phys. Rev. Lett.* **22**, 1238 (1969).

are three-dimensional). This can be seen using the qualitative arguments of Schmid.[16]

Consider the general expression for the diamagnetic susceptibility of an atom with a mean square radius $\langle r^2 \rangle$:

$$\chi = -\tfrac{1}{6}(ne^2/mc^2)\langle r^2 \rangle \qquad (5.7)$$

and make the appropriate identifications

$$n \to |\psi|^2$$

$$e \to 2e$$

$$m \to 2m.$$

The free energy difference is given by the GL expression (neglecting the fourth-order term since we are near T_c)

$$G_s - G_n = A(T)|\psi|^2 V. \qquad (5.8)$$

Combining this with the GL expression for

$$\xi(T) = \frac{\hbar}{(2m|A(T)|)^{1/2}} \qquad (5.9)$$

gives the excess free energy of the superconducting droplet,

$$G_s - G_n = \frac{\hbar^2}{2m\xi(T)^2}|\psi|^2 V \approx k_B T. \qquad (5.10)$$

The right-hand side of (5.10) follows from the assumption that $\delta G \sim T\delta S \sim k_B T$. This gives an expression for $|\psi|^2$ which, when substituted in Eq. (5.7) for n, gives

$$\chi = \frac{\pi^2}{3}\frac{k_B T}{\Phi_0{}^2}\frac{\xi^2(T)\langle r^2 \rangle}{V} \qquad (5.11)$$

where Φ_0 is the flux quantum $= hc/2e$. For a three-dimensional material, $V \sim (4/3)\pi\xi^3$, $\langle r^2 \rangle = \xi^2$, and Eq. (5.11) reduces to

$$\chi_{3d} \sim -10^{-7}\left(\frac{T_c}{T - T_c}\right)^{1/2}. \qquad (5.12)$$

For a two-dimensional superconductor, i.e., thickness $d < \xi$, the fluctuating volume is $\sim \pi\xi^2 d$ so the susceptibility with the field perpendicular to the plane is

[16] A. Schmid, *Phys. Rev.* **180**, 527 (1969).

$$\chi_{2d}{}^{\perp} \sim (\xi/d)\chi_{3d} \sim - \left(\frac{T_c}{T - T_c}\right) \qquad (5.13)$$

while the parallel susceptibility still varies as χ_{3d}.

Lawrence and Doniach[17] have used a model in which the 2d layers are weakly coupled by Josephson tunneling and find near T_c that the 2d diamagnetic Curie–Weiss dependence changes to an anisotropic 3d-like behavior. Such fluctuation behavior was looked for in layered crystals of $TaS_2 \cdot C_5H_5N_5$. These are single crystals in which metallic layers 6 Å thick of TaS_2 are separated by layers of the organic molecule pyridine also 6 Å thick. They become superconducting near 3 K. Initial measurements[18] implied that the diamagnetic fluctuations could be seen to 30 K, or to $T/T_c \sim 10$, which is too large to be explained by any theory. Continuing studies showed,[19] however, that the background Pauli susceptibility was *itself* decreasing due to the formation of charge density waves. True fluctuation susceptibility was determined by measuring its dependence on magnetic field and extends at best to 2 or 3 times T_c.[20]

C. Resistivity Measurements

Even an abbreviated discussion of transport properties would require an entire monograph. We shall therefore omit the Hall effect, thermoelectric effect, etc., and concentrate on how the resistivity behaves in the vicinity of several of the systems we have been considering.

1. Magnetic Systems

As we remarked in Chapter II, the fact that the very short correlation range in magnetic systems gives rise to short range order far above the transition temperature. This results in a Curie-type susceptibility, when magnetic measurements are made, and an extra scattering when resistance measurements are made. Figure V.6 shows the resistivity versus temperature for three ferromagnets. The data on Ni clearly show that the resistivity decreases more rapidly below the Curie point. The Gd data include the effects of lattice expansion and phonons. When these are subtracted, we

[17] W. E. Lawrence and S. Doniach, *Proc. Int. Conf. Low Temp. Phys. 12th, 1970* p. 361 (1971).

[18] T. H. Geballe, A. Menth, F. J. DiSalvo, and F. R. Gamble, *Phys. Rev. Lett.* **27**, 314 (1971).

[19] F. J. DiSalvo, *Proc. Int. Conf. Low Temp. Phys., 13th, 1972*, Vol. 3, p. 417 (1974).

[20] D. Prober, M. R. Beasley, and R. Schwall, *Proc. Int. Conf. Low Temp. Phys., 13th, 1972*, Vol. 3, p. 428 (1974).

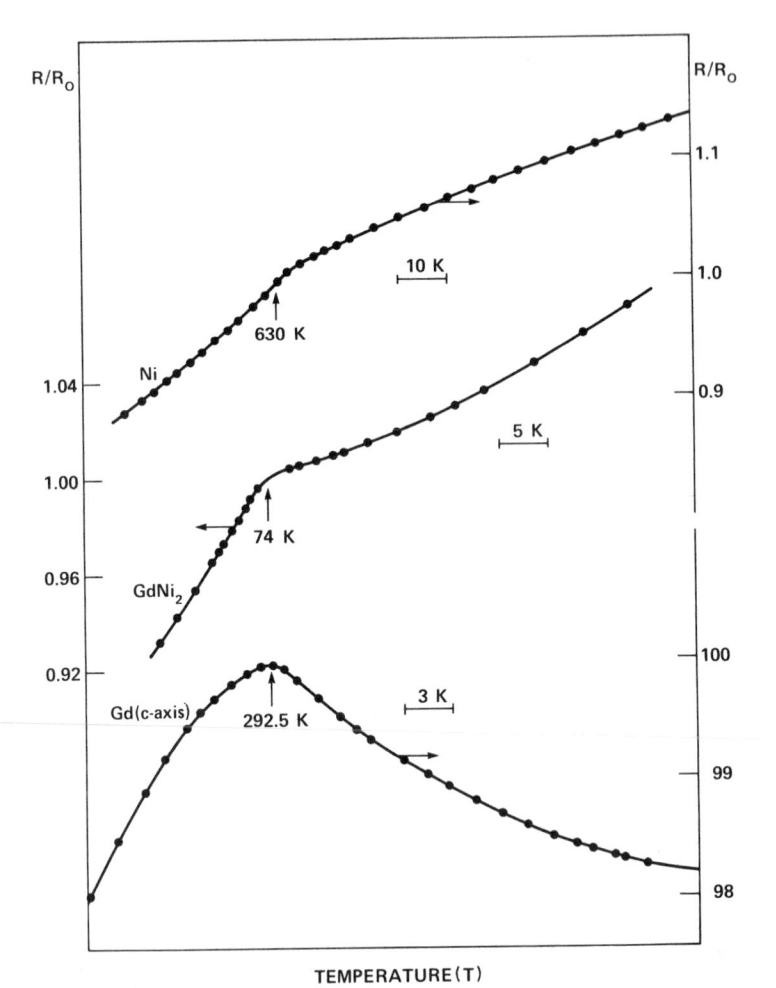

FIG. V.6. Resistivity normalized to its value at the Curie point (indicated by the arrow) versus temperature for three ferromagnets. After Parks.[22]

find that the remaining resistivity also decreases rapidly below the Curie point. This decrease in resistivity is due to the "freezing out" of the exchange scattering of the electrons from the magnetic moments. Assuming an interaction of the form $J\mathbf{S}_i \cdot \mathbf{s}\delta(\mathbf{r} - \mathbf{R}_i)$ between the carrier spin s and a spin $\mathbf{S}_i$ at site $\mathbf{R}_i$, Fisher and Langer[21] have calculated the relaxation time τ in the resistivity expression $\rho = m/ne^2\tau$. For $T \rightarrow T_c^-$, they find that the

[21] M. Fisher and J. Langer, *Phys. Rev. Lett.* **20,** 665 (1968).

scattering is incoherent and decreases with temperature as $|M(T)|^2$. For $T \rightarrow T_c^+$, they argue that the relaxation rate should have the same form as the magnetic energy. Therefore, in this region the resistivity is dominated by short range correlations and its slope should diverge like the heat capacity. Figure V.7[22] shows that this is indeed the case. This relationship between $d\rho/dT$ and C_p also applies to other order–disorder transitions, such as β-brass and the CDW in $2H$-$TaSe_2$.[23] It is interesting to consider the resistivity of a spin glass. This is shown in Fig. V.8a.[24] The arrows indicate the spin–glass freezing point. We see there is no change in the resistivity. This is due to the fact that there is no spatial correlation between the spins either above or below this transition. (We shall return to the chromium data (Fig. V.8b)[25] shortly.)

2. Superconductors

As we mentioned above, superconducting fluctuations also appear, as precursors in the electrical resistivity. Figure V.9, for example, shows the

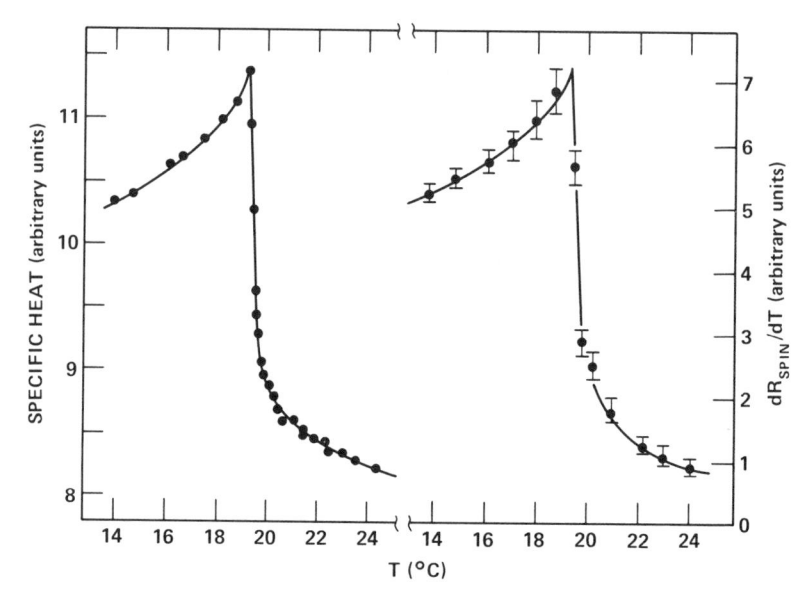

FIG. V.7. Comparison between the specific heat and the temperature derivative of the spin-scattering contribution to the c-axis resistivity of Gd. After Parks.[22]

[22] R. Parks, *AIP Conf. Proc.* **5**, 630 (1971).
[23] R. A. Craven and S. F. Meyer, *Phys. Rev. B* **16**, 4583 (1977).
[24] V. B. Sarkissian and B. R. Coles, *Comments Phys.* **1**, 17 (1976).
[25] D. McWhan and T. M. Rice, *Phys. Rev. Lett.* **19**, 846 (1967).

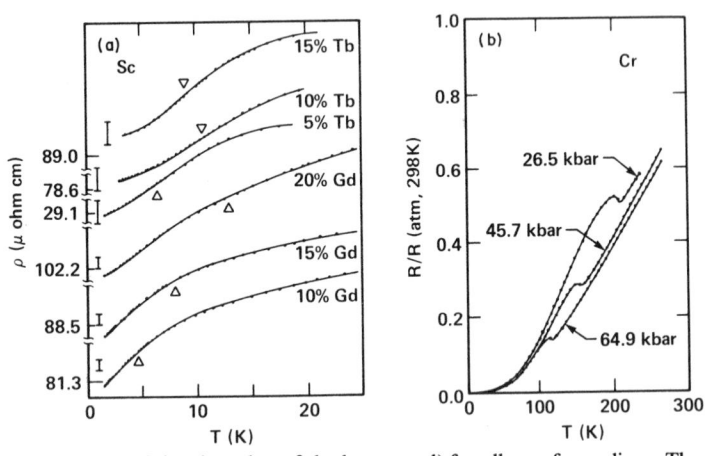

FIG. V.8.   (a) Resistivity (less that of the host metal) for alloys of scandium. The arrows indicate the temperature at which the susceptibility show a cusp. After Sarkissian and Coles.[24] (b) Resistivity of chromium illustrating the pressure dependence of the spin density wave. After McWhan and Rice.[25]

resistivity of an amorphous Bi film. Amorphous films have shorter coherence lengths and hence have larger fluctuation effects than crystalline films (Chapter IX). One might interpret the rounding of the curve as being due to a distribution of transitions into the superconducting state of different regions of an inhomogeneous sample. However, the fit with theory is too good to be ignored. The conductivity is related to the current–current correlation function according to (the Kubo relation)

$$\sigma_{xx}(\omega) = \frac{1}{k_B T} \int_0^\infty \langle j_x(0) j_x(t) \rangle \cos \omega t \, dt. \qquad (5.14)$$

For the current operator we take

$$j = -i(e\hbar/m^*)(\psi^* \nabla \psi - \psi \nabla \psi^*) = (e\hbar/m^*) \sum_{k,q} (2k + q) \, \psi_k^* \psi_{k+q} e^{iq \cdot r}. \qquad (5.15)$$

It is not obvious that one should use the superconducting wave function here. However, it can be shown[26] that close to the transition temperature, this gives the leading contribution to the conductivity. Solving the time-dependent Ginzburg–Landau equation, Eq. (2.106), for $\psi_k(t)$ leads to a

[26] V. Ambegaokar, "Lecture Notes," NATO Advanced Study Institute on Superconductivity, McGill University, Montreal, 1968.

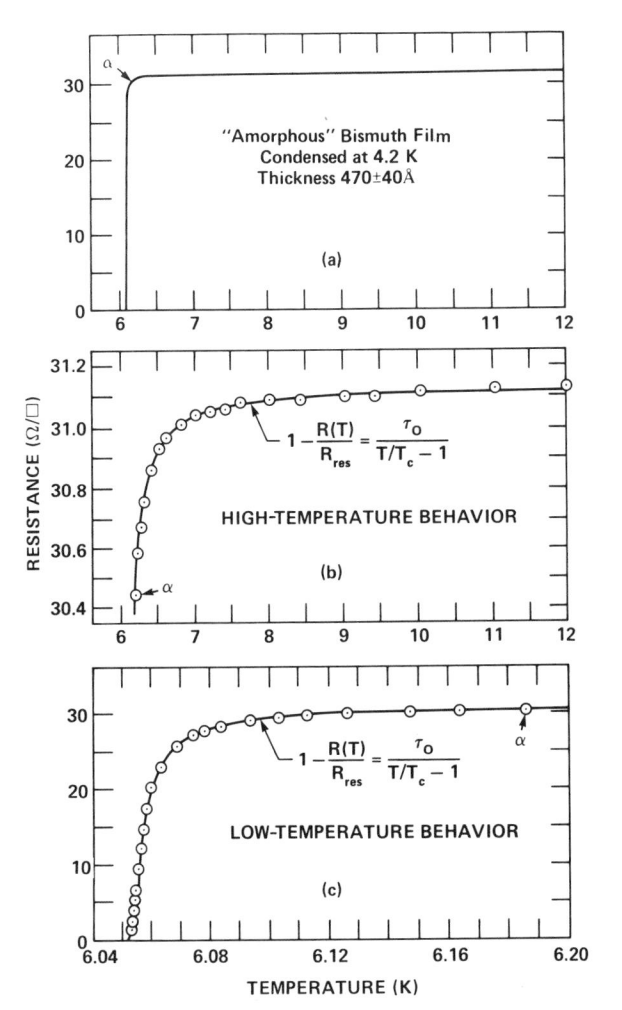

FIG. V.9. Measured resistance per square area of film as a function of temperature for amorphous bismuth. Absolute temperatures are good to ± 0.1 K. Relative temperatures in the vicinity of 6 K are measured to $\pm 2 \times 10^{-4}$ K. The point at 6.185 K labeled "α" is for orientation. After Glover.[14]

fluctuation contribution to the conductivity which in two dimensions (film thickness $d \ll \xi$) takes the form[27]

$$\sigma_{\text{fluc}}(0) = \frac{e^2}{16\hbar d} \frac{T_c}{T - T_c} \tag{5.16}$$

[27] G. Aslamosov and A. I. Larkin, *Phys. Lett. A* **26**, 238 (1968).

where T_c is the only adjustable parameter. Note that this is independent of the mean-free path. The corresponding resistance, $R = 1/\sigma d$, is

$$R(T) = R_{res} \left(1 - \frac{R_{res}e^2}{16\hbar} \frac{T_c}{T - T_c} \right) \qquad (5.17)$$

where R_{res} is the parallel residual resistance. This function is indicated by the solid line in Fig. V.9. This excellent agreement and similar agreement in films of other amorphous elements convinces us that we are indeed observing superconducting fluctuations.

3. Charge Density Waves

Figure V.10[28] shows the resistivity anomalies of the tantalum compounds associated with the formation of two-dimensional charge density waves. The increase in the resistivities of the IT compounds at the ICDW to CCDW transitions is believed to be due to the disappearance of part of the Fermi surface through the electron–hole pairing we discussed in Chapter IV. It is not clear why the resistivity at low temperatures of the 1T sulfide and selenide are so different, since the same superlattice is found in both. Presumably their Fermi surfaces are sufficiently different so that in the case of the sulfide it is completely destroyed by gaps in the commensurate phase. The fact that the resistivity of the 2H compounds *decreases* with the appearance of the charge density wave is also thought to be related to the geometry of the Fermi surface.

Figure V.8b shows that spin density waves have a similar effect on the resistivity, namely, an increase associated with the disappearance of a portion of the Fermi surface.

In addition to the platinum chain compound, KCP, mentioned in Chapter I, *"one-dimensional"* charge density waves have also been observed in the organic salt, tetrathiofulvalene-tetracyanoquinodimethane (TTF-TCNQ). The structure is indicated in the insert of Fig. V.11.[29] Since there are two sets of chains, the ordering is complex. At 54 K the TCNQ chain undergoes a Pierels distortion, giving rise to an incommensurate (along b) structure of the form $2a \times 3.4b \times c$. At 49 K the TTF chains order. From this point down to 38 K interactions between the chains lead to a doubly incommensurate (along a and b) superstructure in which the modulation changes continuously from $2a$ to $4a$. Below 38 K the modulation stays locked at $4a$. The effect of these transitions on the *resistivity* is

[28] F. Di Salvo, "NATO Advanced Study Institute on Electron-Phonon Interactions and Phase Transitions." New York, Plenum Press, 1977. This reference contains references to the original measurements.

[29] S. Etemad, *Phys. Rev. B* **13**, 2254 (1976).

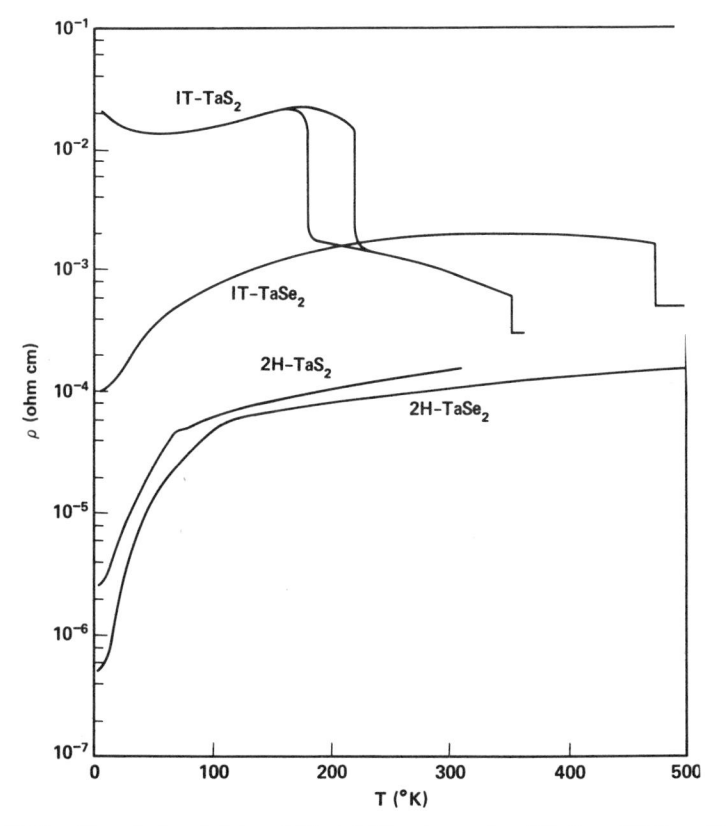

FIG. V.10. Electrical resistivity of 1T and 2H polymorphs of TaS_2 and $TaSe_2$ with current parallel to the layers ($\rho \perp$ c). After DiSalvo.[28]

shown in Fig. V.11. Above the transition region the resistivity has a T^2 behavior. If one plots the *conductivity,* one finds a peak near 58 K which, as we now see, is not a transition temperature. Nevertheless the fact that this conductivity is unusually high for an organic material stimulated enormous research activity on this and related materials. For a review of these quasi-one-dimensional conductors, see Toombs.[30]

D. SCATTERING

1. Electrons and X Rays

Diffraction techniques have played an increasingly important role in the study of ordering phenomena. Electrons which scatter strongly via the

[30] G. A. Toombs, *Phys. Rep. C* **40**, 182 (1978).

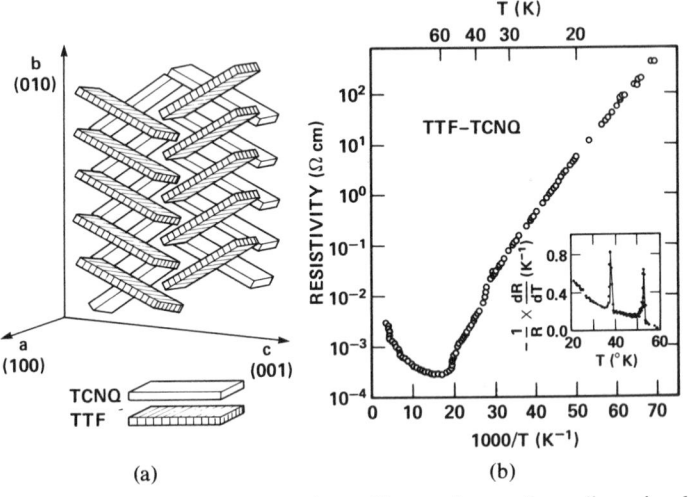

FIG. V.11. (a) Stacking of the molecules making up the quasi-one-dimensional organic metal TTF-TCNQ. (b) Absolute resistivity versus temperature for TTF-TCNQ. Insert shows the logarithmic *derivative* of this resistance. After Etemad[29]

Coulomb interaction are sensitive to charge distributions and small changes thereof. X rays can be employed using high flux monochromatic beams and only require small samples. Both X rays and electrons are sensitive to the geometry of the Brillouin zone.

We have already discussed charge density waves in TaSe₂. These were discovered using electron diffraction. The basal plane electron diffraction pattern of a 200 kV beam transmitted through a crystal of 1T TaSe₂ is shown in Fig. V.12.[31] Not all the visible peaks are on the $l = 0$ plane; spots displaced by $\frac{1}{13}c_0^*$ can still be seen because the sample is thin, causing diffraction peaks to be extended rods in the c_0^* direction. The crystal is thinned to $\leqslant 1000$ Å simply by peeling with cellophane tape. The diffuse rings could be due to the phasons mentioned in Chapter II. The effect of the ICDW–CCDW transition at ~ 473 K on the resistivity can be seen in Fig. V.10.

The uncontrolled intensity variation in Fig. V.12 is an illustration of the difficulty of using electron diffraction techniques for intensity measurements. The strong Coulomb interactions mean that multiple scattering is always important. The relative phases of the three CDW's with respect to each other and the lattice cannot be obtained by electron diffraction. These phases can, however, be determined by other techniques such as core-level shifts of the Ta 4f electrons using X-ray photoemission.[32]

[31] F. J. DiSalvo, *Surface Sci.* **58**, 297 (1976).
[32] See, for example, H. R. Hughes and R. A. Pollak, *Philos. Mag.* [8] **34**, 1029 (1976).

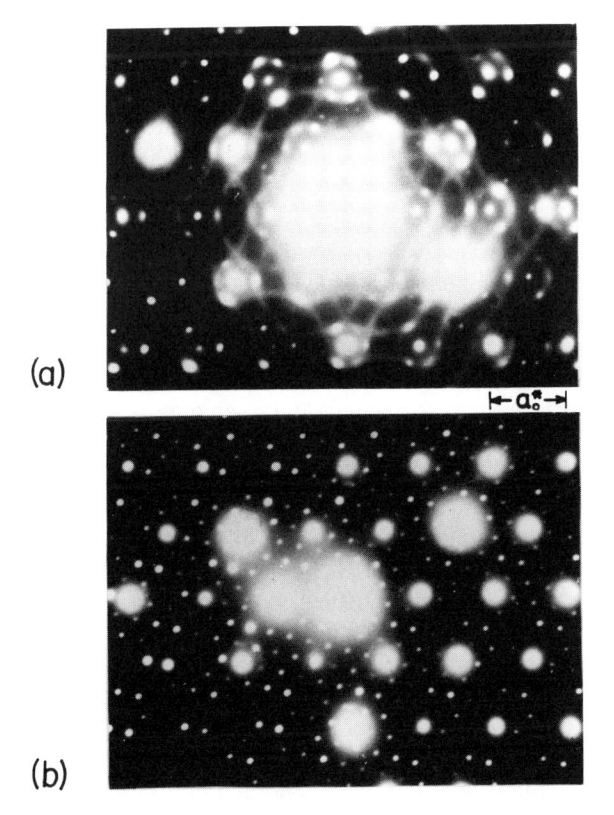

(a)

$\leftarrow a_0^* \rightarrow$

(b)

FIG. V.12. Diffraction patterns are for 1T-TaSe$_2$. (a) (with diffuse rings)—500 K just above T_d = 473 K—that is, in the incommensurate CDW state (the CDW onset temperature is $\approx$ 600 K) Sattelites are the spots that are connected by rings. It can be seen that $\mathbf{q}_0$ is parallel to $\mathbf{a}_0^*$. (b) Room temperature well below T_d = 473 K, the sattelite spots are now rotated by 13°54' from a_0^* directions. They are commensurate with the main peaks as can be seen by following a path three sattelites in one direction and then two more on a second rotation by 120°. After DiSalvo.[31]

2. Neutrons

Neutron scattering is perhaps the most powerful probe of the solid state physicist. Unfortunately, the only high-flux source of thermal neutrons is a nuclear reactor or a proton synchrotron.

We could never hope to cover all the aspects of neutron scattering. We have already mentioned its role in "soft mode" transitions. Here we shall merely mention Bragg scattering in antiferromagnets. The details of neutron scattering are presented in the book by Marshall and Lovesey.[33]

[33] W. Marshall and S. W. Lovesey, "Theory of Thermal Neutron Scattering." Oxford Univ. Press (Clarendon), London and New York, 1971.

In the case of two magnetic sublattices with oppositely directed but equal moments, the coupling of the order parameter to an applied field is generally zero. In this case, we rely on neutron diffraction experiments to measure the order parameter which is the sublattice magnetization. For example, in MnF_2 with the rutile structure, a (100) diffraction peak (forbidden in X ray and neutron nuclear diffraction) appears below the Néel temperature at 70 K. The intensity of this Bragg peak which is due to coherent magnetic scattering of the neutron beam by each sublattice is proportional to the square of the staggered or sublattice magnetization. It follows the expected Brillouin function for $S = 5/2$.

Dysprosium aluminum garnet (DAG) has been widely studied as an example of an Ising antiferromagnet. It also has the feature that the magnetic anisotropy field, H_A, exceeds the exchange field, H_E. This has the consequence that in the presence of an applied field DAG does not show the spin-flop state in which the two sublattice moments rotate into the plane perpendicular to the field and cant along the field, but rather it passes directly into a paramagnetic state. Such a material is referred to as a metamagnet. A low temperatures this field-induced antiferromagnetic to paramagnetic transition is first order, but becomes second order (λ-like) above some temperature T_t. Griffiths[34] has suggested that in addition to the variables, temperature, T, and internal field (applied field corrected for demagnetization), H_i, we also consider the staggered field, H_s, conjugate to the antiferromagnetic order parameter. The resulting three-dimensional phase diagram is shown in Fig. V.13. The interesting feature is that the T-H_i plane A is connected along a line of first-order transitions to two surfaces B and B', sometimes referred to as "wings," extending symmetrically into the regions $H_s \neq 0$ which themselves terminate with increasing temperature in lines of critical points. These two critical lines join the λ-line at T_t which Griffiths calls the *tricritical point*. Blume et al.[35] have shown that when a uniform external field is applied along the [111] axis of DAG, it affects the sublattices unequally, so that a staggered field is also produced. Consequently, the region of experimental accessibility is a sheet which cuts through one of the "wings." The sublattice magnetization is nonzero everywhere on this surface. This has been confirmed by neutron diffraction studies of the staggered magnetization. Between T_N and the tricritical point the staggered magnetization does not go to zero at the apparent phase boundary, but rather exhibits a point of inflection fol-

[34] R. B. Griffiths, *Phys. Rev. Lett.* **24,** 715 (1970).

[35] M. Blume, M. M. Corliss, J. M. Hastings, and E. Schiller, *Phys. Rev. Lett.* **32,** 544 (1974).

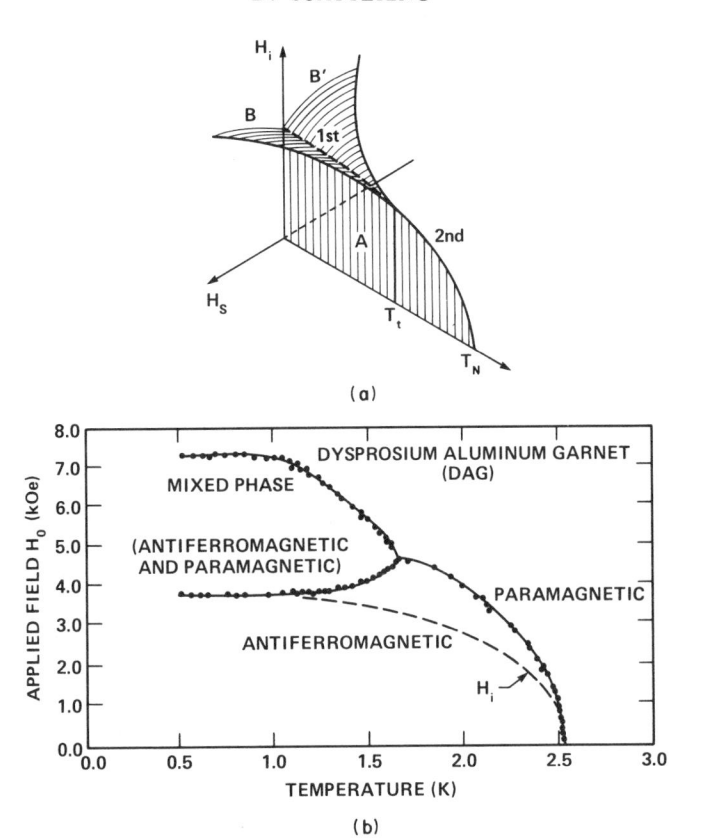

FIG. V.13. (a): Phase diagram of an Ising antiferromagnet in the coordinates temperature (T), internal field (H_i) and staggered field (H_s). (b) Measured phase diagram for dysprosium aluminum garnet in the applied field (H_0) and temperature plane.

lowed by a long tail, and consequently, a persistence of the antiferromagnetic structure into what was thought to be the paramagnetic region.

The basic limitation in neutron scattering is the neutron flux. The High Flux Reactor of the Institute Laue–Langevin (ILL) in Grenoble currently provides the highest flux, approximately 1.2×10^{15} neutrons/cm²/sec. This represents the practical limit for this source imposed by heat transfer and cost. Efforts are now underway to develop pulsed neutron sources. Basically, the scheme is to accelerate protons to say 800 MeV with a synchrotron and then steer the beam to a uranium target where each proton yields about 30 fast neutrons by spallation. These neutrons are then slowed down in a hydrogenous moderator. With such pulsed sources, time-of-flight techniques are used to investigate the elastic and inelastic scattering.

3. Light

The phase diagram shown in Fig. V.13 has also been explored by light scattering.[36] It turns out that if one considers the phase diagram in the T-H_0 plane, where H_0 is the applied field, the first-order line in Fig. V.13 splits into two lines due to demagnetizing effects as shown in the bottom of this figure. These two lines enclose a mixed-phase region in which domains of antiferromagnetic and paramagnetic phases coexist. These domains scatter light strongly whereas the homogeneous phases do not. Therefore, light scattering can be used to locate these phase boundaries. In this way, Giordano and Wolf have been able to map out the shape of the wing critical lines.

Another example of scattering is critical opalescence which is one of the most dramatic aspects of a critical point. We usually associate critical opalescence with the enhanced scattering of light near the liquid–gas critical point. However, the phenomenon is much more general. Whenever we have a probe which couples to the order parameter, we shall find that the scattering of the probe is enhanced when the correlation length becomes comparable to the wavelength of the probe. When the order parameter involves a magnetic dipole, as in ferro- and antiferromagnets, thermal neutrons are an ideal probe because they possess a magnetic moment. A discussion of neutron scattering near the critical point may be found in the review of Als-Nielsen.[37] When the order parameter involves atomic displacements, neutrons are also useful. In particular, inelastic neutron scattering may be used to study the soft phonon modes as we indicated in Chapter I. In cases where the system ends up having made an electric dipole transition light scattering is very useful.

Let us assume that the probe, be it a neutron, photon, etc., couples to some variable of the system at $\mathbf{r}$ and t characterized by an operator $\psi(\mathbf{r}, t)$. The differential scattering cross section is then proportional to

$$\frac{d^2\sigma}{d\omega d\Omega} \sim \int d^3r \int_{-\infty}^{\infty} dt\, e^{i\omega t} e^{i\mathbf{q}\cdot\mathbf{r}} \langle \psi(\mathbf{r}, t)\psi(o, o)\rangle. \tag{5.18}$$

The fluctuation–dissipation theorem tells us this correlation function is proportional to the imaginary part of the susceptibility associated with this quantity, $\chi_\psi''(\mathbf{q}, \omega)$. The coefficient of proportionality involves the thermal factor $[1 - \exp(-\hbar\omega/k_B T)]^{-1}$ which, at high temperatures, varies as ω^{-1}. Therefore, the total scattering cross section is proportional to

[36] N. Giordano and W. P. Wolf, *Phys. Rev. Lett.* **39**, 342 (1977).

[37] J. Als-Nielsen, *in* "Phase Transitions and Critical Phenomena" (C. Domb and M. S. Green, eds.), Vol. 5A, p. 87. New York, 1976.

$$\sigma \sim \int \frac{d\omega \chi_\psi''(\mathbf{q}, \omega)}{\omega} = \chi_\psi'(\mathbf{q}, 0). \tag{5.19}$$

From Eq. (2.61) we see that $\chi(\mathbf{q})$ is proportional to $(q^2 + 1/\xi^2)^{-1}$. In the case of light scattering, the maximum possible momentum transfer is $4\pi/\lambda_{light}$, which is small. Therefore $\chi \sim \xi^2$ and as the correlation length diverges, the total scattering intensity also diverges. This explanation of critical opalescence was first proposed by Ornstein and Zernike in 1914.

It is interesting to note that the dielectric function associated with a phonon is generally represented by a Lorentzian expression

$$\chi(\omega) \sim \frac{1}{\omega_0^2 - \omega^2 + 2i\omega\Gamma}, \tag{5.20}$$

where Γ is the damping. Thus $\chi(0) \sim \omega_0^{-2}$, which shows that a soft mode ($\omega_0 \to 0$) also leads to critical opalescence. If the transition is driven by soft acoustic phonon, then the total scattering associated with acoustic phonons, i.e., *Brillouin* scattering, should diverge. If the transition is driven by a soft optic phonon (at the zone center), then the *Raman* scattering across section should diverge. Figure V.14[38] shows the Brillouin spectrum for $PrAlO_3$ at different temperatures approaching the cooperative Jahn–Teller transition at 151 K. This complements the inelastic neutron results we showed in Fig. I.21.

E. Resonance

We often think of magnetic resonance as a probe of local environments. Indeed, in Chapter VII, we shall indicate its importance in identifying the low-temperature state of magnetic impurities. However, magnetic resonance has also played a role in the study of long range order. In Chapter II, for example, we indicated how Heller and Benedek used NMR to measure the critical exponent β. NMR also provided a crucial test of the BCS theory of superconductivity, as we shall see below.

Magnetic resonance is an enormous subject and we shall not even attempt to review the fundamentals. Rather, we shall indicate a few interesting aspects of *nuclear* magnetic resonance in ordered systems.

Depending upon whether the number of neutrons, N, and protons, Z are even or odd a nucleus can have electric multipole moments Q_l or magnetic multipole moments, M_l. Q_l and M_l vanish unless $l \leq 2I$ where I is the total angular momentum of the nucleus. Thus nuclei with $I = \frac{1}{2}$ have only a magnetic dipole moment but no electric quadrupole moment.

[38] P. Fleury, P. Lazan, and L. Van Uitert, *Phys. Rev. Lett.* **33,** 492 (1974).

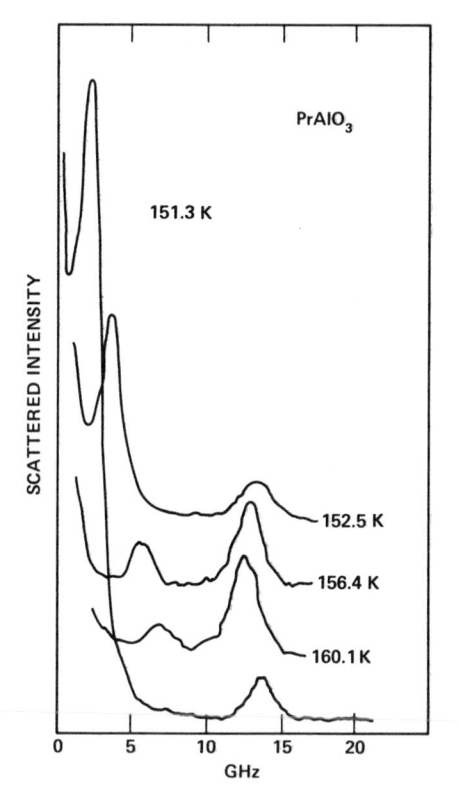

FIG. V.14. Transverse mode Brillouin spectra of PrAlO$_3$ approaching the 151 K Jahn–Teller transition. The soft mode propagates in the [101] direction with a polarization along [101]. The mode at 13 GHz is a TA mode which does not show softening. After Fleury et al.[38]

The magnetic moment interacts with the applied magnetic field plus the local magnetic field arising from the electrons,

$$\mathcal{H}_{\ell} = -\gamma_n \hbar \mathbf{I} \cdot (\mathbf{H}_0 + \mathbf{H}_{\mathrm{loc}}).$$

This local, or hyperfine, field has various sources. Unpaired s-electrons will give a *contact* hyperfine field. Unpaired d-electrons give an *orbital* hyperfine field. Unpaired d-spins also polarize the core s-electrons which then produce a contact hyperfine contribution. This *core polarization* contribution is negative as illustrated for transition metals in Fig. V.15.[39]

[39] A. J. Freeman and R. E. Watson, *in* "Magnetism" (G. T. Rado and H. Suhl, eds.), Vol. 2A, p. 167. Academic Press, New York, 1965.

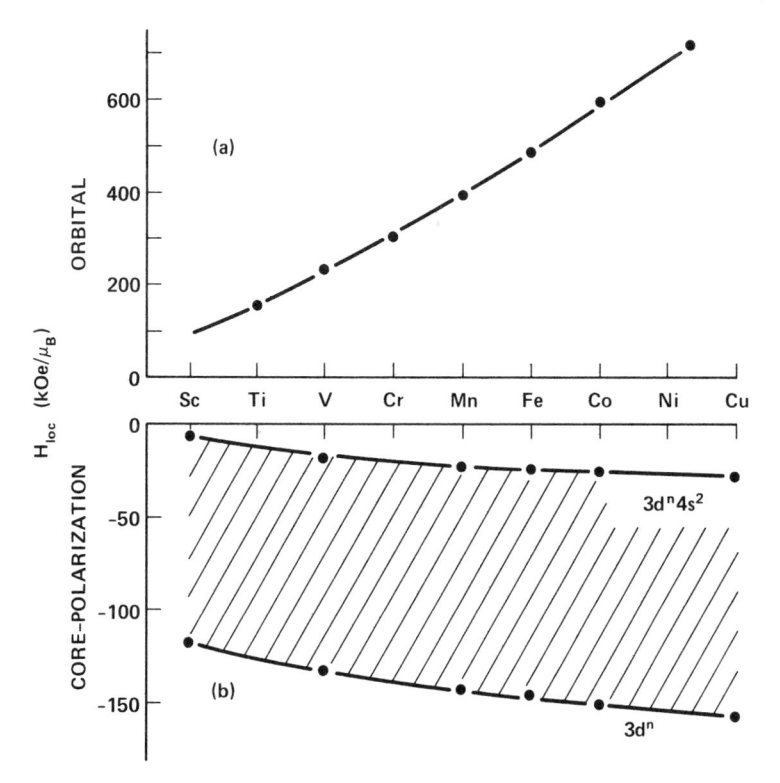

FIG. V.15.   Hartree–Fock hyperfine fields for 3d transition elements. (a) Orbital hyperfine fields for the neutral atoms. (b) Hyperfine field resulting from core polarization for the configurations indicated. After Freeman and Watson.[39]

Since the spin polarization is proportional to the Pauli susceptibility this contribution will, in general, be temperature dependent.

When a nucleus is placed in a metal the conduction electrons also produce contact, orbital dipolar and core polarization hyperfine fields. The resulting displacement of the NMR resonance in the metal relative to its value in a nonmetallic, nonmagnetic environment is the Knight shift, K. Thus, $K = K_s + K_{orb} + K_{core}$ (T). The Knight shift, in conjunction with the susceptibility, can be used to study the d-electron density in metals as we shall illustrate when we discuss the A-15 compounds in Chapter VI.

It is convenient to express the quadrupole moment operator in terms of the angular momentum operators. For example,

$$Q_2{}^o = \frac{eQ}{2I(2I - 1)} [3I_z{}^2 - I(I + 1)] \tag{5.21}$$

where Q is the nuclear quadrupole moment. The quadrupole moment interacts with any electric field gradient present at the nucleus, $V_{\mu\nu}$. Thus

$$\mathcal{H}_Q = \frac{e^2 Qq}{4I(2I-1)} \{[3I_z^2 - I(I+1)] + \eta(I_x^2 - I_y^2)\} \qquad (5.22)$$

where $eq = V_{zz}$ and $\eta = (V_{xx} - V_{yy})/V_{zz}$. When adding $\mathcal{H}_Q$ to $\mathcal{H}_I$ we must keep in mind that the principle axis for the field gradient may not lie along $\mathbf{H}_0$. With this brief introduction to the terminology let us now consider NMR in ordered systems.

1. Ferromagnets

NMR in ferromagnets and antiferromagnets is interesting because of the "rf enhancement" effect. Consider a single domain particle as shown in Fig. V.16a with the magnetization lying along the z-axis in equilibrium. The rf field creates a torque on the magnetization Mh_{rf}. But the resulting transverse component of magnetization, $\delta M \sim M\theta$, experiences a restoring torque from the demagnetizing field, $-4\pi N_z M$, given by $-4\pi N_z M^2\theta$. Therefore $\theta = h_{rf}/4\pi N_z M$ and the magnetization "rocks back and forth" in phase with h_{rf}. This gives an additional rf component with an amplitude $H_{loc}\theta = (H_{loc}/4\pi N_z M) h_{rf}$. If the magnetic anisotropy field H_A exceeds the demagnetization field, then $4\pi N_z M$ is replaced by H_A. This is the case in Co where the rf absorption rate is enhanced by $(1 + H_{loc}/H_A)^2 = 4.7 \times 10^4$.

The situation is even more interesting when the sample is unmagnetized and we have domains as illustrated in Fig. V.16b. The component of the rf

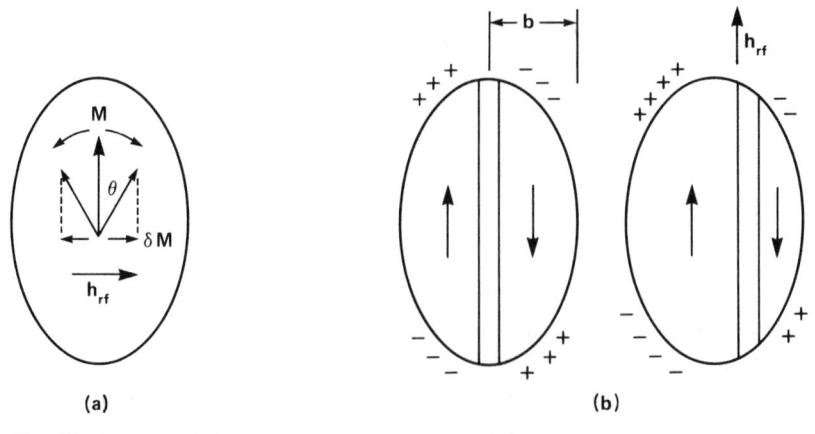

(a)   (b)

FIG. V.16.   (a) and (b) Illustration of the "rf enhancement" effect in ferromagnets.

field which is parallel to a domain wall exerts a torque on the spins within the wall which twists them around moving the wall to the right. As the majority domain grows a demagnetizing field appears which provides an opposing torque. Thus the displacement of the wall Δx is given by

$$4\pi N_z M(\Delta x/2b) = h_{rf}. \tag{5.23}$$

If the thickness of the wall is δ, then the angle $\Delta\theta$ through which a spin turns as the wall moves is $\Delta\theta = (\Delta x/\delta)\pi$. Thus those spins within the wall see an additional rf field given by $H_{loc} \Delta\theta$. Notice that within the domains themselves the rf demagnetizing field just cancels the applied field. Therefore only the nuclei within the domain walls contribute to the resonance.

NMR provides one the flexibility of probing selective ions in a compound. In the ferromagnet MnBi, for example, the hyperfine field at the Bi nucleus is 1.8×10^6 Oe. This implies a large hybridization of the Mn d-electrons with the Bi p-electrons. Since Bi is a very heavy element its spin–orbit interaction is large. Therefore, the large magnetic anisotropy of MnBi very likely derives from the Bi.

2. Superconductors

Fluctuations in the local field lead to relaxation of the nuclear spin. In a metal this relaxation occurs through the scattering of conduction electrons off the nuclei. Let us assume the interaction to be the contact hyperfine interaction which we write as

$$\mathcal{H}_I = A\mathbf{I} \cdot \boldsymbol{\sigma}$$

$$= A \sum_{\mathbf{k},\mathbf{k}'} [I_z(c^{\dagger}_{\mathbf{k}'\uparrow}c_{\mathbf{k}\uparrow} - c^{\dagger}_{\mathbf{k}'\downarrow}c_{\mathbf{k}\downarrow}) + I^+ c^{\dagger}_{\mathbf{k}'\downarrow}c_{\mathbf{k}\uparrow} + I^- c^{\dagger}_{\mathbf{k}'\uparrow}c_{\mathbf{k}\downarrow}]. \tag{5.24}$$

The nuclear relaxation rate, $1/T_1$, is easily calculated by second-order perturbation theory. This involves the square of the matrix element times an energy-conserving delta function summed over all the electronic states. The final result for a normal metal can be understood as follows. Since an electron only spends a time $\hbar/E_F$ at a given nucleus the delta function is broadened by this amount. Furthermore, only those electrons within $k_B T$ of the Fermi level take part in the scattering. Therefore

$$\left.\frac{1}{T_1}\right|_N \sim A^2 \left(\frac{\hbar}{E_F}\right)\left(\frac{k_B T}{E_F}\right). \tag{5.25}$$

Thus $T_1 T$ is expected to be independent of temperature. In fact, since $\mathcal{H}_I$ is also responsible for the Knight shift we have $T_1 T \sim 1/K^2$ which is known as the Korringa relation. Since $N(E_F) \sim 1/E_F$, $T_1 T \sim N(E_F)^2$.

In the superconducting state, we must transform the electron operators in Eq. (5.24) into quasiparticle operators according to Eq. (1.29). The relaxation rate associated with quasiparticle scattering is

$$\left.\frac{1}{T_1}\right|_s = \frac{\pi}{\hbar} A^2 \sum_{\mathbf{k},\mathbf{k'}} \left(1 + \frac{\Delta_\mathbf{k}\Delta_\mathbf{k'}}{E_\mathbf{k}E_\mathbf{k'}}\right) f_\mathbf{k} (1 - f_\mathbf{k'})\delta(E_\mathbf{k'} - E_\mathbf{k} - \hbar\omega). \quad (5.26)$$

The "coherence factor" $(1 + \Delta_\mathbf{k}\Delta_\mathbf{k'}/E_\mathbf{k}E_\mathbf{k'})$ is the result of the quasiparticle transformation. Its importance arises from the fact that it differs from the coherence factor which enters a *non*-pair-breaking scattering process such as acoustic attenuation by a crucial minus sign, i.e., $(1 - \Delta_\mathbf{k}\Delta_\mathbf{k'}/E_\mathbf{k}E_\mathbf{k'})$. In this latter non-pair-breaking case the coherence factor cancels the large density of quasiparticle states generated by opening the gap. Thus a measurement of $(1/T_1)_s/(1/T_1)_N$ constitutes a crucial test of the pairing theory of superconductivity.

Historically, Hebel and Slichter were in the process of measuring $1/T_1$ in aluminum at Illinois while Bardeen, Cooper, and Schrieffer were developing their theory. It was well known by that time that superconductivity was associated with a change in the electronic states at the Fermi level. Since, as we indicated above, these are the states involved in nuclear relaxation, Hebel and Slichter anticipated as early as 1953 a significant difference in the relaxation times between normal and superconducting aluminum.

Experimentally, one is faced with the difficulty presented by the Meissner effect. One might try working with particles small compared with the penetration depth. Hebel and Slichter,[40] however, used the following clever cycling process. The Al resonance was observed in thermal equilibrium in the normal state at a field of about 500 G. The field was then turned off cooling the sample by adiabatic demagnetization making Al superconducting. After a time τ the field was turned back on and the resonance measured in a time short compared with the normal state relaxation time. The decrease of the resonance so observed as a function of τ provided a measure of T_1. They found that $1/T_1$ rose above the normal value upon cooling through T_c. Theoretically, the coherence factor entering $1/T_1)_s$ enhances the peaking in the density of states at the gap edge giving a logarithmic singularity. Fibich[41] showed that broadening of the quasiparticle energies by thermal phonons removes this divergence, leaving one with the moderate increase observed.

[40] L. C. Hebel and C. P. Slichter, *Phys. Rev.* **107**, 901 (1959).
[41] M. Fibich, *Phys. Rev. Lett.* **14**, 561 (1965).

3. NMR and Charge Density Waves

From our introductory discussion, it follows that the NMR associated with a nucleus having an angular momentum I will consist of $2I$ lines given by

$$\hbar\omega_{m\to m-1} = \hbar\omega_0(1 + K_\parallel) + \hbar\omega_Q(m - \tfrac{1}{2})$$

where $K_\parallel$ is the parallel Knight shift and $\hbar\omega_Q = 3e^2qQ/(2I(2I - 1))$.

In the presence of an incommensurate charge density wave the nuclei will experience a distribution of electric field gradients, q, which will cause an inhomogeneous broadening of the quadrupole "satellites." From the size of this broadening we can determine the *amplitude* of the ICDW. Figure V.17 shows the evolution of the ^{93}Nb $m_1 = -\tfrac{3}{2}$ to $M_I = -\tfrac{1}{2}$ resonance in 2H-NbSe$_2$. By assuming that the local change in the electric field gradient is proportional to the change in the charge density, Berthier et al.[42] estimate the amplitude of the charge density wave to be only about 10% of the conduction electron density. It is interesting to note that analysis of X-ray photoemission[43] shows that the amplitude of the commensurate CDW in 1T $-$ TaSe$_2$ has the large value of 1 electron/atom. It is not clear why these systems have such different amplitudes although the 1T transitions all seem to involve a larger fraction of the Fermi surface.

Problem V.2
Suppose the principle axis for the electric field gradient in Eq. (5.22) makes an angle θ with the applied magnetic field. Show that the resonance frequency is then given by

$$\hbar\omega_{m\to m-1} = \hbar\omega_0(1 + K_\parallel) + \hbar\omega_Q(m - \tfrac{1}{2})\left(\frac{3\cos^2\theta - 1}{2}\right).$$

If the probability that the field gradient lies in the direction θ is random, i.e., $P(\theta) = \tfrac{1}{2}$, compute and plot the resonance spectrum $P(\omega)$ for the three possible transitions for $I = \tfrac{3}{2}$. Note that since $-1 < \cos\theta < +1$ the spectrum has cutoffs!

F. TUNNELING

1. Normal Tunneling

An experimental technique that has proven particularly useful in probing the densities of states of both superconductors and ferromagnets is electron tunneling. The concept of tunneling arose with quantum mechanics. It was first applied to the field ionization of atomic hydrogen by

[42] C. Berthier, D. Jêrome, P. Molinié, and J. Rouxel, *Solid State Commun.* **19**, 131 (1976).
[43] G. K. Wertheim, F. DiSalvo, and S. Chiang, *Phys. Rev. B* **13**, 5476 (1976).

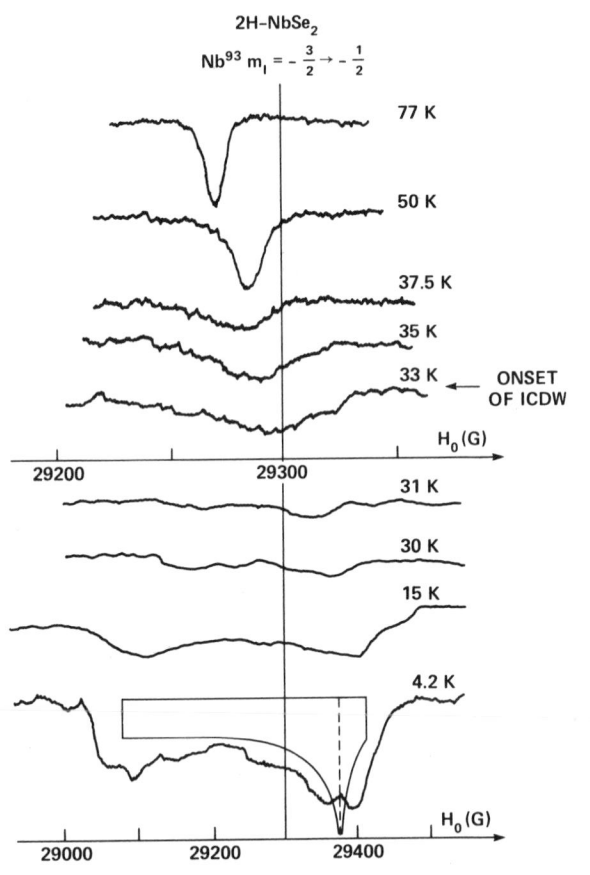

FIG. V.17. Evolution of the resonance between the Nb nuclear levels $m_I = \frac{3}{2}$ to $-\frac{1}{2}$ as a function of temperature. The curve superimposed on the data at the bottom is the calculated lineshape for an electric field gradient distribution associated with a triple charge density wave. The data in the upper half of the figure were obtained by free induction decay, that in the lower half by spin echoes. After Berthier *et al.*[42]

Oppenheimer in 1928. Although there are many ways in which tunneling manifests itself, such as alpha decay, field emission, etc., it was not until Esaki's invention[44] of the tunnel diode in 1957 that tunneling "technology" developed rapidly. The I-V characteristic of an Esaki diode shows the current increasing linearly with voltage above a threshold voltage which is of the order of the band gap; in this region the tunneling is ohmic. In particular, the tunneling current does not reflect any structure in the

[44] L. Esaki, *Phys. Rev.* **109**, 603 (1957).

electron density of states. This insensitivity to the density of states is a general feature of tunneling between normal metals. To see the origin of this let us consider the simple case of a ferromagnet illustrated in Fig. V.18 where the spin-up and spin-down bands are exchange split. The electrons tunnel into a vacuum on the right.

The contribution of an electron in state $\mathbf{k}$ with spin σ to the tunneling current is given by its probability of getting through the barrier, $D_{\mathbf{k}\sigma}$, times the frequency with which it strikes the barrier, $\nu_{\mathbf{k}\sigma}$. Thus the current per unit area is

$$J_\sigma = \sum_{\mathbf{k}} \frac{e}{A} D_{\mathbf{k}\sigma}\nu_{\mathbf{k}\sigma}f_{\mathbf{k}\sigma} \qquad (5.28)$$

where $f_{\mathbf{k}\sigma}$ is the Fermi function and A is the area of the junction.

A common form for the barrier transmission factor is the WKB result

$$D_{\mathbf{k}\sigma} = D(\epsilon_{x\sigma}) = \exp\{-2d\,[2m(V_0 - \epsilon_{x\sigma})/\hbar^2]^{1/2}\} \qquad (5.29)$$

where

$$\epsilon_{x\uparrow} = \hbar^2 k_x^2/2m$$
$$\epsilon_{x\downarrow} = \Delta + \hbar^2 k_x^2/2m \qquad (5.30)$$

where Δ is the exchange splitting. The frequency of striking the barrier is just $v_x/2L = \hbar k_x/2mL = \partial\epsilon_{x\sigma}/\partial k_x 2\hbar L$, where L is the thickness of the

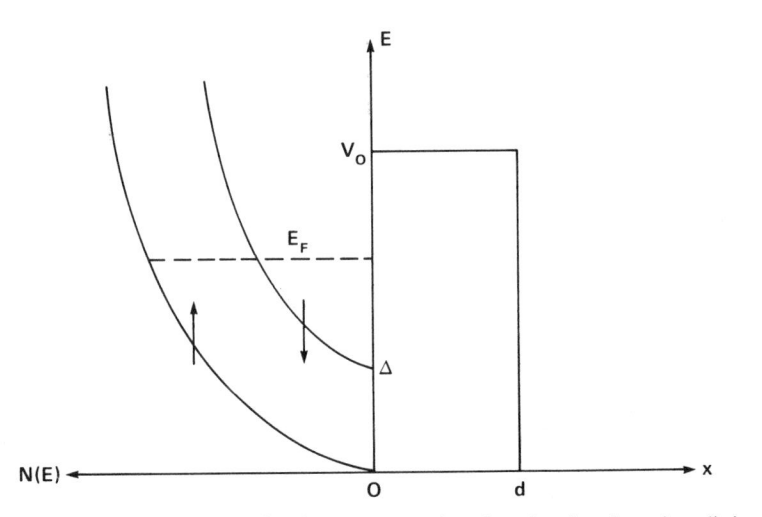

FIG. V.18. Model of a metal–insulator–vacuum junction showing the spin splitting of the metal energy band.

metal. Using these expressions in Eq. (5.28) and converting the sum over the states into an integral gives

$$J_\sigma = \frac{V}{(2\pi)^3} \frac{e}{AL} \frac{1}{2\hbar} \int \frac{d\epsilon_{x\sigma}}{\partial \epsilon_{x\sigma}/\partial k_x} D(\epsilon_{x\sigma}) \int d^2 k_\parallel f(\epsilon^k{}_\sigma) \frac{\partial \epsilon_{x\sigma}}{\partial k_x}. \quad (5.31)$$

From this expression we can easily see the origin of statements on the disappearance of the "density of states." Normally the density of states enters when an integral over momentum coordinates is converted into an integral over energy. However, the presence of the factor $\partial \epsilon_{x\sigma}/\partial k_x$ leads directly to an integral over ϵ_x without an accompanying density of states. One way to avoid this cancellation is to work with a very thin (400 Å) metal film in which the electron states are "box quantized." Under such conditions the concept of a density of states is not valid, and the quantized states can be seen in tunneling.[45] In the "thick film" case there is still an integral involving the momentum coordinate parallel to the plane of the junction. Physically, this factor corresponds to the degeneracy associated with the number of ways an electron can strike the barrier with the same longitudinal momentum mv_x. Notice that if the calculation were carried out in a one-dimensional framework this factor would not become apparent. It is only through this transverse integral that the details of the Fermi surface enter. In fact, the integral over $k_\parallel$ at $T = 0$ K is just the cross-sectional area of the Fermi surface at the height k_x. Thus, the tunneling current is just the average cross-sectional area of the Fermi sphere weighted by the barrier transmission factor.

The cross-sectional area is $\pi k_\parallel^2 = \pi(k_{F\sigma}^2 - k_x^2)$. Therefore,

$$\frac{J_\downarrow}{J_\uparrow} = \frac{\int_\Delta^{\epsilon_F} d\epsilon_x (\epsilon_F - \epsilon_x) \exp\{-2d[2m(V_0 - \epsilon_x)/\hbar^2]^{1/2}\}}{\int_0^{\epsilon_F} d\epsilon_x (\epsilon_F - \epsilon_x) \exp\{-2d[2m(V_0 - \epsilon_x)/\hbar^2]^{1/2}\}}. \quad (5.32)$$

The ratio given by Eq. (5.32) is plotted in Fig. V.19 as a function of the band splitting for a junction characterized by $d = 40$ Å, $V_0 = 5$ eV, and $\epsilon_F = V_0/2$. Also plotted in Fig. V.19 is the ratio of the number of minority spins to majority spins. The difference between these curves is a dramatic illustration of how the geometry of the Fermi surface is suppressed in the tunneling process.

The fact is, however, that experimentally one *does* observe spin-dependent *field emission* from nickel.[46] The polarization of electrons

[45] R. C. Jaklevic, J. Lambe, M. Mikkor, and W. C. Vassell, *Phys. Rev. Lett.* **26**, 88 (1971).
[46] M. Landolt, Y. Yafet, B. Wilkens, and M. Campagna, *Solid State Commun.* **25**, 1141 (1978).

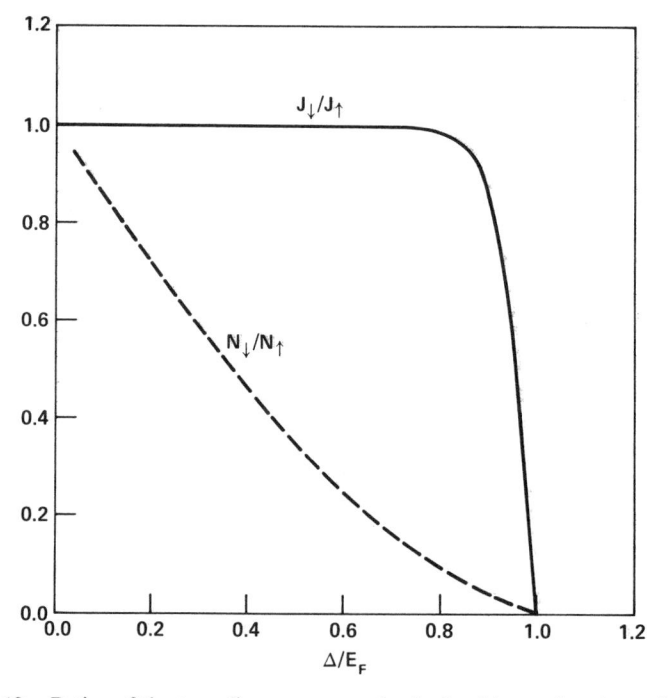

F_IG. V.19.   Ratios of the tunneling currents and spin densities as functions of the band splitting.

emitted from the (100) surface is −3% (i.e., the spin is opposite to the magnetization) while that of electrons from the (110) surface is +5%. This is consistent with the fact that of the three high symmetry directions only the (110) surface has a large proportion of majority d states in the wave function at the Fermi level. This suggests that the tunneling *does* reflect the bulk density of states.

A possible reason for the appearance of the density of states may lie in the use of the WKB approximation [Eq. (5.29)]. If one takes the classical step-potential for a metal and superimposes the linear potential due to the applied electric field, one has a right-triangle barrier. The vertical part of this potential is certainly not smoothly varying as the WKB approximation requires. Therefore, one must treat this boundary exactly, matching the wave function and its logarithmic derivative here. Let us assume that the wave function inside the crystal has the form

$$e^{i\mathbf{k}\cdot\mathbf{r}} + \zeta e^{i(\mathbf{k}-2\mathbf{k}_z)\cdot\mathbf{r}}, \tag{5.33}$$

while outside it is written

$$\zeta e^{i k_\parallel \cdot r} g(z), \tag{5.34}$$

where $g(z)$ is given by the WKB approximation. Assuming the main contribution to the emitted current arises from close to the Fermi level with momenta nearly normal to the surface, the boundary conditions give

$$|\zeta|^2 = 4k_F^2/(\kappa_0^2 + k_F^2), \tag{5.35}$$

where κ_0^2 is proportional to the work function ϕ, i.e., $\kappa_0^2 = 2m\phi/\hbar^2$. The total current is obtained by multiplying $|\zeta|^2$ by the one-dimensional density of states which is proportional to $1/k_F$, leaving a factor $k_F/(\kappa_0^2 + k_F^2)$. Since $k_{F\uparrow} > k_{F\downarrow}$, this current will be spin-dependent. In particular, the polarization becomes $p = (k_{F\uparrow} - k_{F\downarrow})/(k_{F\uparrow} + k_{F\downarrow})$. Had we applied the WKB approximation to the vertical boundary, the corresponding value of $|\zeta|^2$ would have been k_F/κ_0, leading to the disappearance of the dependence of the current on k_F as we saw above. This matching plane method, however, is complicated by hybridization in transition metals. Landolt *et al.*[46] find that the calculated polarization associated with hybridized wave functions is very sensitive to the position of the matching plane.

2. Giaever Tunneling

In 1960 Ivar Giaever,[47] who was studying normal tunneling in oxide junctions, measured the I-V characteristic for an $Al-Al_2O_2-Pb$ junction in which the Pb could be made both normal and superconducting. The result is shown in Fig. V.20. We see that when the Pb is normal the tunneling is ohmic but that the characteristic is much different when the Pb is superconducting. In Fig. V.21 we compare the differential conductance, dI/dV, with the BCS density of states. This similarity suggests that the cancellation of the density of states we discussed above does not occur in the superconducting case. This was further born out by Giaever's result for tunneling between two superconductors, which is shown in Fig. V.22. This is readily understood in terms of the density of states argument sketched out in Fig. V.22(a)–(c).

The appearance of the density of states in the superconducting tunneling was clarified by Cohen, Falicov, and Phillips.[48] The essential point is that the tunneling from one side to the other is via an *electron*, not, for example, a quasiparticle. Thus, if c_q destroys an electron with momentum $\mathbf{q}$ on one side and $c_k^\dagger$ creates one with momentum $\mathbf{k}$ on the other side the tunneling may be represented by a Hamiltonian of the form[49]

[47] I. Giaever, *Phys. Rev. Lett.* **5**, 147; 464 (1960).

[48] M. H. Cohen, L. M. Falicov, and J. C. Phillips, *Phys. Rev. Lett.* **8**, 316 (1962).

[49] J. Bardeen, *Phys. Rev. Lett.* **6**, 57 (1961); **9**, 147 (1962).

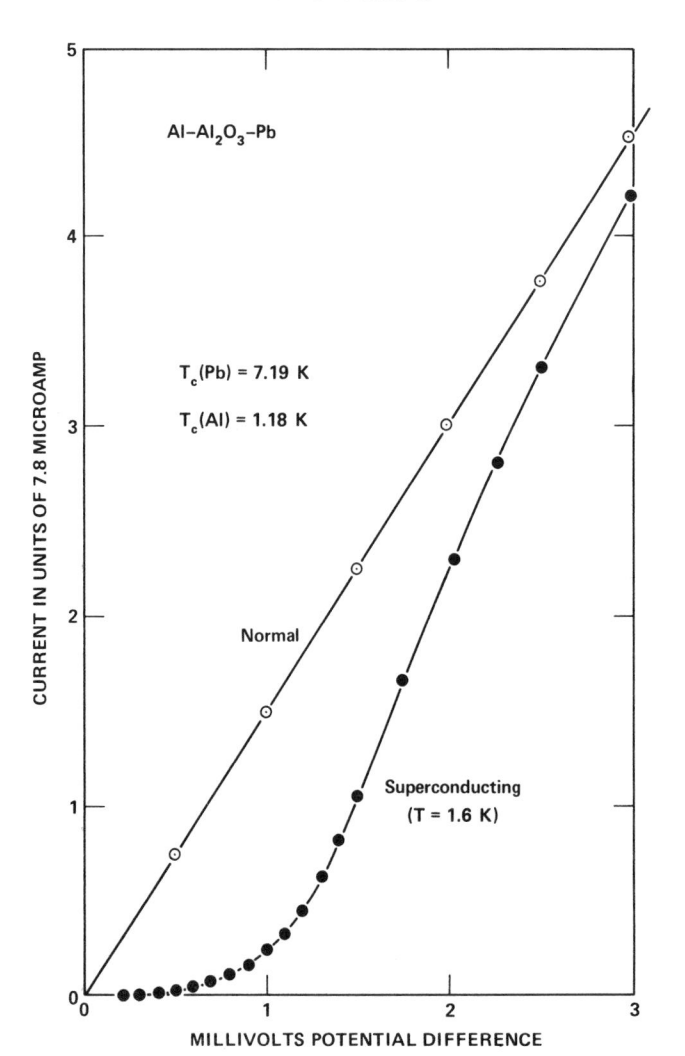

FIG. V.20. I-V characteristics of an Al–Al$_2$O$_3$–Pb junction with the Pb normal and superconducting. After Giaever.[47]

$$\mathcal{H}_T = \sum_{k,q} (T_{kq}c_k^\dagger c_q + T_{qk}c_q^\dagger c_k).$$ (5.36)

The total probability that an electron tunnels from one side, say 1, to the other, 2, is

$$W_{1\to2} = \frac{2\pi}{\hbar} \sum_l |\langle l|\mathcal{H}_T|0\rangle|^2 \delta(E_l - E_0)$$ (5.37)

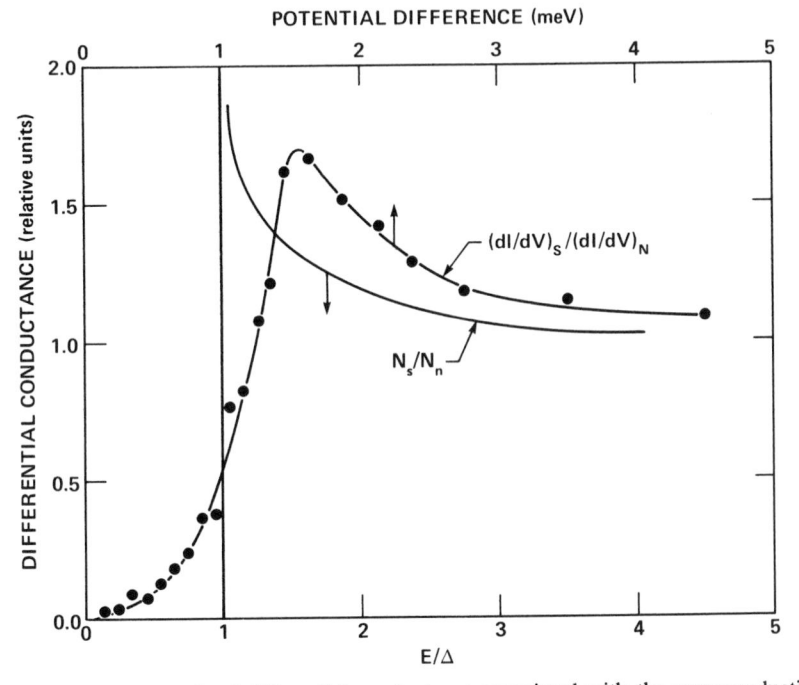

FIG. V.21. Normalized differential conductance associated with the superconducting current in Fig. V.20 compared with the BCS density of states.

where $|0\rangle$ and $|l\rangle$ are products of the eigenstates of the Hamiltonians appropriate for each side. For example, if side 1 is normal the eigenvalues are ϵ_q while those of side 2, the superconducting side, are $E_k = (\epsilon_k^2 + \Delta^2)^{1/2}$. This situation is illustrated in Fig. V.23. In order to evaluate the matrix elements of $\mathscr{H}_T$ between these states the electron operator c_k must be expressed in terms of the quasiparticle operators given in Eq. (1.29)

$$c_{k\uparrow} = u_k \gamma_{k0} + v_k \gamma_{k1}^{\dagger}. \tag{5.38}$$

If one now writes

$$\delta(\epsilon_q + eV - E_k) = \int_{-\infty}^{\infty} dE\, \delta(\epsilon_q - E)\delta(E_k - eV - E) \tag{5.39}$$

and converts the sums over q and k into integrals over ϵ_q and E_k one finds that the tunneling current is proportional to

$$\int_{-\infty}^{\infty} dE |T|^2 N_1(E + eV)N_2(E)[f_1(E + eV) - f_2(E)] \tag{5.40}$$

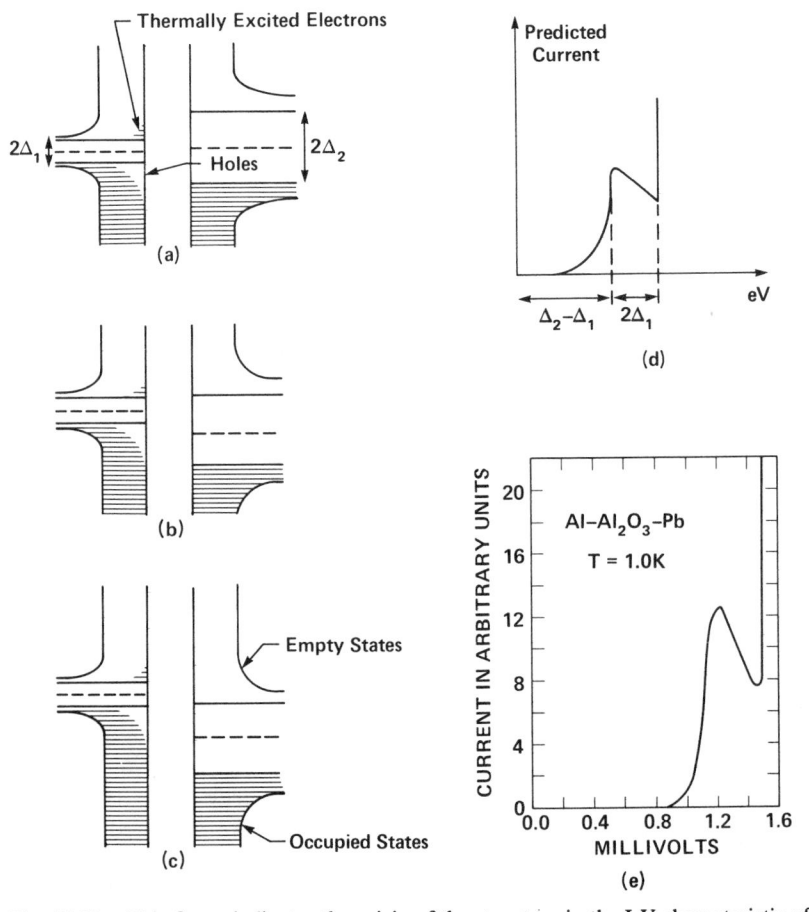

FIG. V.22.   This figure indicates the origin of the structure in the I-V characteristic of a superconductor–insulator–superconductor junction. (a) The two superconductors with no voltage applied. Thermally excited electrons and holes are shown for the smaller gap, while for the larger gap there will be relatively few thermally excited electrons. (b) When a voltage is applied, a current will flow and will increase with voltage, because more and more of the thermally excited electrons in the left-hand superconductor are raised above the forbidden gap in the right-hand superconductor, and can tunnel. When the applied voltage corresponds to half the difference of the two energy gaps, $\Delta_2 - \Delta_1$, it has become energetically possible for all the thermally excited electrons to tunnel through the film. (c) When the voltage is increased further, only the same number of electrons can tunnel, and since they now face a less favorable (lower) density of states, the current will decrease. Finally, when a voltage greater than half the sum of the two energy gaps, $\Delta_2 + \Delta_1$, is applied, the current will increase rapidly because electrons below the gap can begin to follow. This leads to the I-V plot shown in (d). This is to be compared with the experimental data on the same Al–Al$_2$O$_3$–Pb junction used above but with the Al now also superconducting. After Giaever.[47]

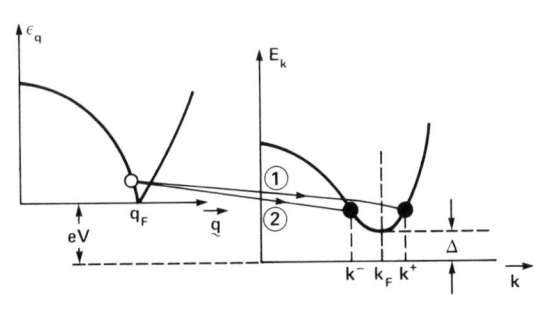

FIG. V.23. Schematic illustration of the electron–hole transitions in metal–superconducting tunneling. The transition labeled 1 is analogous to the process that occurs in metal–metal tunneling. The transition labeled 2, however, corresponds to a process unique to the metal–superconducting case because $1 - n_{k^-} \neq 0$. The total transition probability is proportional to the probability that the states k^+ and k^- are empty, $u_{k^+}{}^2 + u_{k^-}{}^2$, where u_k was defined in Chapter I. Since $u_{k^-}{}^2 = v_{k^+}{}^2$ and $u_k{}^2 + v_k{}^2 = 1$ this illustrates the cancellation of these coherence factors as mentioned in the text.

where $f(E)$ is the particle (or quasiparticle) distribution function. This is the general result. Cohen et al.[48] showed that the BCS phase factors u_k and v_k drop out of the matrix element $|T|^2$ so that it varies only as the reciprocal of the normal metal densities of states. Thus, if side 2 is superconducting $N_2(E) = N_N(E)E(E^2 - \Delta^2)^{-1/2}$ and the current at $T = 0$ will vary as $[(eV)^2 - \Delta^2]^{1/2}$ for $eV > \Delta$.

The fact that this tunneling is sensitive to the superconducting density of states provides us with a powerful probe. The first utilization was to determine the temperature dependence of the energy gap. The data of Townsend and Sutton on Ta shown in Fig. I.12 were obtained by tunneling. As one can see from Fig. V.22, the gap is more accurately determined when both metals are superconducting.

A second important utilization of tunneling was the experimental establishment of the energy dependence of the gap function $\Delta(\omega)$, and thus the verification of Eliashberg's strong-coupling theory as we discussed in Chapter III. In fact, it was noted almost immediately by Giaever that the deviations from the smooth BCS density-of-states curve for Pb-junctions occurred near the zone-boundary acoustic phonon energies for Pb. Extensive measurements by Rowell and McMillan[50] established tunneling as a technique for obtaining the weighted phonon density of states $\alpha^2(\Omega)F(\Omega)$. Figure V.24,[51] for example, shows the deviation in the tunneling density

[50] See, for example, J. M. Rowell and W. L. McMillan, in "Superconductivity" (R. D. Parks, ed.), Chapter 3. Dekker, New York, 1969.

[51] B. Robinson and J. M. Rowell, Toronto Conf. Transition Metal Supercond., 1977 (1978).

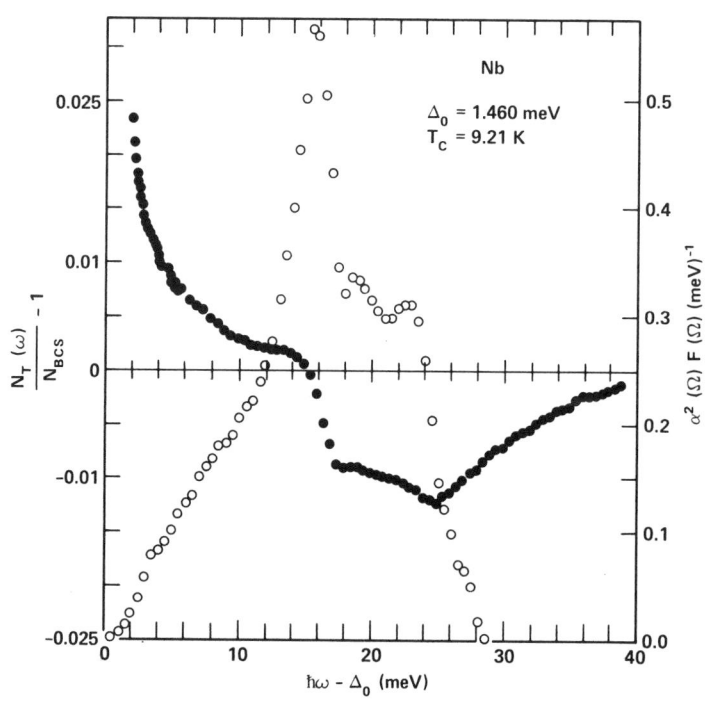

FIG. V.24. The filled circles are the deviation between the measured tunneling density of states for Nb and that predicted by the BCS theory with a gap of $\Delta = 1.46$ meV. The open circles are the values of $\alpha^2(\Omega)F(\Omega)$ calculated by strong-coupling theory (Chapter III) with a Coulomb repulsion of $\mu^* = 0.11$ (after Robinson and Rowell[51]). Notice that α^2F rises much more rapidly at low frequency than, for example, Pb (Fig. III.6).

of states for Nb from that calculated for a BCS density of states and the value of $\alpha^2(\Omega)F(\Omega)$ necessary to produce this deviation.

While such experiments are easy to carry out, one must use a number of criteria to be selective before a junction can be expected to give reliable, nonspurious results. A number of these are summarized by Rowell.[52] It is particularily important to establish that any nontunneling conductance is small. This is done experimentally by showing that the conductance at zero bias is less than 1% of that in the normal state for transition metals (or $\leq 10^{-3}$ for Pb or Hg). If this be true, then, one can use standard techniques to differentiate the I-V curve and obtain a reliable tunneling density of states. Transition metals have oxides of intermediate valence which are frequently conducting that form at the junction. They

[52] J. M. Rowell, In "Tunneling Phenomena in Solids" (E. Burstein, ed.), Chapter 20. Plenum, New York, 1969.

are probably responsible for the great difficulty that has been found in preparing good transition metal junctions. If one attempts to analyze a poor junction, unreasonably low, or even negative values for μ^* are obtained. Niobium is a particularly bad actor. The junction for which the data of Fig. V.24 were taken was made by depositing the Nb in very high vacuum ($\sim 10^{-11}$ torr) at 4 K onto a previously prepared Al–Al$_2$O$_3$ substrate. Under such circumstances there is little possibility for any oxygen to combine with Nb at the junction interface.

It is worth pointing out that it now also appears possible to determine $\alpha^2 F$ by measuring the voltage derivative of the resistance of tiny point contacts between *normal* metals at low temperatures.[53] This derivative turns out to be proportional to the reciprocal electron–phonon scattering time which, in turn, is proportional to $\alpha^2 F$. Measurements on point contacts between noble metals give structure that can be correlated with the phonon density of states obtained from neutron scattering.

3. Spin-Dependent Tunneling

In the presence of a magnetic field the quasiparticle energies become $E_{\uparrow,\downarrow} = (\epsilon^2 + \Delta^2)^{1/2} \pm \mu H$. The resulting density of states splits as shown in the left portion of Fig. V.25. This splitting manifests itself in the tunneling conductance as observed by Meservey, Tedrow, and Fulde[54] in 1970.

This assumes, of course, that σ_z is a good quantum number. In the presence of strong spin–orbit coupling, the spin states become admixed and the field dependence of the tunneling will be reduced. It is curious to note, however, that even in gallium, which is expected to have a large spin-orbit interaction, such a Zeeman splitting is observed. Tedrow and Meservey[55] used this splitting to measure the spin-dependence of the tunneling between superconducting Al and ferromagnetic Ni. The resulting conductance is shown in Fig. V.26. The asymmetry of this conductance may be understood by considering Fig. V.25. As the Al voltage increases, the Ni Fermi surface is displaced upward. Assuming there is no flipping of electron spins during tunneling we expect first a peak (b, Fig. V.26) at $eV = \Delta - \mu H$ because of the coincidence of the spin-down Al density-of-states peak with the Fermi surface of the Ni. The corresponding negative voltage peak (a), $eV = -(\Delta - \mu H)$, will be smaller in magnitude because of the smaller spin-up density of states at the Fermi surface assumed for Ni. At low temperatures the normalized conductance reduces

[53] A. G. Jansen, F. Mueller, and P. Wyder, *Phys. Rev.* **16**, 1325 (1977).
[54] R. Meservy, P. M. Tedrow, and P. Fulde, *Phys. Rev. Lett.* **25**, 1270 (1970).
[55] P. M. Tedrow and R. Meservey, *Phys. Rev. Lett.* **26**, 192 (1971).

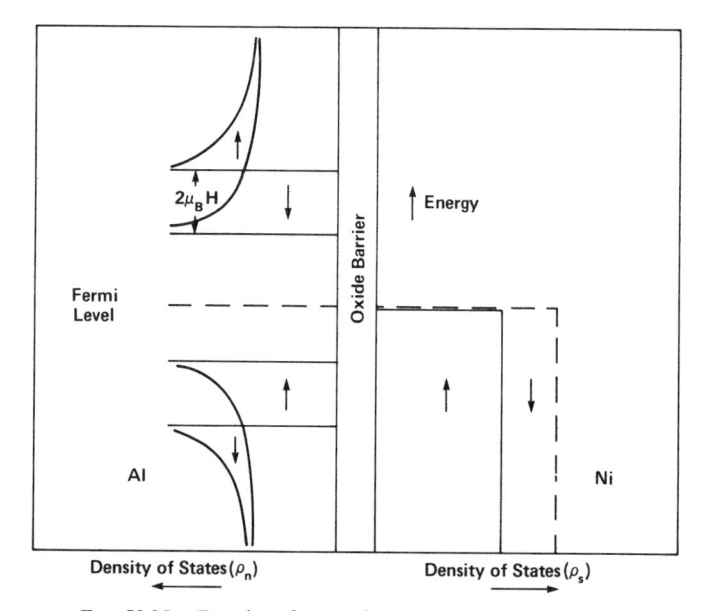

FIG. V.25. Density of states for an Al–Al$_2$O$_3$–Ni junction.

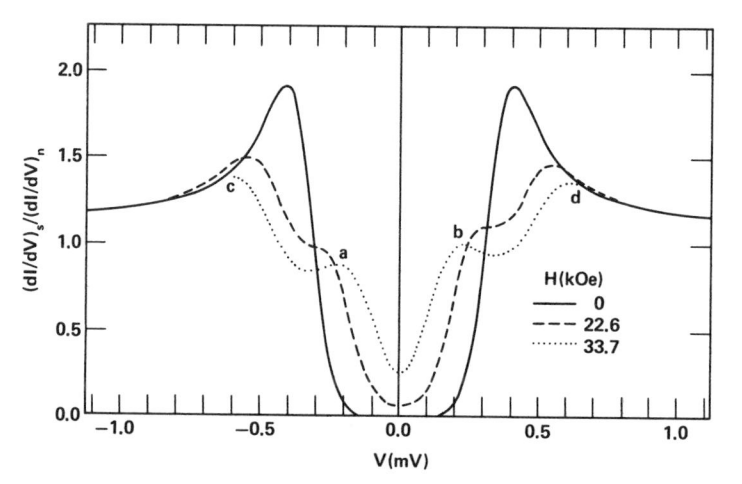

FIG. V.26. Experimental results for the normalized conductance as a function of applied voltage for an Al Al$_2$O$_3$–Ni tunneling junction for different values of the applied field H (kOe). After Tedrow and Meservey.[55]

to

$$\sigma = [x\rho_s(eV + \mu H) + (1 - x)\rho_s(eV - \mu H)]$$

where ρ_s is the superconducting density of states and $x = \rho_{n\downarrow}/$ $(\rho_{n\downarrow} + \rho_{n\uparrow})$ is the fraction of tunneling electrons in the Ni whose moments are parallel to the applied field. The ratio of the peaks, σ_b/σ_a, is just $x/(1 - x)$. From this ratio we obtain the value of the effective carrier polarization in the Ni,

$$\rho = (\rho_{n\downarrow} - \rho_{n\uparrow})/(\rho_{n\downarrow} + \rho_{n\uparrow}) = 2x + 1. \qquad (5.41)$$

The result is $\rho = +7.5\%$, which is not inconsistent with the field emission results mentioned above considering the polycrystalline nature of the Ni. The energy region probed by tunneling is within 1 meV of E_F. Therefore, according to the band structure of Ni (Fig. IV.14), we would have expected to detect primarily minority spins. The likely explanation is that the tunneling current consists predominantly of s-electrons which have been polarized by the d-electrons. This idea is also supported by tunneling measurements[56] on a wide compositional range of alloys. As Fig. V.27 shows, the spin polarization follows the saturation magnetization of the alloy.

4. Josephson Tunneling

a. *Theory:* Let us now consider the tunneling of Cooper pairs. This occurs whenever we have two superconductors connected by a "weak link." This usually takes the form of two superconductors separated by a thin (approximately 50 Å) oxide, although it can also be a *short* ($L < \xi$) microbridge. The tunneling Hamiltonian is still given by Eq. (5.36). Since this describes the tunneling of individual electrons, the pair current must involve this Hamiltonian in second order. In dealing with pairs it is convenient to introduce the operators

$$\sigma_k = c_k^\dagger c_{-k}^\dagger$$

$$\sigma_{z,k} = \tfrac{1}{2}(1 - c_k^\dagger c_k - c_{-k}^\dagger c_{-k}). \qquad (5.42)$$

These commute like spin operators and are therefore referred to as pseudospin operators. If we now apply a canonical transformation to Eq. (5.36) which removes the single particle tunneling to first order, we are left with pair tunneling contributions that take the form[57]

[56] R. Meservey, D. Paraskevopoulos, and P. M. Tedrow, *Phys. Rev. Lett.* **37**, 858 (1976).
[57] P. R. Wallace and M. J. Stavn, *Can. J. Phys.* **43**, 411 (1965).

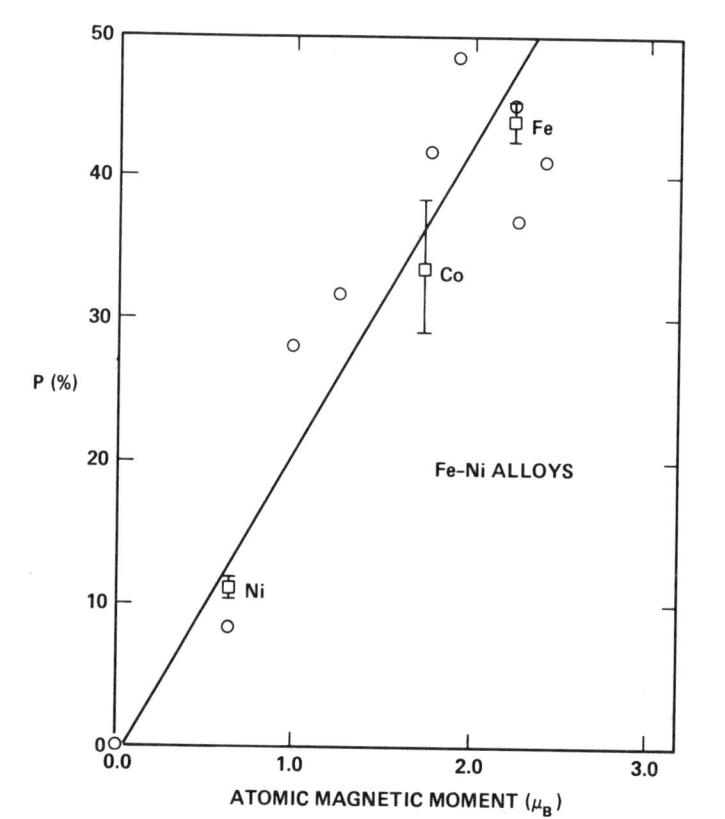

F$_{\text{IG}}$. V.27.   Measured polarization of Fe–Ni alloys (circles) and the three pure elements (squares). The straight line is a least-square fit of the alloy data. After Meservey, Paraskevopoulos, and Tedrow.[56]

$$\mathscr{H}_T = g \sum_{\mathbf{k},\mathbf{q}} (\sigma_\mathbf{k}^\dagger \sigma_\mathbf{q} + \sigma_\mathbf{q}^\dagger \sigma_\mathbf{k}). \tag{5.43}$$

For simplicity we have assumed that the coefficient g, which is proportional to $|T_{\mathbf{kq}}|^2$, is independent of $\mathbf{k}$ and $\mathbf{q}$. The tunneling current is given by $\mathscr{J} = -2e\dot{N}_L$ where N_L is the number of pairs in the left-hand superconductor:

$$N_L = \frac{1}{2} \sum_\mathbf{k} (c_\mathbf{k}^\dagger c_\mathbf{k} + c_{-\mathbf{k}}^\dagger c_{-\mathbf{k}}) = \sum_\mathbf{k} \left(\frac{1}{2} - \sigma_{z,\mathbf{k}}\right). \tag{5.44}$$

Therefore

$$\mathscr{J} = -2e\dot{N}_L = -2e(i/\hbar)[\mathscr{H}_T, N_L]. \tag{5.45}$$

Carrying out the commutation,

$$\mathcal{J} = -2e(i/\hbar)g \sum_{k,q} (\sigma_k^\dagger \sigma_q - \sigma_q^\dagger \sigma_k). \qquad (5.46)$$

If the coupling between the two superconductors is weak, the wave function is a product of the form $|\phi_L\rangle|\phi_R\rangle$ where $|\phi\rangle$ is the BCS function (1.34). Notice that this may be written as

$$|\phi\rangle = \prod_k (u_k + |v_k|e^{i\phi}\sigma_k)|0\rangle.$$

Using the fact that $\sigma_k^\dagger \sigma_k = 1$, the average of Eq. (5.46) is easily found.

$$J = \langle \phi_L|\langle \phi_R|\mathcal{J}|\phi_L\rangle|\phi_R\rangle = J_0 \sin(\phi_L - \phi_R) \qquad (5.47)$$

where

$$J_0 = (4eg/\hbar) \sum_{k,q} (u_k|v_k|)_L (u_q|v_q|)_R. \qquad (5.48)$$

Thus we have the amazing result that if we connect two such weakly coupled superconductors by an ideal current generator a current J will flow in the absence of any voltage. This is the *dc Josephson effect*. The sign and magnitude of the current depend upon the relative phases of the superconducting wave functions.

Problem V.3

Combine Eq. (1.36) with Eq. (5.47) to calculate the work done in establishing a phase difference $\Delta\phi$ between two superconductors. Since the work done on a system in an isothermal process is equal to the change in free energy, what is the equilibrium phase difference between the two superconductors?

b. Interference Phenomena: Since this current depends upon the relative phases of the wave functions, it should not be surprising that this leads to an interesting magnetic field dependence. Consider, for example, the cross section of a junction shown in Fig. V.28a. Let us apply a magnetic field H along the y-axis. The current density is given by

$$\mathbf{j}_s = \frac{e^*\hbar}{2m^*i} (\psi^*\nabla\psi - \psi\nabla\psi^*) - \frac{e^2}{m^*c} \psi^*\psi\mathbf{A}. \qquad (5.49)$$

Let us assume that $d \ll \xi(T)$. This means that $|\psi|$ does not change over the region of interest. Therefore we write $\psi = |\psi| \exp(i\phi(\mathbf{r}))$ and the current density becomes

$$\mathbf{j}_s = \frac{2e}{m^*} |\psi|^2 \left(\hbar\nabla\phi - \frac{2e}{c} \mathbf{A}\right) = 2e|\psi|^2\mathbf{v}_s \qquad (5.50)$$

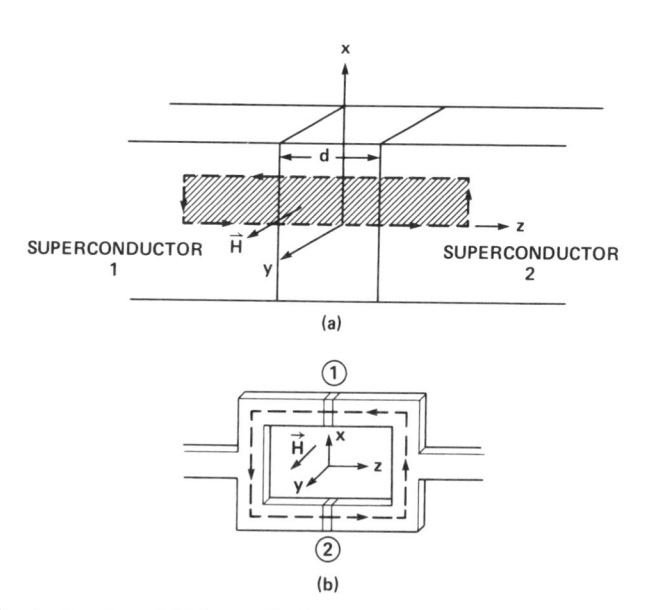

FIG. V.28.   (a) Junction of thickness d in the presence of a field **H**. (b) Two junctions connected in parallel.

or

$$\nabla\phi = \frac{m^*\mathbf{v}_s}{\hbar} + \frac{2\pi}{\Phi_0}\mathbf{A}. \qquad (5.51)$$

Here $\mathbf{v}_s$ is the velocity of the supercurrent that screens the field in the superconducting regions. This velocity lies in the x-direction. For the vector potential, we choose a gauge such that **A** lies in the x-y plane. Then Eq. (5.51) tells us that ϕ is independent of z. Therefore, it is convenient to solve Eq. (5.51) deep inside the superconducting regions where $v_s = 0$ and the vector potential has some constant value. Therefore, Eq. (5.51) becomes

$$\frac{d\phi(x, y)}{dx} = \frac{2\pi}{\Phi_0} A_x(\infty)$$

$$\frac{d\phi(x, y)}{dy} = \frac{2\pi}{\Phi_0} A_y(\infty). \qquad (5.52)$$

In these equations $\Phi_0 = hc/2e$. Integrating these equations,

$$\phi(x, y) = \phi(0, 0) + (2\pi/\Phi_0)[A_x(\infty)x + A_y(\infty)y]. \qquad (5.53)$$

The phase difference across the junction is then

$$\phi_1(x, y) - \phi_2(x, y) = \phi_1(0, 0) - \phi_2(0, 0)$$
$$+ (2\pi/\Phi_0)[A_{1x}(-\infty) - A_{2x}(\infty)]x + (2\pi/\Phi_0)[A_{1y}(-\infty) - A_{2y}(\infty)]y. \quad (5.54)$$

The term $[A_{1x}(-\infty) - A_{2x}(\infty)]x$ is just the flux passing through the shaded area in Fig. V.28. To see this we note that $A_z = 0$ so

$$[-A_{1x}(-\infty) + A_{2x}(\infty)]x = \oint \mathbf{A} \cdot d\mathbf{l} = \iint \nabla \times \mathbf{A} \cdot d\mathbf{S}$$
$$= \iint \mathbf{H} \cdot d\mathbf{S} = \Phi(x). \quad (5.55)$$

The penetration depth λ is defined such that

$$\Phi(x) = \iint H_y(x', z')dx' \, dz' = (d + 2\lambda) \int_0^x H_y(x') \, dx'. \quad (5.56)$$

Therefore the phase difference in the presence of a field is

$$\Delta\phi(x, y) = \Delta\phi(0, 0) + \frac{2\pi(d + 2\lambda)}{\Phi_0} \left[\int_0^x H_y(x') \, dx' - \int_0^y H_x(y') \, dy' \right]. \quad (5.57)$$

If we now assume that the tunneling current is small, then its magnetic field may be neglected. In this case the only field is the applied field in the y-direction. Thus, $H_y = H_0$, $H_x = 0$. If the width of the junction is Y and its height X, then the total current flowing through the junction, from Eq. (5.47), is

$$I = J_0 Y \int_{-X/2}^{X/2} dx \sin \Delta\phi(x, o)$$
$$= J_0 XY \sin[\phi_1(o, o) - \phi_2(o, o)] \frac{\sin(\pi\Phi/\Phi_0)}{\pi\Phi/\Phi_0} \quad (5.58)$$

where $\Phi \equiv H_0 X(2\lambda + d)$. In Fig. V.29[58] we show the current through a Sn–SnO–Sn junction as a function of the field. It shows the pattern predicted by Eq. (5.58).

If we had not assumed the tunneling current to be small, then this current also contributes a magnetic field. If there is a time dependence there will also be a displacement current. Thus the total field is given by the Maxwell equation

$$\frac{\partial H_x}{\partial y} - \frac{\partial H_y}{\partial x} = -\frac{4\pi}{c} J + \frac{\epsilon}{c} \frac{\partial E}{\partial t}. \quad (5.59)$$

[58] B. Josephson, *Phys. Lett.* **1**, 251(1962).

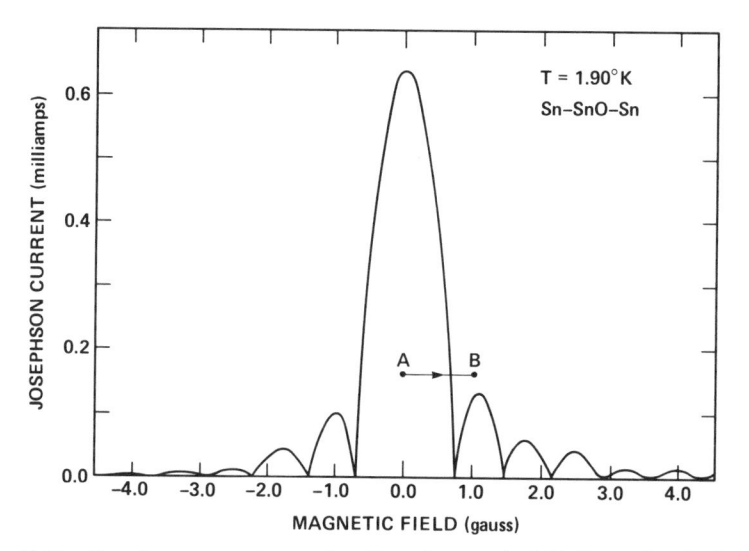

FIG. V.29. Josephson current vs. a function of magnetic field illustrating the interference effect described by Eq. (5.48). After Josephson.[58]

From Eq. (5.57)

$$\frac{\partial \Delta\phi}{\partial x} = \frac{2\pi(d + 2\lambda)}{\Phi_0} H_y$$

$$\frac{\partial \Delta\phi}{\partial y} = -\frac{2\pi(d + 2\lambda)}{\Phi_0} H_x.$$
(5.60)

Therefore, differentiating Eq. (5.60), using Eq. (5.47) and the so-called *ac Josephson relation* $E = V/d = (\hbar/2ed)\partial\Delta\phi/\partial t$ from Eq. (1.36), Eq. (5.60) becomes

$$\frac{\partial^2 \Delta\phi}{\partial x^2} + \frac{\partial^2 \Delta\phi}{\partial y^2} - \frac{1}{\bar{c}^2}\frac{\partial^2 \Delta\phi}{\partial t^2} = \frac{1}{\lambda_J^2} \sin \Delta\phi$$
(5.61)

where

$$\lambda_J = \left[\frac{c\Phi_0}{8\pi^2 J_0(d + 2\lambda)}\right]^{1/2} \quad \text{and} \quad \bar{c} = \left(\frac{1}{1 + 2\lambda/d}\right)^{1/2}\frac{c}{\epsilon^{1/2}}.$$
(5.62)

This equation, known as the *sine-Gordon equation*, has a solution corresponding to a localized traveling wave solution which is called a *soliton*. In this particular context it corresponds to a single flux quantum moving across the junction. We shall encounter this same equation in several different systems in Chapter VIII. It is one of the beauties of physics that the same equations govern such diverse situations!

Let us now consider what happens when we connect two Josephson junctions in parallel, as shown in Fig. V.28b. If the phase difference across junction 1 is $\Delta\phi_1$ and that across junction 2 is $\Delta\phi_2$, then the current through this device is

$$I = I_0(\sin \Delta\phi_1 + \sin \Delta\phi_2).$$
(5.63)

By Eqs. (5.53)–(5.58) these phase differences may be related to the total flux Φ through the area indicated by the dashed line in Fig. V.28b,

$$\Delta\phi_1 - \Delta\phi_2 = \frac{2\pi\Phi}{\Phi_0}. \tag{5.64}$$

Therefore

$$I_{max} = 2I_0 \left| \cos\frac{\pi\Phi}{\Phi_0} \right|. \tag{5.65}$$

This has the same form as the two-slit interference pattern in optics. Thus two Josephson junctions in parallel are referred to as a Superconducting QUantum Interference Device (SQUID). Notice that for a loop with an area of the order of 0.1 cm² the interference fringe interval corresponds to a field of the order of 10^{-6}G. Since it is possible to read the interference pattern to about a thousandth of a fringe, this means we can resolve field changes of the order of 10^{-9} G. This is the origin of the enormous sensitivity of a SQUID magnetometer mentioned at the beginning of this chapter.

c. Application: Not long ago superconductivity was considered the domain of the low-temperature research physicist. This is no longer the case. Cryogenic engineering has now advanced to the stage where low temperatures can be economically maintained. Superconducting transmission lines are being actively developed. Similarly, Josephson junctions are being developed for use in information processing systems which require high speeds and low-power dissipation. The basic switching mechanism that enables a Josephson junction to function as a logic element is easily seen from Fig. V.30 and the I-V characteristic of the junction. Consider the geometry shown in Fig. V.30a. Let us assume that the current in the control circuit biases us at point A in Fig. V.29. At this point, the junction is capable of passing a current I_{max}^A with zero voltage. Consider, for example, that the current is I_g^A. If the current in the control circuit is sufficient (this turns out to be 2.5 mA) to increase the field by an amount associated with one flux quantum, then the junction moves to point B in Fig. V.29. Since this maximum possible Josephson current is now less than I_g^A, the junction must move to the single-particle tunneling curve. Where the junction establishes itself on this curve depends upon the load resistance R. As shown in Fig. V.30b, this creates a voltage V_g across the junction, which in turn produces a current $I_R = V_g/R$ in the load. This current acts as a control to other circuits. Normally the device remains in this voltage state even after the control current is removed. The current through the device must be reduced to zero in order to return the junction to the zero voltage state. This resetting is called a "latching" operation.

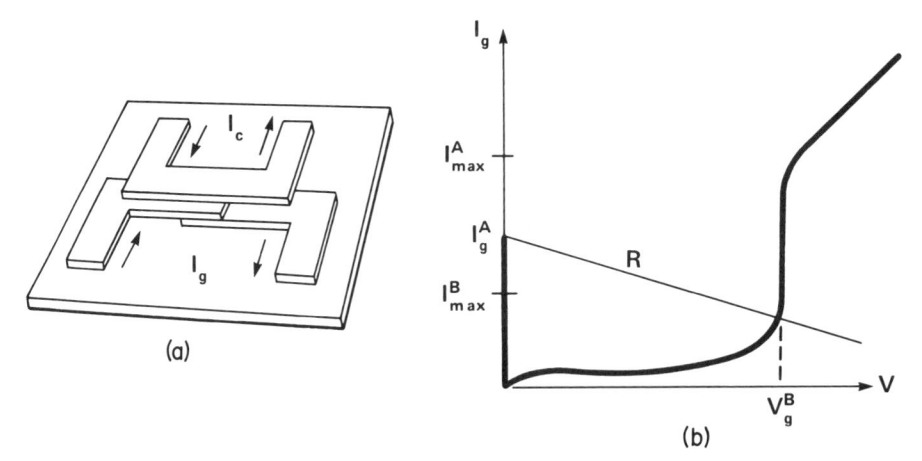

FIG. V.30. (a) Geometry of a Josephson tunnel junction. The tunneling takes place through a thin oxide layer between the stripes carrying the current I_g. The control current I_c generates a magnetic field at this junction. (b) I-V characteristic of such a junction with an external load line labeled R.

Complex latching logic and memory circuits have been built using Josephson junctions.

Although Josephson devices are touted as being superior to semiconductor devices in that they have a lower power-delay product, it is interesting to note that because the thermal conductivity decreases linearly with temperature at low temperatures, a "one-electron" field effect transistor (FET) at room temperature would have the *same* power-delay product as a Josephson junction at 4 K. The point is, however, that the dimensions of a one-electron FET would be in 100's of Angstroms whereas the one flux quantum Josephson device has dimensions of order microns. Thus, for the same power-delay product the Josephson device is more easily fabricated. This ratio of characteristic lengths reflects the ratio of the London penetration depth to the Debye screening length.

VI. Systematics of Long Range Order

In this chapter we shall survey the occurrence of superconductivity and magnetism. It is our intention that reality will provide the "experience" to complement the theoretical concepts introduced earlier and provide a more complete understanding.

The periodic table is a meaningful vehicle for organizing data. It was developed by Mendeleev in the 1860's using regularities of chemical behavior which derive from valence electron effects which are of the order 1 eV or less. We now know that the periodicity displayed by the elements is a result of the Pauli exclusion principle. Mendeleev might have been surprised to find that as the frontiers of science have widened, periodicity manifests itself all the way up to the GeV range where it is useful in predicting new "particles," and down to the millivolt range where it is useful in predicting the occurrence of superconductivity. The periodic table is useful in building one's intuition as to what might occur by presenting what is known to exist in an organized form.

Crystal structures are not essential for long range order as we shall see in Chapter IX. However, a recurring theme in this chapter will be the important interplay between long range order and structure. An element can frequently exist in different crystal structures (polymorphs) which have almost the same internal energy. The free energy difference between two polymorphs, $\Delta G = \Delta E - T\Delta S$, is temperature dependent through the entropy term primarily, although the differences in heat capacity also play a role in shifting the sign of ΔE.

Vibrational entropy is always an important factor in determining structure. Metallic close-packed elements in the alkaline, alkaline earth, and rare earth groups, as well as the lighter members of the transition metals—a total of 26 elements—crystallize from the melt as bcc polymorphs. As first pointed out by Zener,[1] there is more vibrational entropy in the bcc phases which have lower Debye temperatures than their close-packed counterparts. Grimvall and Ebbsjo[2] have confirmed Zener's ideas by demonstrating with calculations using a large variety of intera-

---

[1] C. Zener, "Elasticity and Anelasticity of Metals," p. 32. Univ. of Chicago Press, Chicago, Illinois, 1948.

[2] G. Grimvall and I. Ebbsjo, *Phys. Scr.* **12**, 168 (1975).

tomic potentials that there is invariably a small structural dependence in the Debye temperature with θ_{bcc} being somewhat lower than θ_{fcc} or θ_{hcp}.

Magnetic entropy is of importance when there are other possible electronic configurations with lower spin entropy but with greater binding energy. Plutonium holds the record—it exists in six different polymorphs below its melting point. Iron itself exists over the temperature range 916°C to 1394°C in a weak or nonmagnetic fcc structure as shown in Fig. VI.1.[3]

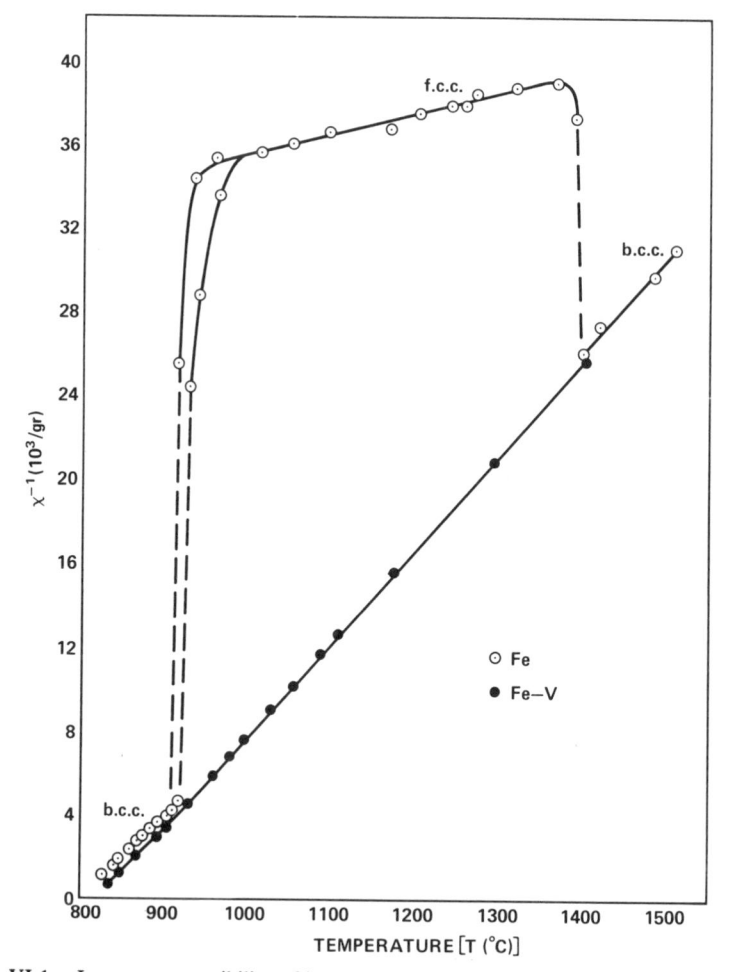

FIG. VI.1. Inverse susceptibility of iron and iron containing 5% vanadium. After Sucksmith and Pearce.[3]

[3] W. Sucksmith and R. R. Pearce, *Proc. R. Soc. London, Ser. A* **167**, 189 (1938).

It is in equilibrium at both temperatures with bcc phases [low-temperature (α) and high-temperature (δ) polymorphs]. The latter phases both obey the same Curie–Weiss law.

Problem VI.1

Use the information shown in Fig. VI.1 to compare the properties of bcc and fcc iron. Explain the transitions at 916 K (bcc → fcc) and 1394 K (fcc → bcc) in terms of plausible changes in magnetic, electronic, and vibrational entropies. Describe an experiment which would determine the importance of the electronic term.

Finally, it should be noted that there can be a close trade-off between the degrees of freedom associated with itinerant electrons and the internal energy—resulting in the occurrence of metal–insulator transitions. A well known case is that of grey Sn, a covalently bonded, tetrahedrally coordinated semimetal with zero band overlap (or equivalently a semiconductor with zero-gap) which exists in equilibrium with the metallic, 4 + 2 coordinated (meaning two neighbors at one distance and four others at a slightly different distance) white tin phase at 313°C. Such transitions are only accidentally found at atmospheric pressure, but undoubtedly occur among the metalloids along some isobars (as, for example, Ge at ~100 Kbar (see Fig. I.3). They are also found in some three-component phases based upon compounds such as V_2O_3 and SmS. These are systems where one can adjust composition and hence the electron density by chemical substitution, such as Cr in V_2O_3 and Gd in SmS.

A. SUPERCONDUCTIVITY

1. Introduction

It is not evident *a priori* that superconductivity, which involves energies of the order of millivolts, should reflect the systematics of the periodic table. However, such behavior was found by Matthias[4] in surveying the occurrence of superconductivity in the 1950's. Particularly among the transition metal elements and alloys superconducting transition temperatures were found to be a periodic function of the valence electron-per-atom ratio (e/a). This also appears to be true for the limited number of nontransition metal solid solutions which are available for investigation.

The BCS theory does not emphasize special effects which might arise from short range or chemical aspects. It utilizes the normal density of states and permits an interpretation of relative variations in T_c as a function of composition. However, it is not apparent why niobium has the

[4] B. T. Matthias, *Prog. Low Temp. Phys.* **2**, 138 (1957).

highest transition temperature among the elements, and is a major constituent in the highest temperature superconducting compounds. Thus, chemical considerations must be important in addition to the electron-per-atom ratio or density of states considerations. As noted in Chapter III, Hopfield, and subsequently Gaspari, Gyorffy, and others, have formulated the BCS theory in a way in which the local atomic potential appears explicitly.

Superconducting transitions are found in metallic elements as low as 0.016 K for tungsten up through 9.3 K for niobium. The transitions therefore span a broader range of temperature than do the melting points of the elements themselves. There are of the order of 1000 compounds in which superconductivity has been found to occur. The distribution in temperature is given in Fig. VI.2. All those above 17 K contain niobium and those above 18 K occur in the A15 structure commonly known as β-tungsten (or, more correctly, as the Cr_3Si type). The distribution curve of T_c's rises exponentially below 10 K. For $T_c = \theta \exp(-1/N(0)V)$ where θ has the known moderate distribution of Debye temperatures ranging around 300, this suggests $N(0)V$ is fairly sharply peaked around 0.3. Most metals other than magnetic ones become superconducting below 10 K. The handful that exist above 15 K presumably have some special characteristic that results in the high value of $N(0)V$.

2. Polyvalent Nontransition Metals

Empirically, if there is sufficient density of electrons, a metallic element will be found superconducting unless magnetic scattering or internal fields interfere. The covalently bonded, low-coordination semiconductors and metalloids (Si, Ge, P, As, Sb, Bi, Se, and Te), of groups IV, V, and VI, are all superconducting if their coordination is increased so they become metallic. Iodine, which becomes metallic under pressure, is expected to be superconducting. The more highly coordinated, "collapsed" (i.e., reduced in volume), metallic phases can be produced in the laboratory either by high pressure or by quenching from the vapor phase by condensation onto substrates held at low temperatures (see Chapter IX).

A sufficient condition for the occurrence of superconductivity is that there be at least three conduction electrons per atom available. For elements with filled d-bands below the Fermi level such as Zn, Cd, and Hg there need only be two electrons per atom. Beryllium ($1s^2 2s^2$) is an exceptional case. It is the only alkaline earth that is superconducting at atmospheric pressure. The fact that it has a detectable, although very low T_c (0.026 K), may be due to its light mass. One cannot expect that the transitions for the other alkaline earths will even remotely scale as their respec-

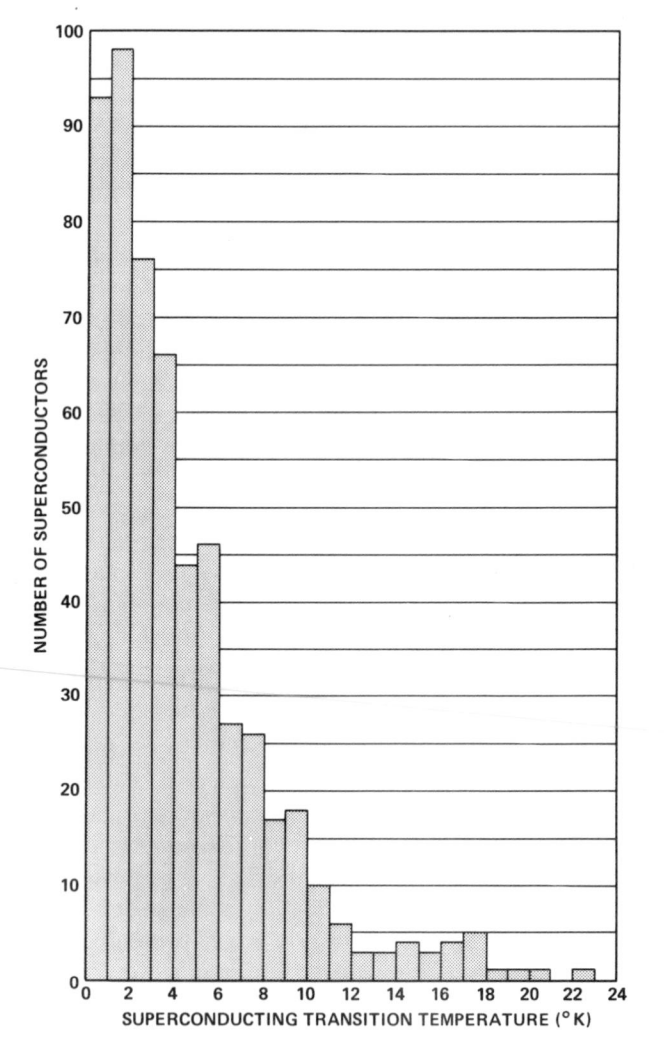

FIG. VI.2. Distribution of superconductors versus transition temperature. The highest transition as of 1978 is Nb_3Ge at $T_c = 23$ K.

tive Debye temperatures because slight changes in the effective inter-action, $N(0)V = \lambda - \mu^*$ (here, the weak coupling formulation is obviously adequate), will cause T_c to drop many powers of e. Therefore, theoretical predictions[5] that Mg, and indeed all the nonmagnetic metals, should become superconducting are impossible to disprove by direct

---

[5] P. B. Allen and M. L. Cohen, *Phys. Rev.* **187**, 525 (1969).

experiment. Beryllium, in a disordered state, prepared by quench-condensing the vapor onto a liquid helium-cooled substrate, has $T_c \sim$ 9 K. The mechanism responsible for this more than two orders of magnitude increase in T_c has not been established. It could be due to an increase in the density of states at the Fermi level which is near a minimum in the alkaline earth metals. On the other hand, the observed surprisingly high T_c might be spurious, due to the formation of filaments of a second phase compound formed by Be with some refractory metal present in the evaporator. Compounds such as $Be_{13}W$ are known to be superconducting. The low-temperature films are usually investigated by measurement of their electrical properties which are subject to the ambiguities concerning the possible role of filaments and other inhomogeneities that have been discussed in Chapter V. More refined investigations such as tunneling would be of interest for Be.

Only under special conditions do monovalent elements become superconducting. Cesium does so under high pressure. Presumably this is related to the substantial d-character that is mixed into the 6s conduction band of Cs in its collapsed phase. The noble metals with filled d-bands have not been found to be superconducting. However, efforts to extrapolate to the pure elements by working with superconducting solid solutions suggest that pure Cu, Ag, and Au may eventually become superconducting at very low temperatures.

The variation of T_c with chemical sequence is not easily established for nontransition metals. Nevertheless, where λ (Eq. 3.35), is not too small, regularities of T_c and of course the superconducting energy gaps are found. If e/a is held constant, as, for example with the white tin polymorphs of the group IV elements under pressure, then we have the following transition temperatures:

Element	T_c
Si	7.1
Ge	5.4
Sn	3.7

Here the decrease of T_c correlates with the increase of mass, or the decrease of the Debye temperature.

Dynes and Rowell[6] have studied the system Tl-Pb-Bi, which is one of the few cases among the nontransition metals where the mass can be held almost constant, and the variation of T_c with e/a can be studied in a con-

[6] R. C. Dynes and J. Rowell, *Phys. Rev. B* **11**, 1884 (1975).

tinuous fashion. It is possible to make good tunnel junctions throughout the fcc Tl-Pb phase and even beyond into the Bi-region up to $e/a = 4.4$. T_c reaches 9 K, which appears to be the limit for nontransition element phases. T_c is indeed, found to be a smoothly increasing function of e/a over this range as can be seen in Fig. VI.3.

3. The Transition Metals and Their Solid Solutions

The variation of T_c among the alloys of Table VI.2 is typical of the d-elements and near neighbor alloys as shown in Fig. VI.4.[7] This figure shows that T_c is a strongly varying function of the filling of the d-band and is a good illustration of Matthias' rule which correlates T_c with e/a. His rule predicts, for example, that the solution of any metal to the right of Zr in Zr initially will raise T_c. The position of the maxima at 4.7 and 6.5 were established by Hulm and Blaugher.[8] It is not surprising that the technologically important alloys are those of Nb and Ti, although, of course, other properties than T_c are also important.

The transition metals are more amenable to study by chemical means than the nontransition metals because there are many more regions of

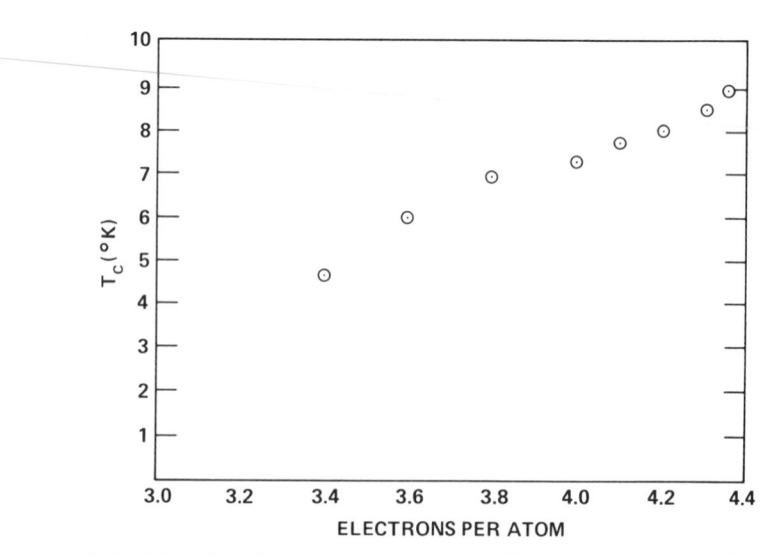

FIG. VI.3. Variation of transition temperature with electron-to-atom ratio in the Tl-Pb-Bi alloy family. After Dynes and Rowell.[6]

[7] G. Gladstone, M. A. Jensen, and J. R. Schrieffer, in "Superconductivity" (R. D. Parks, ed.), p. 665. Dekker, New York, 1969.

[8] J. K. Hulm and R. D. Blaugher, *Phys. Rev.* **123**, 1569 (1961).

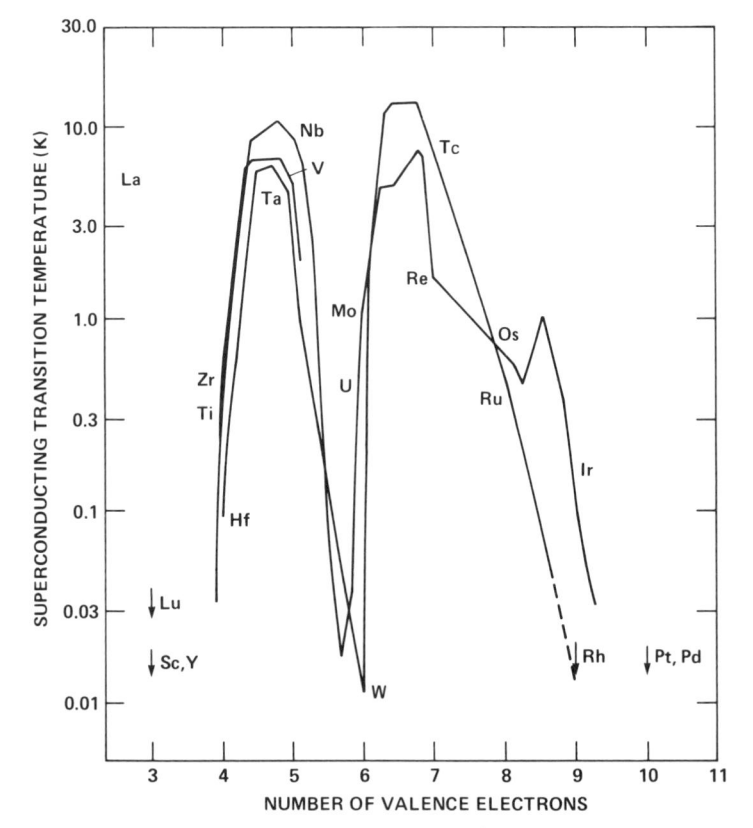

FIG. VI.4. Variation of T_c in the transition metals. After Gladstone et al.[7]

solid solubility where the average Z can be varied over wide ranges. One frequently can measure a number of parameters of the system and obtain correlations which give insight into the microscopic interactions.

The transition metals can have ratios of $T_c/\theta_D \gtrsim 0.05$ which implies they fall in the strong-coupling regime. In Chapter III we outlined the procedure which McMillan used to construct the analytic expression

$$T_c = \frac{\langle \Omega \rangle}{1.20} \exp \left(\frac{-1.04(1 + \lambda)}{\lambda - \mu^* - 0.62 \lambda \mu^*} \right) \tag{6.1}$$

where $\langle \Omega \rangle$ is the average phonon frequency,

$$\langle \Omega \rangle \equiv \frac{\displaystyle\int \alpha^2 F \, d\Omega}{\displaystyle\int \frac{\alpha^2 F}{\Omega} \, d\Omega} . \tag{6.2}$$

TABLE VI.1[a]

EMPIRICAL VALUES OF THE COULOMB PSEUDOPOTENTIAL μ^* FOUND FROM
THE ISOTOPE SHIFT α, T_c, AND θ_D USING EQ. (3.45)

Metal	α	T_c (K)	θ_D (K)	μ^*
Zr	0.00 ± 0.05	0.55	290	0.17
Mo	0.37 ± 0.04	0.92	460	0.09
Re	0.38	1.69	415	0.10
Ru	0.0 ± 0.15	0.49	550	0.15
Os	0.21	0.65	500	0.12

[a] After McMillan.[9]

The electron–phonon coupling constant λ is also determined by the phonon spectrum according to Eq. (3.20),

$$\lambda = 2 \int_0^\infty \frac{d\Omega \alpha^2 F}{\Omega}. \tag{6.3}$$

If one has tunneling data available, then $\langle \Omega \rangle$ and λ can be computed directly. Unfortunately, this is the case for only a few metals. In the absence of such data λ can be determined from T_c using Eq. (6.1).

The value of μ^* is determined from the isotope shift coefficient [Eq. (3.45)] neglecting the strong-coupling correction. In Table VI.1,[9] we indicate the values of μ^* obtained this way for several transition metals. McMillan noted that these values of μ^* all lie around 0.13. He therefore takes this as the value for *all* the bcc 4d transition metal alloys. In Table VI.2, we list the resulting values of λ determined from T_c using Eq. (6.1).

As we indicated in Chapter III, once λ is known we can calculate the bare density of electronic states $N(0)$ from the measured heat capacity. This is also tabulated in Table VI.2.

In cases where tunneling data are available we have a check on the McMillan equation. In Table VI.3,[10,11] we list several metals, most of which are not transition metals, for which $\alpha^2 F$ and therefore λ have been measured directly. With these values of λ the transition temperatures may be calculated from Eq. (6.1) and compared with the observed values. The correlation is quite good as the last two columns show.

4. Intermediate Phase Compounds

A large number of binary compounds have structures and characteristics that are intermediate between the end members. Such materials are

[9] W. L. McMillan, *Phys. Rev.* **167**, 331 (1968).

[10] R. C. Dynes, *Solid State Commun.* **10**, 615 (1972).

[11] B. Robinson and J. M. Rowell, *Toronto Conf. Transition Metal Supercond., 1977* (M. J. G. Lee, J. M. Perz, and E. Fawcett, eds.), p. 666 (1978).

TABLE VI.2[a]

EMPIRICAL VALUES OF λ AND $N(0)$ FOUND FROM T_c, θ_D, AND γ
FOR THE BCC 4D TRANSITION-METAL ALLOYS

Alloy	Percent second metal	T_c (K)	θ_D (K)	γ (mJ/mole K^2)	λ	$N(0)$ (states/eV/atom)
ZrNb	50	9.3	238	8.3	0.88	0.93
	75	10.8	246	8.9	0.93	0.98
NbMo	15	5.85	265	6.3	0.70	0.79
	40	0.60	371	2.87	0.41	0.43
	60	0.05	429	1.62	0.31	0.26
	70	0.016	442	1.46	0.29	0.24
	80	0.095	461	1.49	0.33	0.24
	90	0.30	487	1.67	0.36	0.26
MoRe	5	1.5	450	2.2	0.45	0.32
	10	2.9	440	2.6	0.51	0.36
	20	8.5	420	3.8	0.68	0.48
	30	10.8	395	4.1	0.76	0.49
	40	12.6	340	4.4	0.86	0.50
	50	11.5	320	4.4	0.85	0.50
MoTc	50	12.6	300	4.6	0.91	0.51
ZrRh	3	3.1	244	3.62	0.59	0.48
	4	3.8	226	3.83	0.64	0.50
	5	4.8	210	5.08	0.70	0.63
	6	5.75	196	6.80	0.78	0.81
	7	5.95	192	7.36	0.80	0.87

[a] After McMillan.[9]

referred to as "intermediate phases" and contain most of the interesting superconducting materials. In many cases the properties are simply an average over the periodic table. This is the "rigid-band" approximation which roughly applies when the end members are not too dissimilar. Consider, for example, the materials listed below. We see that MoPd has the same structure as Ru and is also superconducting.

Element/compound	Mo	Ru	MoPd	Pd
Valence electrons per atom	6	8	8	10
Structure	bcc	hcp	hcp	fcc
T_c	0.9	0.5	3.5	0

TABLE VI.3[a]

A COMPARISON OF THE EXPERIMENTAL AND CALCULATED T_c's FOR
METALS AND ALLOYS UTILIZING THE McMILLAN EQUATION (6.1)

Material	θ_D (meV)	$\langle \Omega \rangle$ (meV)	λ	μ^*	$T_{c_{cal}}$ (K)	$T_{c_{exp}}$ (K)
Pb	9.05	5.20	1.55	0.131	6.48	7.19
In	9.65	6.91	0.834	0.125	3.44	3.40
Sn	17.2	9.50	0.72	0.111	3.77	3.75
Hg	6.19	3.3	1.60	0.11	4.51	4.19
Tl	6.80	4.98	0.78	0.127	2.10	2.33
Ta	22.2	12.0	0.69	0.111	4.26	4.48
Nb[b]		12.84	0.98	0.11	9.6	9.21
Amorphous Ga		5.47	2.25	0.17	8.48	8.56
Amorphous Bi		2.86	2.46	0.105	5.38	6.11

[a] After Dynes,[10] except where noted.

[b] Robinson and Rowell.[11]

When a given filling of the conduction band or, equivalently, the number of valence electrons per atom, determines the structure, the compound is referred to as an "electron-compound." Binary intermediate phases are, for the most part, "electron-compounds," and it is not surprising that their superconductivity follows the Matthias' valency dependence, very much like the elements and solid solutions given in Fig. VI.4. These phases comprise the bulk of those materials shown in Fig. VI.2 which have transitions which are exponentially distributed in the temperature decade 1–10 K. We shall not discuss these low-temperature superconductors, but refer the interested readers to Roberts compilation[12] for a complete listing with references. Rather, we shall focus on those intermediate phase materials showing high T_c's.

a. *Structure:* High T_c's seem to be intimately related to structure. Therefore let us first consider what governs the structure of these phases. This is evidently determined by the considerations mentioned in the Introduction. A regular sequence of simple structures is followed as the d-band is filled: bcc up to the half-filled band, then hcp, and, finally for the noble metals, fcc.

Complex structures with large unit cells are found particularly between A atoms with less than half filled d-shells and B atoms which can be non-transition metals, or transition metals which have greater than half filled d-shells. The B atoms are "electropositive" and can transfer electrons to the A atoms which can thus gain intraatomic (Hund's rule) configuration

[12] B. W. Roberts, *J. Phys. Chem. Ref. Data* **5**, 581 (1974).

stability. Metallic Mn can act both ways, i.e., it is amphoteric in its polymorphs. α-Mn, which has 58 atoms per unit cell, provides different sites for Mn-ions with different configurations. The extra stability of those ions which end up with half-filled d-shells evidently provides the driving force for transferring the electrons. The great stability of the interstitial rocksalt structures, which we discuss below, which have the highest known melting points (in excess of 4000°) is probably also due in part to charge transfer.

When considering compound formation between transition metal A atoms ($<\frac{1}{2}$ filled d-shell) and B atoms, the relative size can also be an important factor in determining which structure becomes the ground state. Efficient packing of the atoms is favored. Empirical radius ratio limits have been found for most structures. For instance, in the A-15 structure the radii of the A and B elements do *not* ordinarily differ by more than 15%. At high temperatures, vibrational entropy favors the bcc structure and also the other simple closed packed structures with respect to more complex ground state phases. Thus, A_3B compounds (with A=Nb, V; B=Au, Pt, Ir) are bcc at high temperatures and A-15 (β-tungsten) at low temperatures.

A speculative model which might be applicable to the formation of some complex structures involves the formation of a commersurate charge density wave. The charge density wave in 1T TaS_2, for example, has an amplitude corresponding to the transfer of almost 1 full electron. Thus, the Ta^{4+} disproportionates into Ta^{3+} and Ta^{5+}. Here the driving force is the Fermi surface instability.

The sequence in which the complex phases form as the d-band is filled is discussed in many places.[13] There are phases with ideally fixed stoichiometries; A_3B (β-tungsten), A_2B (Laves phases and derivatives), AB(CsCl), AB_3 (ordered Cu_3Au) and AB_5. The latter include the very hard magnetic compounds such as $SmCo_5$.

In addition, there are phases which occur over a wide range of stoichiometry including the commonly occurring tetragonal σ-phase, and the above-mentioned α-Mn phase. The purpose here is not to overwhelm the reader with specifics, but rather to illustrate the wealth of homogenous phases that are encountered. The delicate balance between the one-electron energies, intraatomic configuration energy and vibrational entropy leaves plenty of room for variety.

Even when a particular structure does not normally form it may be metastable (i.e., it represents a local minimum in configuration space) and

[13] See, for example, M. V. Nevit, *in* "Electronic Structure and Alloy Chemistry of Transition Elements" (P. A. Beck, ed.), p. 101. Wiley, New York, 1963.

may be prepared if some particular path of synthesis, such as epitaxial growth, favors its presence. Such metastable phases are interesting because they often exhibit strong electron–phonon coupling.

b. NaCl Compounds—Carbides, Nitrides and Oxides: The oldest class of superconducting compounds is composed of compounds of the sodium chloride (NaCl) family, such as NbC,[14] where a transition metal is combined in one-to-one proportion with carbon, nitrogen, oxygen or boron. These compounds not only exhibit high T_c's but also admit a wide range of solid solutions. Superconductivity occurs most strongly in those compounds which have 9 (e.g., NbC, ZrN) and 10 (e.g., NbN, WC) valence electrons per molecule. It occurs weakly in those which have 11(NbO), except for the strong superconductivity in PdH which will be discussed shortly. These compounds are sometimes referred to as interstitials in recognition of the (not always) small anion.

Vacancies can occur on both anion and cation sites, typically simultaneously. For example, TiO exists with roughly 15% vacancies on each site. When the compound is synthesized at high pressures the vacancies are squeezed out and T_c increases from less than 1 K to greater than 3 K. Figure VI.5[15] shows that generally T_c is markedly reduced on either side of the stoichiometric composition. TiO, an exception, has its max T_c ($\sim$3.2 K) with $\sim$6% of the cation sites vacant. NbN which forms with 9% anion vacancies has surprisingly little change in T_c when the vacancies are filled, by N-ion implantation. Figure VI.6 shows a somewhat unusual behavior for the dependence of T_c upon the solution of two binary (called pseudobinary) phases with a common constituent. Usually the intermediate superconductivity is lower than a linear interpolation between the end members. Here a maximum T_c is found for some intermediate composition. Furthermore, the maximum T_c is found whether the substitution is on the anion or cation site. In the case of Nb(CN) the anion vacancies are almost filled at the 30% C composition where $T_c^{max} = 18$ K is found and for (NbTi) $T_c^{max} \sim 18$ K is found. [The reader studying Fig. VI.6 may be disappointed to know that solid solutions of (NbTi) and (NbC)N have been studied and T_c^{max} is still $\sim$18 K.]

The microscopic reasons for the high transition temperatures in these materials are not obvious. In going from Nb to NbN the transition temperature is doubled, while the density of states, $N(0)$ is decreased. Thus some combination of the other parameters in the expression for λ,

[14] NbC ($T_c = 11$ K) was one of the earliest superconducting compounds discovered [W. Meissner and H. Franz, *Naturwissenschaffen* **18**, 418 (1930)].

[15] J. K. Hulm and R. D. Blaugher, *AIP Conf. Proc.* **4** (1972).

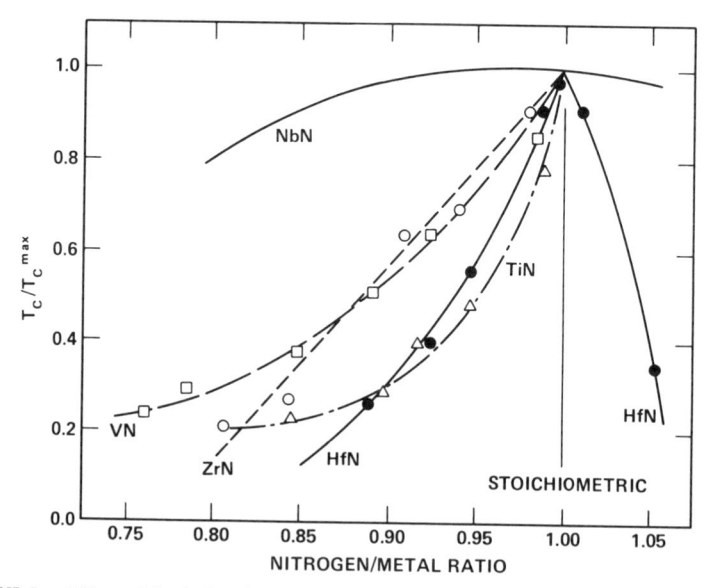

FIG. VI.5. Effect of deviation from stoichiometry on T_c for various nitrides. After Hulm and Blaugher.[15]

$$\lambda = \frac{N(0)\langle I^2 \rangle}{M\langle \Omega^2 \rangle} \tag{6.4}$$

must increase. Smith[16] has shown by inelastic neutron scattering that the phonon dispersion curves of superconducting carbides have anomalous dips in contrast to the nonsuperconducting ones as illustrated in Fig. VI.7.

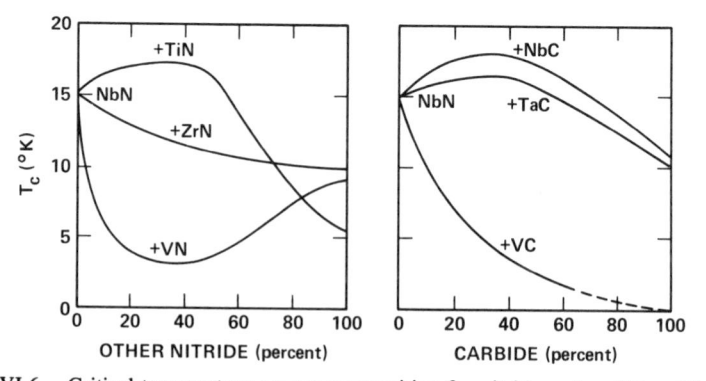

FIG. VI.6. Critical temperature versus composition for nitride and carbide additions to NbN. After Hulm and Blaugher.[15]

[16] H. G. Smith, *AIP Conf. Proc.* **4,** 321 (1972).

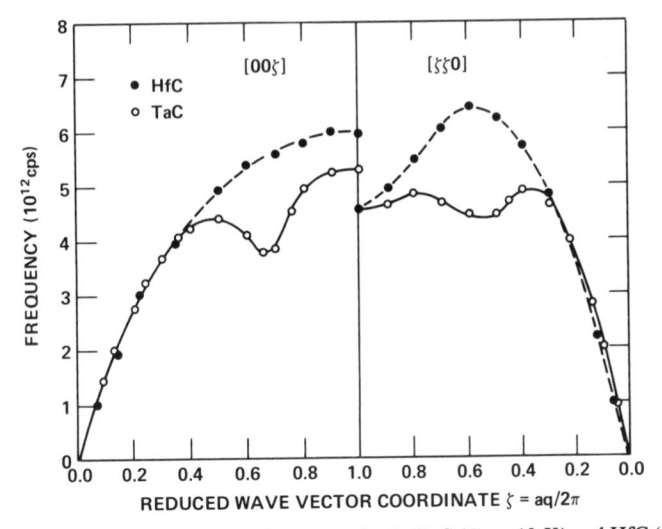

FIG. VI.7.   Longitudinal acoustic phonon modes in TaC ($T_c \simeq 10$ K) and HfC (nonsuperconductivity). After Smith.[16]

In this connection, it is interesting to note that Varma and Weber,[17] using the tight binding approach we mentioned in Chapter III, have calculated the phonon dispersion curves for 4d bcc metals and alloys. They find phonon anomalies similar to those shown in Fig. VI.7. The origin of these anomalies is due to interband contributions to the electron–phonon coupling "constant" $g_{k\mu,k'\mu'}$, where μ is a band index. That is, they are not directly related to the Fermi surface as were the Kohn anomalies we discussed in connection with charge density wave formation.

It is interesting to note that superconductivity is also found in low-carrier concentration *semiconducting* NaCl phases of nontransition elements such as SnTe and InTe. Interestingly, the T_c vs. carrier concentration relation found for those two compounds in the carrier decades 10^{20}–10^{22} per cm³ extrapolates to the high T_c NbN-based phases in a smooth way as can be seen in Fig. VI.8. The occurrence of superconductivity in low-carrier density (i.e., semiconductors–semimetallic) phases has been reviewed by Hulm.[18]

Palladium Hydride: Finally, let us consider PdH. Palladium and hydrogen also form a so-called interstitial system with the NaCl structure. It has the composition $PdH_{0.63}$ when in equilibrium with H_2 at atmospheric pressure. The vacancies on the H sublattice order at temperatures below

[17] C. M. Varma and W. Weber, *Phys. Rev. Lett.* **39**, 1094 (1977).

[18] J. K. Hulm, *Proc. 2nd. Symp. Spring Supercond. Symp., 1969* NRL Report 6972 (1969).

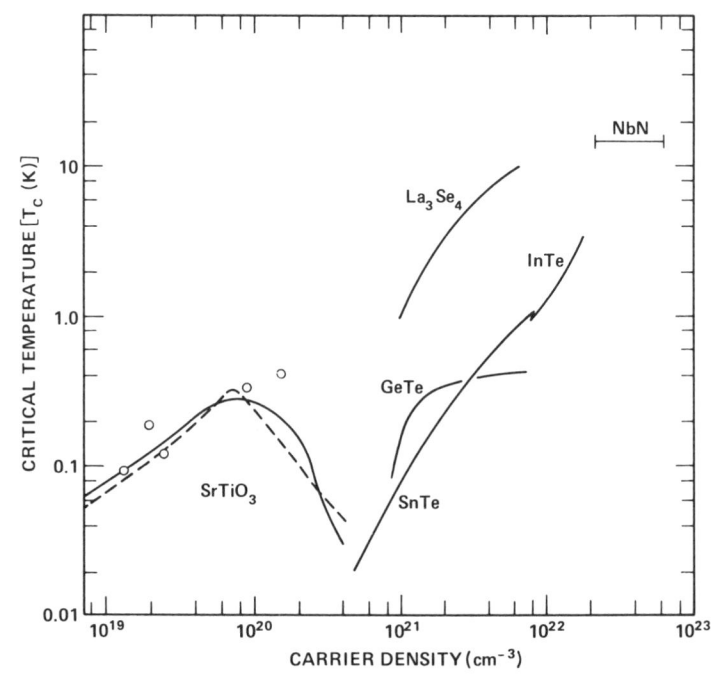

FIG. VI.8. Superconducting transition temperatures for various semiconductors as a function of carrier density. The three sets of data for $SrTiO_3$ correspond to three different methods for generating carriers. The curious maximum shown in this figure for $SrTiO_3$, which was originally believed to be due to a band structure effect, is more likely due to the presence of paramagnetic (pair-breaking) scattering defects (see Chapter VII).[18]

liquid nitrogen. Superconductivity has been found[19] for H concentrations greater than 0.8 with T_c reaching a maximum for what is believed to be the stoichiometric PdH composition. Due to the large difference in mass between H and Pd the expression for λ, Eq. (6.4), can be decoupled into an acoustic component consisting chiefly of Pd vibrations and an optic branch consisting of H vibrations.

$$\lambda = \frac{N(0)\langle I^2 \rangle_{Pd}}{M\langle \Omega^2 \rangle_{Pd}} + \frac{N(0)\langle I^2 \rangle_{H}}{M_H\langle \Omega^2 \rangle_{H}}. \tag{6.5}$$

The optical mode frequencies are low and have a large zero point amplitude. Of particular interest is the *reverse* isotope effect found by Stritzker and Buckel;[20] T_c for PdH = 9.5 K whereas PdD = 11.6 K. There is good

[19] T. Skoskiewicz, *Phys. Status Solidi A* **11**, K123 (1972).
[20] B. Stritzker and W. Buckel, *Z. Phys.* **257**, 1 (1972).

evidence that the contribution to λ is dominated by the optic mode both from tunneling experiments[21] and model calculations.[22] The cause of the reverse isotope effect is thus rooted in the optical modes; evidently connected with the larger zero-point motion of H relative to D.[23]

Buckel and Stritzker[24] have also found that T_c's of up to 17 K can be reached by implantation of H and D into Ag-Pd alloys. Other Pd-based alloys and implantations of other ions lead to superconductivity, but not so high as in the (PdAg) H system.

It is clear that hydrides, under special circumstances where the hydrogen can couple to the conduction electrons, have the potential of being very high temperature superconductors.[25]

The special conditions which obtain for the PdH system, unfortunately have not been found generally. In PdH the hydrogen s-band is partly occupied, and thus neither a simple "protonic" nor "hydrogenic" model is applicable.

c. A-15 Structures: Another class of intermediate phase compounds are those with the A-15 structure. Since this class comprises all the highest temperature superconductors, they have been studied by most of the techniques known to solid state physics. The relationship between the A-15 structure and the unusual behavior frequently found in both lattice and electronic properties is particularly challenging. The three highest transition temperature superconductors, Nb_3Ge ($T_c = 23$ K), Nb_3Ga

---

[21] P. J. Silverman and C. V. Briscoe, *Phys. Lett. A* **53**, 221 (1975).

[22] See, for example, D. A. Papaconstantopoulos and B. M. Klein, *Phys. Rev. Lett.* **35**, 110 (1975).

[23] The effect of the zero-point motion of the hydrogen is discussed qualitatively by R. Miller and C. Satterthwaite [*Phys. Rev. Lett.* **34**, 144 (1975)]. Other superconducting compounds with large mass differences leading to simple optical modes have been studied, but no negative isotope effect has been found. E. Bucher and C. Palmy [*Phys. Lett. A* **24**, 340 (1967)] have studied the isotopic mass dependence of T_c in $Be_{22}Mo$, ($T_c = 2.5$ K) where the contributions to λ may be approximately decoupled. They find a decrease in T_c as the mass of Mo is increased from 92 to 100 atomic units and conclude that the coupling to the low frequency Mo-modes is five times stronger than to the Be-modes. Distinct contributions of optical modes to the low temperature heat capacity in the superconducting alkali metal (hexagonal) tungsten bronzes have also been found [C. N. King, J. A. Benda, R. L. Greene, and T. H. Geballe, *Proc. Int. Conf. Low Temp. Phys., 13th, 1972* Vol. 3, p. 411 (1974)], but with no direct evidence for contributions to λ.

[24] W. Buckel and B. Stritzker, *Phys. Lett. A* **43**, 403 (1973).

[25] The only other superconducting metal hydrides are the thorium hydrides [C. B. Satterthwaite and I. L. Toepke, *Phys. Rev. Lett.* **25**, 741 (1970)] Th_4H_{15} ($T_c = 8.6$ K) and Th_4D_{15} ($T_c = 8.6$ K), which have a cubic lattice with 16 Th and 60 H atoms per unit cell. Again the contribution of the hydrogen modes is believed to be important, but there is also the possibility of some contribution of the 5f levels of Th, either to λ or μ^*.

$(T_c = 20$ K), and to a lesser extent Nb_3Al $(T_c = 18.8$ K) are metastable at the stoichiometric composition. They disproportionate into a low T_c Nb-rich A-15 phase (presumably with Nb atoms on metalloid sites), and a metalloid-rich phase.

When Nb_3Ge is grown from the melt under nearly equilibrium conditions, the A-15 phase turns out to have the composition near Nb_4Ge with $T_c = 6.3$ K. To maintain the metastable stoichiometric phase one must use various metallurgical "tricks" that permit nonequilibrium states which show a high degree of spatial order. Vapor phase condensation and rapid quench techniques are particularly useful.

Nb_3Sn $(T_c = 18$ K) and V_3Si $(T_c = 17$ K) can be grown in the A-15 phase as stoichiometric single crystals, and it is for these and other stable stoichiometric compounds that lattice and electronic anomalies have been studied in great detail.[26]

The special structure-related properties of the cubic A-15 structure are based upon the three sets of orthogonal noninteresting chains of A atoms (i.e., Nb atoms) running along the [100] directions shown in Fig. VI.9. Along each chain there is an important screw axis governing the disposition of the B atoms. The nearest neighbor distances along the chain are 10–15% less than nearest neighbor distances in the corresponding pure A elements.

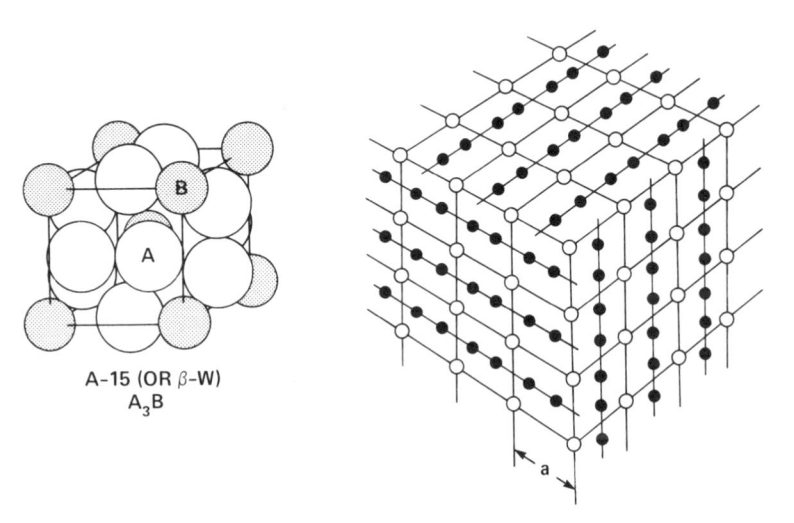

A-15 (OR β-W)
A_3B

FIG. VI.9. (a) Unit cell of the cubic A-15 (β-tungsten or Cr_3Si) structure. (b) Extended arrangement of atoms in the stoichiometric A-15 structure emphasing the three orthogonal, nonintersecting linear chains of A atoms.

[26] See, for example, L. R. Testardi, *Rev. Mod. Phys.* **47**, 637 (1975).

Geller has introduced a set of self-consistent A-15 radii based upon contact between A and B site atoms, which have been brought up to date recently,[27] and are useful in predicting the lattice constants of ordered stoichiometric A-15 phases.

One example of the usefulness in having a set of self-consistent radii is illustrated by Nb_3Si. According to Geller's radii, Nb_3Si should have a lattice constant of ~5.06 Å, if it could be formed as an ordered stoichiometric, A-15 phase. It then is expected to have a superconducting transition temperature substantially higher than Nb_3Ge on the basis of its position in the periodic table. The few times Nb_3Si has been synthesized with an A-15 structure at all (rather than in the related Ti_3P structure in which it normally forms) it had lattice constants much larger than predicted from the Geller radii. Nb_3Ge films, on the other hand, have the lattice constant predicted by the Geller radii, when prepared as stoichiometric high T_c films, and one that is much larger otherwise. Thus, it is assumed that stoichiometric Nb_3Si has not yet been synthesized. The unit cell of $Nb_3Si(Ti_3P)$ is related to the A-15 phase by the doubling of two of the cubic axes, but there is no simple relation between the atomic sites.

There is a variety of empirical evidence that, when B-sites are occupied by transition metals, the special properties of the A-15 phase, including the high T_c's, are lost. First we have the results just mentioned, that excess Nb (presumably on B-sites) lowers T_c. Further, of the more than 40 known A-15 compounds, the majority have transition metals on the B-sites and are ordinary metals. Finally, radiation damage experiments using fast neutrons (14 MeV), or 2 MeV helium atoms, with fluences in excess of 10^{19} cm^{-2} result in large reductions of T_c. The initially high T_c's can be restored by moderate (800 °C) annealing. Intensity measurements of X ray and neutron Bragg peaks show that a significant amount of anti-site (A atoms on B sites and vice-versa) disorder exists in the radiation damaged material. The long range order parameter (i.e., the fraction of A atoms on A sites) decreases in irradiated Nb_3Al and thermally disordered V_3Ga.[28] The T_c depression could also be due to other defects which might stiffen the lattice, reduce peaks in the density of states, or even remove gap anisotropy (see introduction to Chapter VII). All these mechanisms have been proposed. It is not likely that a quantitative understanding will be achieved until a satisfactory microscopic model for the A-15 compounds is established.

It is perhaps not surprising that the strong-coupling which must exist manifests itself in normal state properties. There are anomalies in both

[27] G. R. Johnson and D. H. Douglass, *J. Low Temp. Phys.* **14**, 565 (1974).

[28] A. R. Sweedler and D. E. Cox, *Phys. Rev. B* **12**, 147 (1975); R. Flükiger, J. L. Staudenmann, and P. Fischer, *J. Less Common Met.* **50**, 253 (1976).

electronic and lattice properties of Nb_3Sn and V_3Si, which have been prepared as single crystals, and in other A-15 compounds as well. For example, both the Knight shifts on the A and B atoms and the static magnetic susceptibility are markedly temperature dependent for V_3Ga, V_3Si and Nb_3Sn unlike the expected behavior for a Pauli temperature-independent spin susceptibility. For V_3X the dependence is stronger for the higher T_c compounds (Fig. VI.10a).[29] Analysis of the Knight shift as a function of susceptibility (with temperature as the implicit parameter) shows that there is almost complete pairing of the d-electrons[30] when the compounds becomes superconducting. However, not all the high temperature compounds show such behavior. For example, the susceptibilities of Nb_3Al ($T_c = 18.9$) and Nb_3Ge ($T_c = 23.2$) are not usually temperature dependent.

Related behavior is also seen in the linear heat capacity coefficients which are a measure of the phonon-enhanced density of states. Those compounds with large coefficients have the temperature dependent properties. For example, the coefficient per transition metal atom of Nb_3Sn is approximately twice as great as that of niobium, and for V_3Ga it is nearly

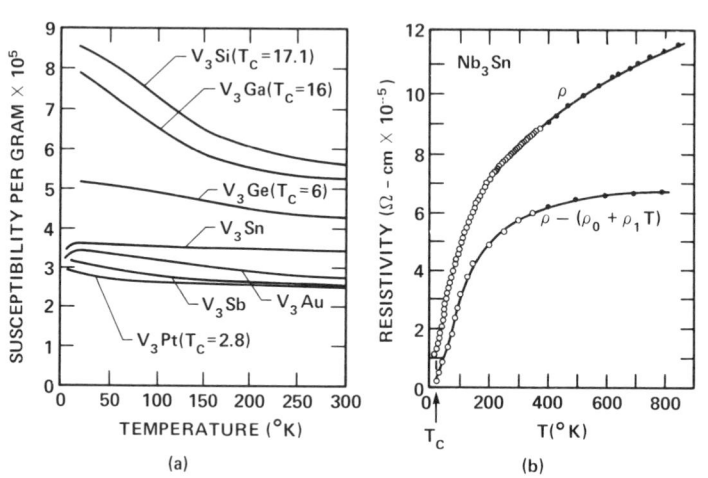

FIG. VI.10. (a) Temperature dependence of the magnetic susceptibility of several V_3X compounds. Note that the temperature dependence occurs for the high T_c compounds. (The T_c's for V_3Sn, V_3Au, and V_3Sb are all less than 1 K). After Williams and Sherwood.[29] (b) The temperature dependence of the electrical resistivity of Nb_3Sn. The lower solid curve is the $\rho_2 \exp(-T_0/T)$ contribution. After Woodward and Cody.[32]

[29] H. J. Williams and R. C. Sherwood, *Bull. Amer. Phys. Soc.* [2] **5**, 430 (1960).
[30] A. M. Clogston, S. C. Gossard, V. Jaccarino, and Y. Yafet, *Phys. Rev. Lett.* **9**, 262 (1962).

three times as great as it is for vanadium metal.[31] Again, however, the coefficients of Nb_3Al and Nb_3Ge have the values expected for typical d-band metals such as elemental niobium.

Anomalous temperature dependences of resistivity have also been observed in the A-15 compounds. Woodward and Cody[32] found that the resistivity of Nb_3Sn could be fit to an accuracy of 1% between T_c and 850 K by $\rho = \rho_0 + \rho_1 T + \rho_2 \exp(-T_0/T)$ where ρ_0, ρ_1, and ρ_2 are constants independent of temperature and $T_0 = 85$ K. The exponential term makes an appreciable contribution as Fig. VI.10b[32] shows. Later the resistivity at high temperature was found to saturate at a value on the order of 150 $\mu\Omega$-cm.[33] It was also found[34] that the low-temperature resistivity could be fit rather well to a simpler T^2 law, but over the smaller temperature range from T_c to 50 K. The low-temperature resistivity of V_3Si ($T_c = 17$ K) was also found to follow a T^2 law. Such a T^2 dependence is often an indication of electron–electron scattering. However, in this case the magnitude observed is much larger than can be accounted for by this means. Ordinarily, electron–phonon scattering gives a T^5 dependence. This, however, is based on a Debye phonon spectrum. Webb et al.,[34] have shown that if one uses a more realistic phonon density of states the resistivity does indeed show a T^2 behavior. Furthermore, Allen et al.[35] have shown that the saturation behavior at high temperatures is an indication that the mean free path, $\Lambda = \hbar \langle v(E_F)^2 \rangle^{1/2}/2\pi\lambda k_B T$, has become comparable with the interatomic spacing. The average Fermi velocity, $\langle v^2 \rangle^{1/2}$, was calculated from a self-consistent band structure; λ was determined from T_c. For Nb_3Ge, $\lambda \sim 1.8$ and the resulting mean free path was found to be 5 Å at 300 K. Values of ρ approaching 150 $\mu\Omega$-cm are found for other A-15 compounds. This is also believed to be a consequence of Λ saturating near the interatomic spacing.[33]

Another interesting anomaly is the cubic-to-tetragonal distortion that starts a few degrees above the superconducting transition.[36] Such transitions have been found in single crystals of Nb_3Sn and even in polycrystalline samples of other A-15 compounds. This is the "martensitic" transition we mentioned in Chapter II. Despite its closeness in temperature to the superconducting transition, it is now generally believed that this structural transition is not directly related to T_c but rather that both are the re-

[31] M. Weger and I. B. Goldberg, *Solid State Phys.* **28**, (1) (1973); A Junod, J. L. Staudenmann, J. Müller, and P. Spitzli, *J. Low Temp. Phys.* **5**, 25 (1971).

[32] D. W. Woodward and G. D. Cody, *Phys. Rev. A* **136**, 166 (1964).

[33] Z. Fisk and G. Webb, *Phys. Rev. Lett.* **36**, 1084 (1976).

[34] G. W. Webb, Z. Fisk, J. J. Engelhardt, and S. D. Bader, *Phys. Rev. B* **15**, 2624, (1977).

[35] P. B. Allen, W. E. Pickett, K.-M. Ho, and M. L. Cohen, *Phys. Rev. Lett.* **40**, 1532 (1978).

[36] B. W. Batterman and C. S. Barrett, *Phys. Rev.* **149**, 296 (1966).

sult of a more fundamental mechanism. The point is that the structural transition is associated with a soft phonon at the zone center. Since the phonon density of states is very small near $\Omega = 0$, this will not have much effect on T_c as expected from the result of Bergmann and Rainer (ref. 33, ch. III). The elastic constant $(c_{11} - c_{12})$ softens with decreasing temperature and approaches zero at the martensitic transition.[37] In contrast the bulk modulus and c_{44} are relatively temperature independent.

The lattice parameters of Nb_3Sn as a function of T are shown in Fig. VI.11.[38] In V_3Si, the martensitic distortion is reversed; the a axis becomes smaller. Vieland and Wickland[39] have found that about 8% aluminum substituted for tin is just sufficient to suppress the martensitic transition. At that concentration T_c reaches a maximum = 18.6 K. The martensitic distortion is easily suppressed by impurities or imperfections of one sort or another that have not yet been well studied. It is necessary to investigate each individual defect carefully to find out whether it endangers the transition or not. Most data indicate that the T_c of the tetragonal phase is a few tenths of a degree lower than the related cubic phase. Presumably

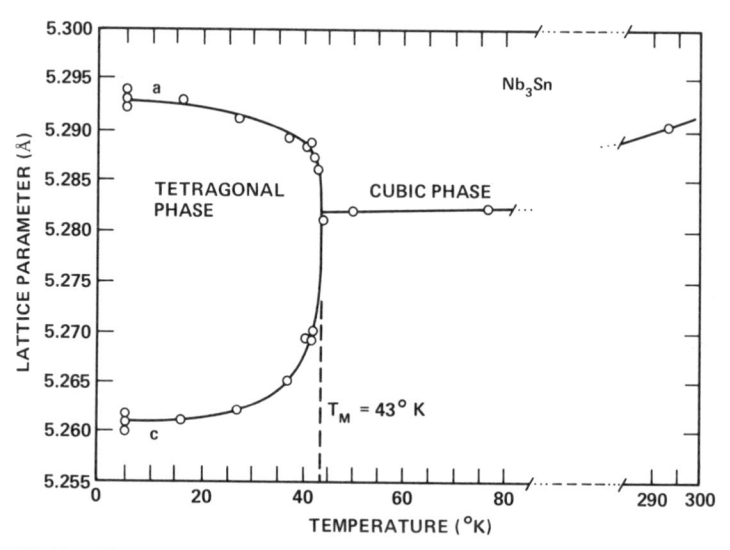

FIG. VI.11. The c and a lattice parameters of Nb_3Sn vs. T showing the cubic-to-tetragonal transformation at $T_m = 43$ K. After Mailfert et al.[38]

[37] L. R. Testardi and T. B. Bateman, *Phys. Rev.* **154**, 402 (1967).
[38] R. Mailfert, B. W. Batterman, and J. J. Hanak, *Phys. Lett. A* **24**, 315 (1967).
[39] L. J. Vieland and A. W. Wickland, *Phys. Lett. A* **34**, 43 (1971).

only a small portion of the Fermi surface is eliminated by the martensitic transition.

In both V_3Si and Nb_3Sn any discontinuous volume change at T_m is extremely small. Further, there is no evidence of latent heat even though the heat capacity associated with its transition, as shown in Fig. VI.12,[40] showed no evidence of hysteresis over a wide range of temperature variation. In view of the argument presented in Section II.A.6b, the structural transition must be a *very* weak first-order transition.

The acoustic phonon spectrum is also strongly temperature dependent. In Fig. VI.13,[41] the marked softening of the elastic moduli with decreasing temperature (i.e., the decrease in slope of the solid lines) is found to continue far out into the zone, although at a reduced level, as determined by inelastic neutron scattering. The additional 21 optical modes must also be considered. Schweiss *et al.*,[41a] using time-of-flight neutron spectroscopy, which is applicable to polycrystalline samples, find that the marked softening occurs over the whole spectrum.

There does not exist a microscopic explanation of these anomalous properties. With 36 or 38 electrons per unit cell this should not be surprising. However, it does not mean to say that there have not been numerous models proposed. Most of these models exploit the one-

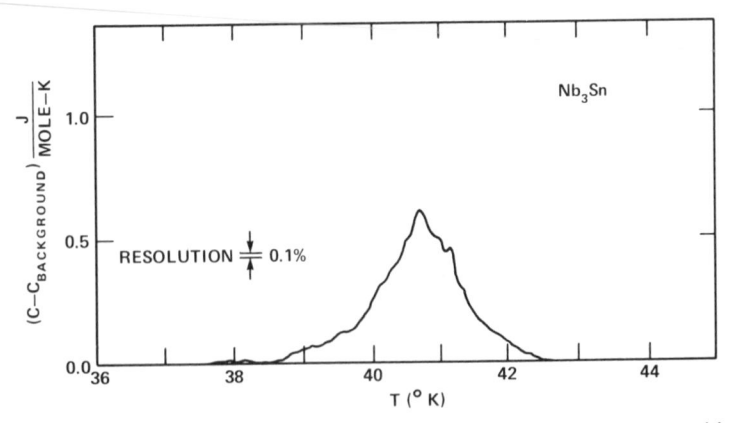

FIG. VI.12. Nb_3Sn heat capacity measurements through the martensitic transition. No evidence for first-order or time-dependent effects at the martensitic transition were found. After Harper.[40]

[40] J. M. E. Harper, unpublished Ph.D. dissertation, Stanford University, Stanford, California (1975).

[41] G. Shirane and J. Axe, *Phys. Rev. Lett.* **27**, 1803 (1971).

[41a] B. P. Schweiss, B. Renker, E. Schneider, and W. Reichardt, in "Superconductivity in d- and f-Band Metals" (D. Douglas, ed.) p. 189. Plenum, New York, 1976.

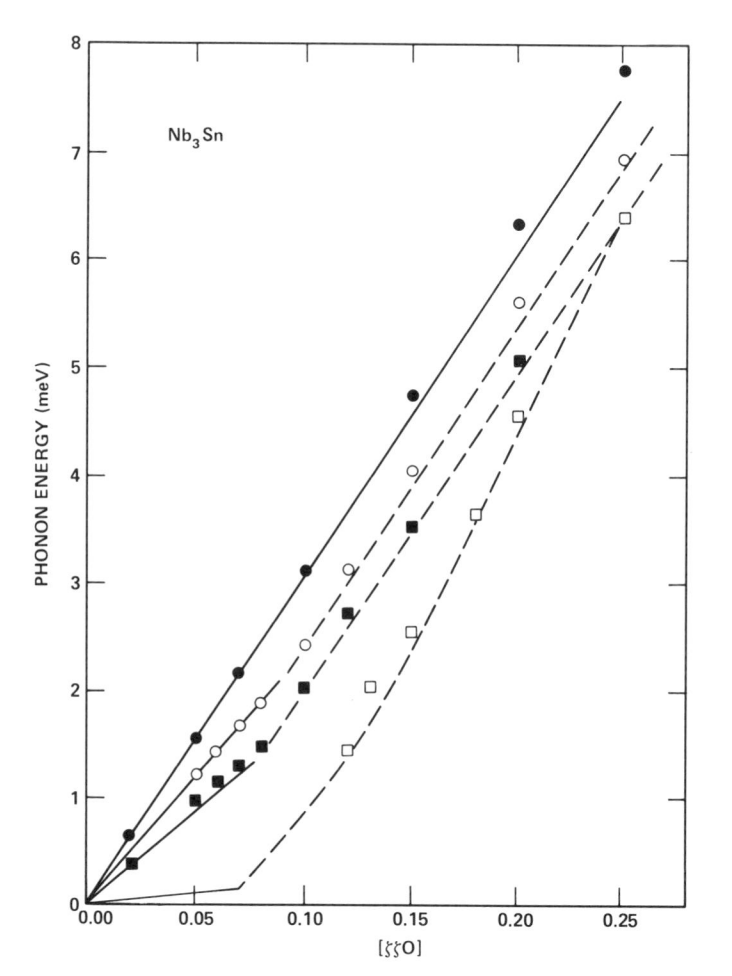

FIG. VI.13.   Acoustic phonon dispersion curves for $q \| [110]$ waves with $[110]$ polarization in Nb_3Sn. $q = (\zeta, \zeta, 0)2\pi/a = 1.19$ Å. Note the mode softening on cooling: 259 K; 80 K; 46 K. After Shirane and Axe.[41]

dimensional character of the structure shown in Fig. VI.9. The first such model was due to Labbé and Friedel.[42] They take a one-dimensional tight binding d-band density of states with characteristic square root singularities as shown in Fig. VI.14a and assume that the Fermi level lies within several meV of this singularity. In the tetragonal phase, the degeneracies are lifted as shown in Fig. VI.14b. The shifting of one of the singularities to lower energy allows the total electronic energy to decrease. Thus, the

[42] J. Labbé and J. J. Friedel, *J. Phys. Radium* **27**, 708 (1966).

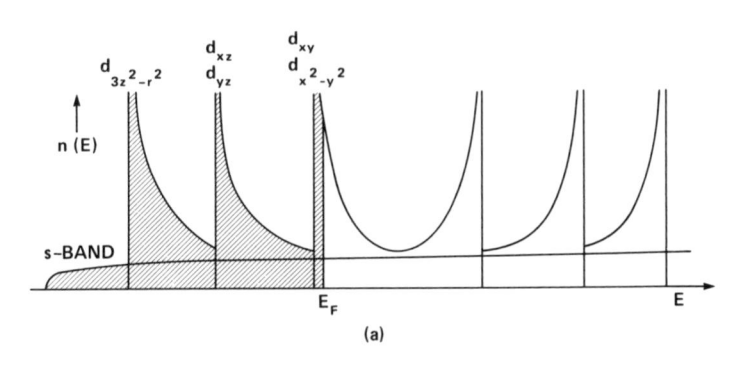

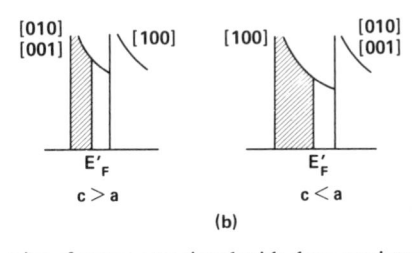

FIG. VI.14.   Density of states associated with three noninteracting linear chains.

martensitic transition in this model is the result of the Jahn–Teller effect. Band calculations by Mattheiss,[43] however, show that interchain coupling is sufficient to wash out fine structure in the density of states of this order. This suggests that the one-dimensional aspect alone is not sufficient. In this respect it is interesting to note that a subsequent self-consistent pseudopotential band calculation[44] shows that the Fermi level does indeed lie in a peak in the density of states of width 0.07 eV. The appearance of sharp peaks in this calculation is the result of a large number of d-bands lying within a 10 eV interval plus the restriction that no band crossings are allowed off symmetry planes.

Gor'kov[45] has suggested that the martensitic transition may be associated with the formation of a charge density wave that results in a pairing of the transition metal atoms. This model assumes that the Fermi level lies near the X-point in the band structure. This is not found in either of the band calculations.[43,44] Bhatt and McMillan[46] have developed a phenomenological Landau theory based on Gor'kov's idea of a CDW-driven transi-

[43] L. F. Mattheiss, *Phys. Rev. B* **12**, 2161 (1975).

[44] K.-M. Ho, W. E. Pickett, and M. L. Cohen, to be published.

[45] L. P. Gor'kov, *JETP Lett. (Engl. Transl.)* **17**, 379 (1973).

[46] R. N. Bhatt and W. L. McMillan, *Phys. Rev. B* **14**, 1007 (1974).

tion without making any inappropriate assumptions about the band structure. This theory is successful in explaining many of the experimental results.

5. Ternary Compounds

In structures where there are more than two sites within the unit cell it is possible to find long range order involving primarily only one or two of these sites. Such systems enable us to probe the spatial extent of the order parameter within the unit cell by making chemical substitutions on those sites not directly involved in the ordering. Ternary compounds offer a number of examples.

Consider the *silver oxide clathrates*, $Ag_6O_8 \cdot AgX$ where $X = NO_3^-$, BF_4^-, or HF_2^-. The unit cell is composed of four Ag_6O_8 polyhedra, or cages, each enclosing an X^- ion (hence the term clathrate which comes from the Greek word "enclose") and connected by Ag^+ ions.

The unit cell is roughly the same size whether a linear (HF_2^-), planar (NO_3^-) or spherical ion (BF_4^-) is within the cage. The compounds are superconducting with T_c's all about 1 K.[47] The Ag positions comprising the cage polyhedra are all equivalent; thus if these six Ag ions are to balance the charge of the eight oxygens they are in what is referred to as a mixed valence state. The weak interaction between the X^- and the conduction band electrons is evident by only small kinks in the resistivity of the NO_3 compound believed to be associated with alignment of the planar NO_3^- ion. Similarly, a marked increase in the linear heat capacity coefficient for the HF_2^- compound is believed to be due to the six equivalent states within the cage available to that linear anion rather than electronic in origin.[48]

Superconductivity is also found in the noncubic *tungsten bronzes* with the general formula A_xWO_3, $0 < x < \frac{1}{3}$, if A is an alkali metal, and $0 < x < 0.16$ if A is an alkaline earth metal.[49] Knight shift measurements support a model in which the conduction band is formed from the p and d orbitals of the WO_3.

Spatial separation of the superconducting and magnetic order within the unit cell of the intercalated layered *transition metal dichalcogenides* is easy to visualize. $2H\text{-}TaS_2$, which undergoes a charge density wave (CDW) transition at 80 K, becomes superconducting at ~ 0.15 K. If

[47] M. B. Robin, K. Andres, T. H. Geballe, N. A. Kuebler, and D. B. McWhan, *Phys. Rev. Lett.* **17**, 917 (1966).

[48] Mary M. Conway, N. E. Phillips, T. H. Geballe, and N. A. Kuebler, *J. Phys. Chem. Solids* **31**, 2673 (1970).

[49] A. R. Sweedler, Ch. J. Raub, and B. T. Matthias, *Phys. Lett.* **15**, 108 (1965).

nitrogen-containing organic molecules such as amines are intercalated (inserted) between the covalently bonded layers, the CDW ordering is suppressed, the Fermi surface remains intact, and the T_c rises above 3 K. Figure VI.15 shows that T_c increases with the spatial density of nitrogen bonds linking the alkylamines to the metallic TaS_2 layers.[50] When there are less than eight carbons in the chain, the amines lie flat; and fixed interlayer spacing corresponds to the thickness of the amine. For 12 or more carbons in the amine chain the interlayer spacing corresponds to a bilayer of amine molecules stacked perpendicular to the layers, and T_c remains constant at 3 K even for the octadecylamine compound where the interlayer spacing is 57 Å. Studies of the interaction and ordering between the layers have not been definitive largely because of the difficulty in obtaining perfect enough crystals. Pyridine can be much more easily intercalcated and relatively well-formed crystals of $TaS_2 \cdot (C_6H_5N)_{1/2}$ with the TaS_2 layers separated by about 6 Å. The critical field slopes for the field aligned parallel to the layers is larger than for any known material reaching as high as 200 kOe per degree.[50a]

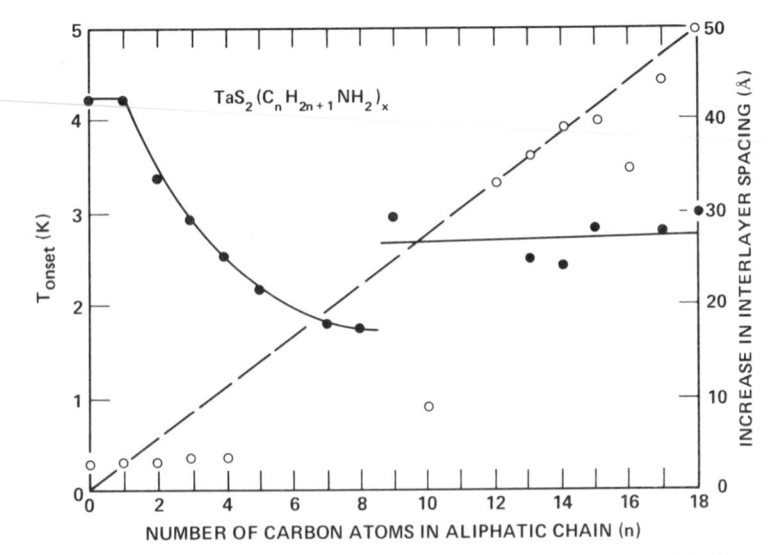

FIG. VI.15.   The onset temperature of superconductivity (solid dots) and the increase in the interlayer spacing (open circles) as a function of the number of carbon atoms in the n-alkyl amine intercalates in TaS_2. After Gamble et al.[50]

[50] F. R. Gamble, J. H. Osiecki, M. Cais, R. Pisharody, J. F. Di Salvo, and T. H. Geballe, Science 174, 493 (1971).
[50a] D. Prober, R. Schwall, and M. R. Beaseley, to be published.

Two ternary phases with an interesting combination of magnetic and superconducting properties have recently been discovered. The *Chevrel compounds*,[51] so named for their discoverer, have the formula $M_yMo_6X_8$ where M is divalent or trivalent and X is a chalcogen. The Mo_6X_8 units occur as clusters with Mo–Mo bonds that are shorter than in elemental Mo. The sites for the M atom between the clusters are not rigidly specified, and the stoichiometry can vary from $y = 3.6$ for Cu to ~1 for the rare earths and heavy metals. The Mo_6X_8 unit can form stable binary compounds with no M-site occupancy, or pseudobinary compounds with Br or I partially substituted for X, as in $Mo_6Se_6I_2$. The crystal structure is rhombohedral-hexagonal with the rhombohedral angle close to 90°.

The T_c's range up to 15 K and there is no obvious correlation between the M atom and T_c. This can be understood if the conduction band is assumed to arise from the 4d electrons of the Mo clusters. The M atom then influences T_c indirectly by its effect on the phonon structure, and by charge transfer to the cluster. It is obvious that there must be overlap between clusters since the binary Mo_6Se_8 is metallic, but the insertion of M will increase the intercluster distances and thus reduce the amount of overlap and the width of the d-band. Hydrostatic pressure studies[52] show, apart from some low-pressure irregularities, a marked decrease in T_c with increasing P, as might be expected on the basis of this argument.

Both the Chevrel phases and a second class of ternaries with the formula RE · Rh_4B_4 exhibit an interesting interplay between magnetic ordering on RE sites and superconductivity. We defer further discussion of the rich ordering phenomena observed in these structures until the discussion of the coexistence of magnetism and superconductivity in Chapter VII.

6. Superconducting Polymers—$(SN)_x$

The polymer formed from chains of alternating sulfur and nitrogen atoms, $(SN)_x$ is not a new material; it was first reported prior to 1910. Greene, Street, and Suter[53] discovered superconductivity in the polymer at 0.3 K after Walatka and coworkers[54] observed the high electrical conductivity, unusual for any polymer, persisted down to 4 K.

$(SN)_x$ is prepared by the solid state polymerization of S_2N_2 which is itself formed by the thermal decomposition of S_4N_4. The polymerization

[51] See the review by Ø. Fischer (*Appl. Phys.* **16**, 1 (1978)) for a more detailed discussion.

[52] R. N. Shelton, *in* "Superconductivity in d- and f-Band Metals" (D. Douglas, ed.), p. 137. Plenum, New York, 1976.

[53] R. L. Greene, G. B. Street, and L. J. Suter, *Phys. Rev. Lett.* **34**, 577 (1975).

[54] V. V. Walatka, M. M. Labes, and J. H. Perlstein, *Phys. Rev. Lett.* **31**, 1139 (1973).

process must be carried out very slowly at 0 °C for two days followed by two weeks at room temperature in order to obtain good quality crystals. Scanning electron microscope pictures show the crystals are made of oriented bundles of fibers. These are quite evident in the poorer quality crystals, as in twinning, and the presence of several percent hydrogen. The defects and fibrous morphology lead to difficulty in determining the crystal structures.[55] The structure consists of chains of SN which are almost planar, lying in the (102) plane of the monoclinic lattice. Comparison of the shortest interchain bond lengths with corresponding Van der Waal distances shows the interaction between chains is weak, but finite.

Each SN molecule with eleven valence electrons can contribute one electron to the conduction band. A purely one-dimensional system would be expected to show a Peierls transition. The widely studied compounds TTF-TCNQ and KCP mentioned in earlier chapters are examples of quasi-one-dimensional conductors that undergo transitions at low temperatures believed to result from the Peierls mechanism. The fact that no dimerization or metal–insulator transition is found in (SN_x) is understandable if interchain interactions are important. Evidence for interchain interaction comes from transport and optical measurements. Thus the polymer is really an anisotropic three-dimensional conductor. dc conductivity measurements are not reliable and apt to be too low either in the chain direction due to defects, or in the perpendicular direction because of the fibrous morphology. The optical (Drude) conductivity determined from reflectivity measurements is less subject to spurious extrinsic effects, particularly the high resistance defects. Indeed at room temperature $\sigma_{opt}^{\parallel}$ along the chain is $\sim 2.5 \times 10^{-4} \, \Omega^{-1} cm^{-1}$ almost an order of magnitude higher than $\sigma_{dc}^{\parallel}$. Similarly $\sigma_{opt}^{\perp}$ is $380 \, \Omega^{-1} cm^{-1}$, a factor of 40 greater than $\sigma_{dc}^{\perp}$. Although it is metallic, it is strongly damped due to the fiber boundaries.

There are other properties of $(SN)_x$ which, like the conductivity, are inconsistent probably because of spurious effects. H_{c2} is more anisotropic than band structure would indicate. Also $H_{c2\parallel}$ is not strongly dependent upon $\sigma_{dc}^{\parallel}$ (4.2 K) as it should be. Therefore at present the observed properties may result at least partly from imperfections and extrinsic effects. In fact, all highly anisotropic single crystals particularly the intercalated layered metal dichalcogenides as well as the $(SN)_x$ polymers can be expected to have their intrinsic properties masked by defects. As the quality of the crystals is improved, further understanding of the ultimate intrinsic properties should be possible.

[55] See, for example, the review of R. L. Greene and G. B. Street, *in* "Chemistry and Physics of One-Dimensional Metals" (H. J. Keller, ed.), p. 167. Plenum, New York, 1977.

B. MAGNETISM

With only a few exceptions, ferromagnetism is only observed in two regions of the periodic table: in the last half of the first transitional series, and among the rare earths.

1. Transition Metals

In Fig. VI.16[56] we show the ionic radii of the divalent transition metal ions. As we move across the series d-orbitals become more localized. This has two effects: it means that the d-bands of the metals will be narrower at the right end; it also means that the average Coulomb energy, U, will be larger at the right end. Thus Ti and V are Pauli paramagnets with a relatively temperature-independent susceptibility. *Scandium,* however, with only one d-electron shows a large susceptibility. The origin of this is found in its generalized susceptibility $\chi(\mathbf{q})$ shown in Fig. VI.17.[57] We see that it shows a maximum at $(0,0,0.31)2\pi/c$, indicating an incipient spin-density wave. Overhauser[58] suggested that in such cases a small amount of a magnetic impurity should catalyze the appearance of the SDW. However, if one adds Gd or Tb to scandium one does not see evidence of the opening of gaps on the Fermi surface. Rather, one finds spin glass behavior, i.e., a cusp in the low field susceptibility with no anomalies in the resistivity (see Fig. V.8).

As we noted in Chapter IV, *chromium* has unique magnetic properties.

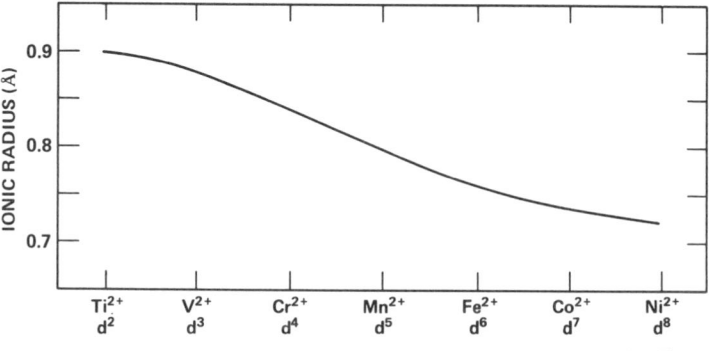

FIG. VI.16. Ionic radii of the transition metal ions. After Pauling.[56]

[56] L. Pauling, "Nature of the Chemical Bond," 3rd ed. Cornell Univ. Press, Ithaca, New York, 1960.

[57] J. Rath, R. P. Gupta, and A. J. Freeman, *AIP Conf. Proc.* **24,** 327 (1974).

[58] A. W. Overhauser, *Phys. Rev. Lett.* **3,** 414 (1959).

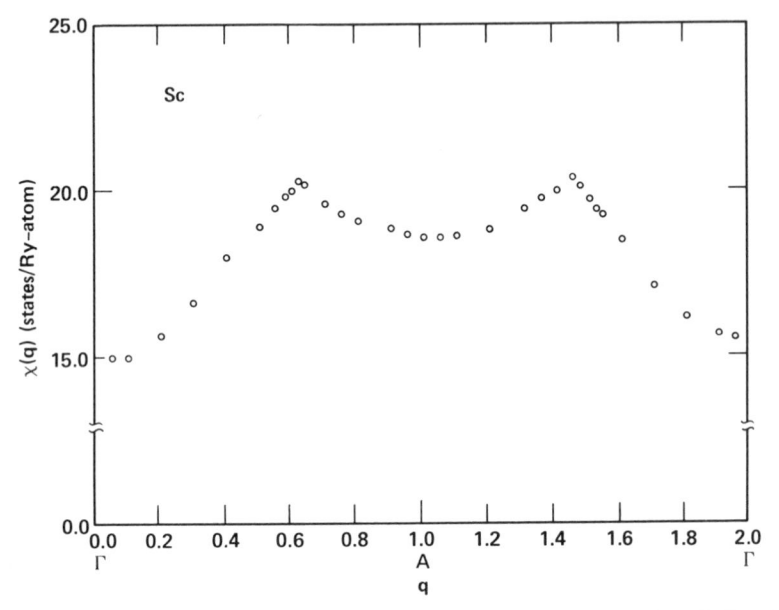

FIG. VI.17. Generalized susceptibility of scandium metal along the Γ-A-Γ direction. After Rath *et al.*[57]

Above 310 K there is no evidence of the existence of paramagnetic local moments, while below this temperature, Cr is antiferromagnetic with a spin polarization whose period is incommensurate with the lattice. This spin-density wave is transverse; the direction of the magnetization is at right angles to the wavevector. At 120 K there is a spin–flop transition at which the magnetization rotates to become parallel to the Q-vector which remains along a cube axis, i.e., a longitudinal SDW. The behavior is summarized in Fig. VI.18.

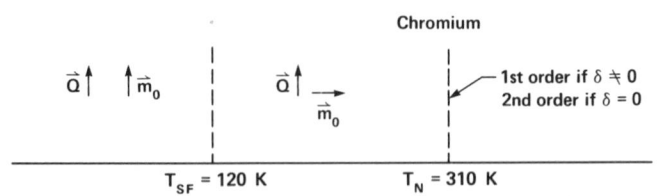

FIG. VI.18. Summary of the magnetic transitions in chromium. Here Q corresponds to the wave vector of the spin-density wave and m_0 its amplitude. δ is a measure of the incommensurability.

Problem VI.2

The Clausius–Clapeyron equation, $dP/dT = \Delta S/\Delta V$, may be obtained by differentiating the equilibrium condition $G_1(T, P, H) = G_2(T, P, H)$ with respect to temperature and using Eq. (1.9) with $\Delta H = 0$. Derive a similar relation for dH/dT in terms of the changes in entropy and magnetic susceptibility $\Delta\chi = \chi_1 - \chi_2$. The field at which spin-flop occurs in chromium is experimentally found to satisfy

$$H_{SF}^2 = 6.76 \times 10^{10} (1 - T/T_{SF}) \text{ gauss}^2$$

while $\Delta\chi = 1.44 \times 10^7$ emu/cm^3.

a. What is the latent heat associated with this transition?

b. The pressure dependence of T_{SF} is given by $(1/T_{SF})(\partial T_{SF}/\partial P) = -4.63 \times 10^{-11}$ dynes^{-1} cm^2. What is the change in volume at the transition?

Manganese exists in four allotropic forms, the highest naturally being bcc (δ-Mn) which is antiferromagnetic with a moment of 1 μ_B. The α-phase, stable at room temperature and below, is cubic with 58 atoms per unit cell. There are three inequivalent sites with different electronic configurations which order antiferromagnetically at 100 K. (This tendency to form metallic structures with different configurations at different sites extends to the other members of column VII, Tc and Re.)

Iron, cobalt, and *nickel* and their solid solutions are the common 3d ferromagnets with Curie temperatures well above room temperature. The magnetic properties of the elements are listed in Table VI.4.

2. Alloys

Slater[59] noted that the magnetic properties of 3d solid solutions, particularly their moments, can be averaged over the periodic table and plotted in a meaningful way as a function of the filling of the d band. If the components are fairly near-neighbors of the periodic system, then the averaging is just like that for the superconducting 4d and 5d solid solutions (Fig.

TABLE VI.4[a]

PHYSICAL PROPERTIES OF THE TRANSITION METAL FERROMAGNETS

		n_B	T_c (K)	p_{eff}	θ_p (K)	γ (mJ/mole-K^2)
Fe		2.216	1043	3.20	1101	5
Co	hcp	1.716				
	fcc	1.75	1394	3.15	1408	5
Ni		0.616	631	1.61	650	7.3

[a] The effective moment, p_{eff}, is obtained from the Curie constant, C, by $C = N\mu_B^2 p_{eff}^2/3k_B$. γ is the linear coefficient of the specific heat.

[59] J. C. Slater, *J. Appl. Phys.* **8**, 385 (1937).

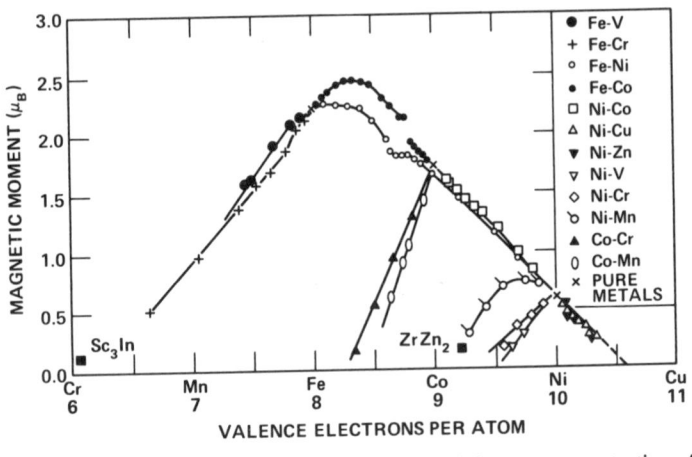

FIG. VI.19.   Saturation magnetization as a function of electron concentration. After Bo-zorth.[60]

VI.4). Figure VI.19[60] illustrates this behavior. One should also note the nonintegral numbers of Bohr magnetons in contrast to the integral number of spins found for the rare earth magnetic metals discussed below. If the pairs of atoms have too much charge contrast, then states become localized and the simple averaging fails as is evident in the figure.

The Slater–Pauling curve may be understood by considering the d-electrons to be exchange split as illustrated in Fig. VI.20. The structure shown is a rough approximation to those obtained by detailed band calculations as discussed in Chapter IV. In the rigid-band approximation, one assumes that the only effect of alloying is to shift the Fermi level. The majority spin band in nickel is completely full (in recognition of which it is sometimes referred to as "strong" meaning further splitting cannot increase the magnetism). The maximum in μ_B between cobalt and iron reflects the composition where the majority band is just completely full. Further removal of electrons as one moves across the compositional axis towards iron results in depletion of both majority and minority bands.

Pauling[61] offered an alternative explanation based on the idea that 2.56 d-orbitals hybridize with s- and p-orbitals to form nonmagnetic bonding orbitals. The remaining 2.44 d-orbitals fill according to Hund's rule to give the magnetic moments.

There is no doubt that the magnetic electrons partake in the transport

[60] R. Bozorth, "Ferromagnetism," p. 441. Van Nostrand-Reinhold, Princeton, New Jersey, 1951.

[61] L. Pauling, *Phys. Rev.* **54**, 899 (1938).

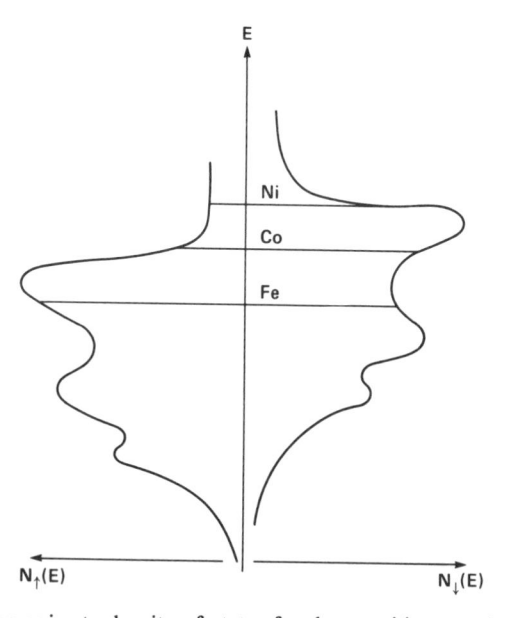

FIG. VI.20. Approximate density of states for the transition metal alloys. The Fermi levels for Fe, Co, and Ni are indicated.

properties of the transition metal ferromagnets, i.e., that they are itinerant. Nevertheless, there are some physical properties where a localized model is a reasonable and tractable approximation. The situation that pertains is of course somewhere between completely localized and free as is strikingly seen in the neutron diffraction obtained distribution of magnetism in bcc *iron* by Shull and Mook (Fig. VI.21).[62] This figure emphasizes the complexity of the real space distribution of spin density. Similar neutron investigations for *nickel* allow Mook[63] to apportion the moment, in μ_B, as follows:

$$
\begin{aligned}
\text{3d spin } (n_\uparrow - n_\downarrow) &= +0.656 \\
\text{3d orbital} &= +0.055 \\
\text{4s polarization} &= \underline{-0.105} \\
&\ +0.606
\end{aligned}
$$

The moment density is quite asymmetric about the lattice sites. About 80% of the 3d magnetic electrons occupy t_{2g} orbitals.

[62] C. Shull and H. Mook, *Phys. Rev. Lett.* **16**, 184 (1966).
[63] H. A. Mook, *Phys. Rev.* **148**, 495 (1966).

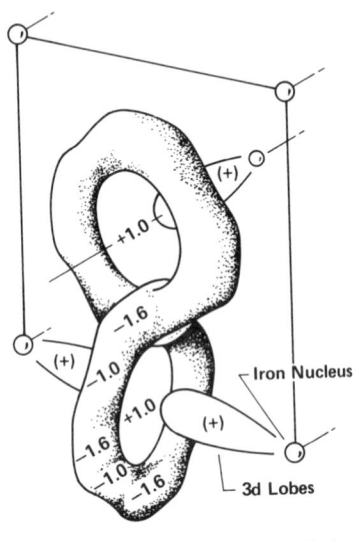

FIG. VI.21. The spin density of the interstitial regions deduced from neutron diffraction measurements. The magnetization is given in kilogauss. After Shull and Mook.[62]

For hexagonal *cobalt* Moon[64] finds

$$
\begin{aligned}
\text{3d spin } (n_\uparrow - n_\downarrow) &= +1.86 \\
\text{3d orbital} &= +0.13 \\
\text{4s polarization} &= \underline{-0.28} \\
&\ +1.71
\end{aligned}
$$

In this case, the magnetic moment looks like an almost spherical distribution of positive moment localized around each atomic site decreasing to a negative level in the region between atoms.

3. Transition Metal Compounds

a. The Heusler "Alloys": The Heusler alloys, Cu_2MnX, with X = Al, Ga, In, or Sn, are of interest because they are ferromagnetic while none of the component elements are. The structure is based on the CsCl structure with the Cu atoms on one sublattice and the Mn and X atoms ordered alternately on the other. The magnetization corresponds to a moment of $4\mu_B$ associated with the Mn. Since the nearest neighbor Mn distances are large (4.2 Å) the exchange is through the indirect RKKY mechanism. The Curie temperatures run as high as 600 K for Cu_2MnAl.

---

[64] R. M. Moon, *Phys. Rev. A* **136**, 195 (1964).

The Cu may be replaced by Ni, Pd and Co. It is interesting to note that in the cases of Ni_2MnX and Pd_2MnX neutron scattering shows that the moment is still confined to the Mn sites. In Co_2MnSn the Co contributes only 0.75 μ_B, but raises the Curie temperature substantially (830 °C) above that of the corresponding Ni_2MnSn ($T_c = 334$ °C).

b. *Special Compounds:* $SrRuO_3$, Sc_3In, and $ZrZn_2$ are metals in which ferromagnetism is unexpectedly found. With the arbitrary assignment of all the 10 d-shell electrons for Zn and In, the latter two fall on the Slater–Pauling curve as indicated in Fig. VI.19. The Laves phase compound $ZrZn_2$ is of particular interest because of the rapid rise in T_c where Zr is replaced by isoelectronic Ti. Neutron diffraction studies show the magnetism is concentrated in the Zr-Zr bonds, but is not localized.

c. *Transition Metal Halides:* The list of nonmetallic magnetic compounds is enormous. The halides constitute an interesting class because they contain many of the "lower dimensional" magnetic systems. Representative examples are shown in Table VI.5.

$CsCuCl_3$ is the only known example of a spin $\frac{1}{2}$ Heisenberg ferromagnetic chain. It is not a particularly good example for there appears to be a large antiferromagnetic interchain interaction. One-dimensional antiferromagnets occur more readily in nature. $(CH_3)_4NMnCl_3$, abbreviated TMMC, is a well-studied example. The magnetic moments of the Mn ions are coupled by superchange through the Cl ions in the $MnCl_3$ complexes. These $MnCl_3$ chains are magnetically insulated by large intervening $(CH_3)_4N$ complexes. For a discussion of the magnetic properties of one-

TABLE VI.5
REPRESENTATIVE MATERIALS WHICH BEHAVE LIKE LOWER
DIMENSIONAL MAGNETIC SYSTEMS

	Material	S	J	T_c	Hamiltonian
One-dimensional ferromagnets	$CsCuCl_3$	$\frac{1}{2}$	?	10.4	Heisenberg
	$CoCl_2-2NC_5H_5$	$\frac{1}{2}$	9.5	3.5	Ising
	$CsNiF_3$	1	?	2.6	XY
One-dimensional antiferromagnets	$(CH_3)_4NMnCl_3$ (TMMC)	$\frac{5}{2}$	-6.5	0.84	Heisenberg
	$CsCoCl_3$	$\frac{1}{2}$	-100	8	Ising
Two-dimensional ferromagnets	K_2CuF_4	$\frac{1}{2}$	11.2	6.5	Heisenberg
	$FeCl_2$	1	3.4	23	Ising
	$CoCl_2$	$\frac{1}{2}$	~10	24.7	XY
Two-dimensional antiferromagnets	K_2NiF_4	1	-56	97	Heisenberg
	K_2CoF_4	$\frac{1}{2}$	-97	107	Ising

and two-dimensional systems we refer the reader to the article by de Jongh and Miedema.[65]

In three dimensions the halides are all antiferromagnetic to our knowledge. This is a result of the inherent antiferromagnetic nature of superexchange as we mentioned in Chapter IV (see Fig. IV.3).

4. Rare Earths

The rare earth metals have similar chemical properties, which makes their separation difficult, but exhibit a large variety of magnetic long range order. The origin of this may be seen by considering the electron distribution of a typical rare earth atom such as gadolinium (Fig. VI.22).[66] The neutral atom has the configuration $(Xe)4f^7 5d^1 6s^2$. In the crystal the $5d^1 6s^2$ electrons form metallic bands with the result that we are left with a $4f^7$ trivalent ion imbedded in a sea of nearly free electrons. Notice, however,

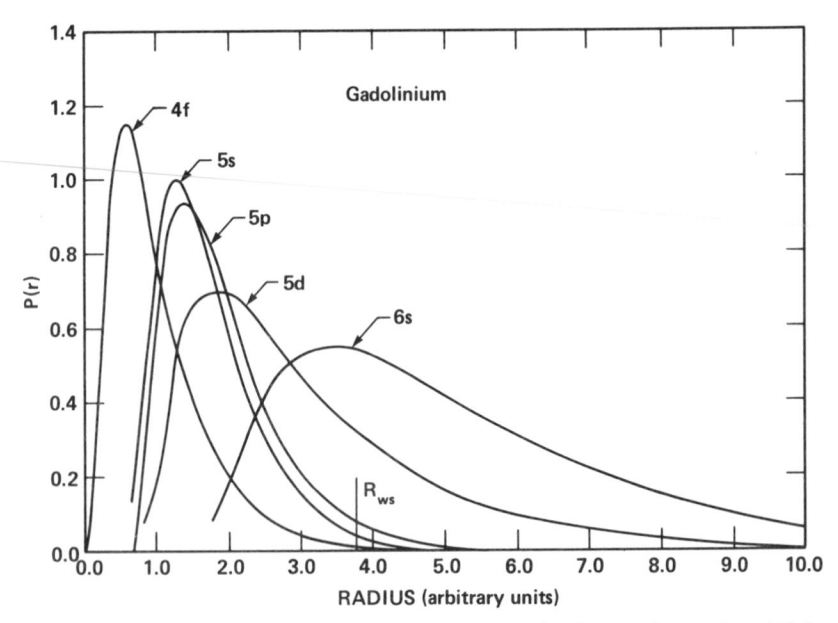

FIG. VI.22. The relative outer radial extent of the atomic electrons in atomic gadolinium. Not shown are the various orthogonality oscillations of the wave functions at small radial distances. R_{ws} refers to the Wigner–Seitz radius appropriate for Gd in its metallic state. After Freeman.[66]

[65] L. J. de Jongh and A. R. Miedema, Adv. Phys. 23, 1 (1974).

[66] A. J. Freeman, in "Magnetic Properties of Rare Earth Metals" (R. J. Elliott, ed.), p. 248. Plenum, New York, 1972.

that the 4f electrons lie inside the $5s^2 5p^6$ shells of the Xe core. This shields the 4f electrons from their crystalline environment. Consequently the intraatomic Coulomb interaction is the largest interaction which means that, to lowest order, the state of the rare earth ion may be characterized by its total angular momentum, J. In the case of Gd the ground state is $^8S_{7/2}$.

These ionic moments are then coupled via the indirect exchange mechanism discussed in Chapter IV. As we found there, this exchange is proportional to $\Sigma_q \chi(\mathbf{q}) \exp(i\mathbf{q} \cdot \mathbf{r})$ where $\chi(\mathbf{q})$ is the generalized magnetic susceptibility of the electron sea. Thus the exchange and, hence, the magnetic order, is governed by the Fermi surface.

Since the 4f's are not completely shielded by the $5s^2 5p^6$ shells, they experience forces which reflect the symmetry of the crystalline environment. In the light rare earths, where the 4f's are more extended, these crystal fields are comparable to the exchange. In Pr this prevents the appearance of long range order by removing all the degeneracy of the ground state. In the heavy rare earths this crystalline anisotropy is small, but leads to a variety of different spin structures, such as helices, cones, etc. Furthermore, since the temperature dependence of the exchange and the anisotropy are very different, as we already noticed in Chapter IV, one obtains different spin structures as a function of temperature. These are summarized in Fig. VI.23.

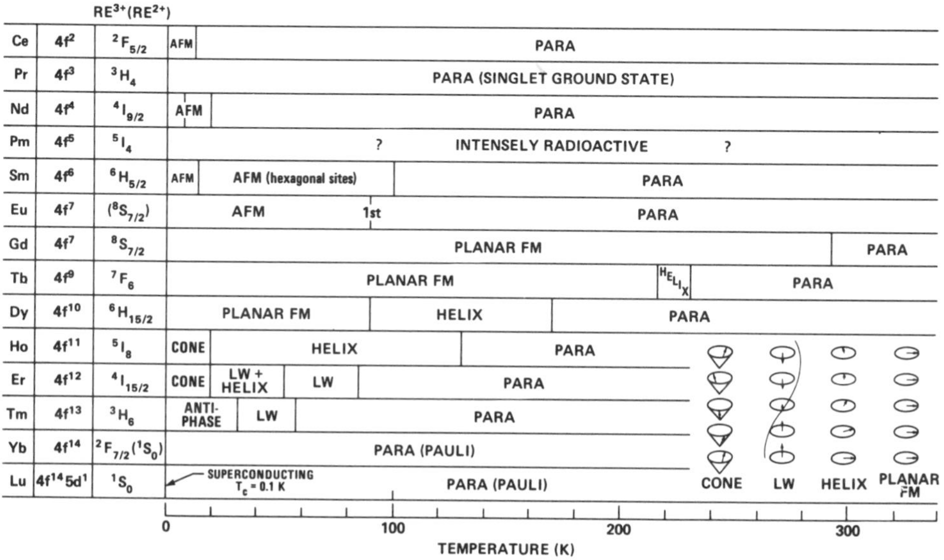

FIG. VI.23. Summary of the magnetic properties of the rare earth metals.

For a survey of the properties of the rare earth metals, we recommend the collection edited by Elliott.[67]

5. Special Oxides

a. *Spinels:* Spinel is the name of the mineral $MgAl_2O_4$. However, there exists a large class of ferrimagnetic compounds with the same structure which are now commonly referred to as the spinel ferrites. This structure is shown in Fig. VI.24. The smallest cubic unit cell consists of eight molecular units, i.e., 32 oxygen ions. The cations occupy interstitial positions in this cubic oxygen lattice, of which there are two types. In one, called the A site, the cation is surrounded by four oxygen ions located at the corners of a tetrahedron. In the other, called the B site, the cation is surrounded by six oxygen ions at the vertices of an octahedron. Eight A sites and 16 B sites are occupied per unit cell. If these A sites are occupied by the divalent ions and the 16 B sites by the trivalent ions the structure is called a *normal* spinel. If, however, the B sites are occupied half by divalent and half by trivalent ions distributed at random, with the other trivalent ions in the A sites, the structure is called an *inverse* spinel.

The spinel ferrites play a prominent role in the history of magnetism. We tend to take the existence of magnetic insulators for granted. The con-

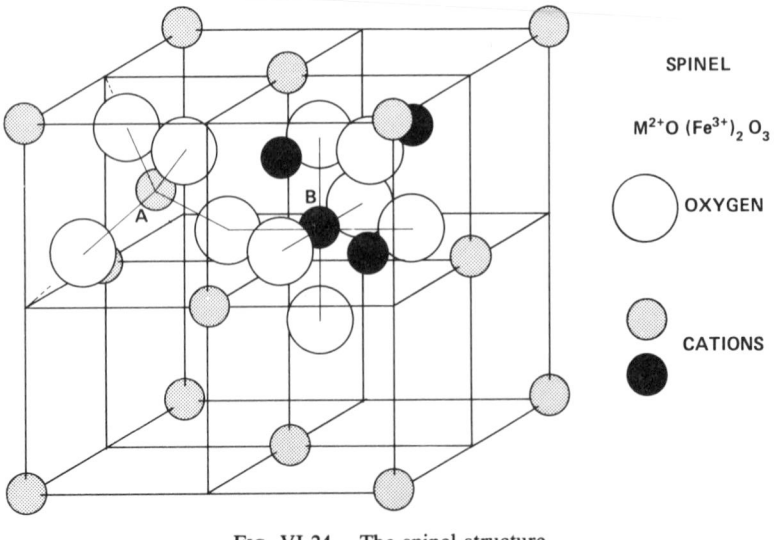

SPINEL

$M^{2+}O (Fe^{3+})_2 O_3$

OXYGEN

CATIONS

FIG. VI.24. The spinel structure.

[67] R. J. Elliott, ed., "Magnetic Properties of Rare Earth Metals." Plenum, New York, 1972.

cept was first patented in 1909 (in Germany), but did not succeed because of the lack of a theoretical base and a supporting technology. Ferrites were eventually developed largely as a result of Snoek's work at Philips during the earlier 1940's on spinels and Néel's 1948 theory of ferrimagnetism (for which he shared the 1970 Nobel Prize in Physics).

A familiar example of an inverse spinel is Fe_3O_4, or $(Fe^{3+})_A(Fe^{2+}Fe^{3+})_BO_4$ known to the ancients as magnetite. Another is the material commonly found in digital magnetic recording tapes, γ-Fe_2O_3, or $(Fe^{3+})_A(Fe^{3+}_{5/3}\square_{1/3})_BO_4$ where $\square$ represents a vacancy.

The advantage of the spinels lies in their ability to be manipulated chemically to give specific magnetic or electrical properties. Suppose, for example, one wants a high permeability. The initial permeability, due to rotation of domains, is proportional to M_s^2/K, so a high permeability may be obtained either by a high saturation magnetization M_s or by a low anisotropy. As we saw in Chapter IV, the anisotropy constant K falls rapidly with increasing temperature producing a maximum in the permeability near the Curie temperature. A high permeability ferrite therefore requires either a high value of M_s or a Curie temperature just above room temperature. In Fig. VI.25,[68] we show how the magnetization of the mixed zinc ferrites varies with composition. Pure $ZnFe_2O_4$ is a normal spinel which is antiferromagnetic below 9.5 K. This variation in moment is due to the fact that the divalent Zn ions go into an A-site displacing an Fe ion into a B-site to replace the vacated M ion. Above 50% the antiparallelism between the diminishing number of Fe_A ions and the Fe_B ions cannot be maintained against the increasing antiferromagnetism within the B ion lattice. Thus, we find 50% Mn-Zn ferrites used in transformers and high-quality inductors.

The spinels also encompass a class of materials involving chalcogenides. One of these, $CdCr_2Se_4$, is the only known high-mobility (7 cm^2/V-sec) ferromagnetic ($T_c = 130$ K) semiconductor. $CdCr_2S_4$ is also ferromagnetic ($T_c = 85$ K) but its mobility is much lower. The copper spinels $CuCr_2S_4$ and $CuCr_2Se_4$ are ferromagnetic, but show metallic p-type conduction attributed to holes in a broad valence band. Ferromagnetic semiconductors offer the opportunity to study the interaction between the magnetic degrees of freedom and carriers. Balberg and Pinch,[69] for example, have observed anomalies in the longitudinal magnetoresistance of p-type $CdCr_2Se_4$ which they attribute to the transfer of energy from the drifting carriers to spin waves.

[68] C. Guilland, *J. Phys. Radium* **12**, 239 (1951).
[69] I. Balberg and H. L. Pinch, *Phys. Rev. Lett.* **28**, 909 (1972).

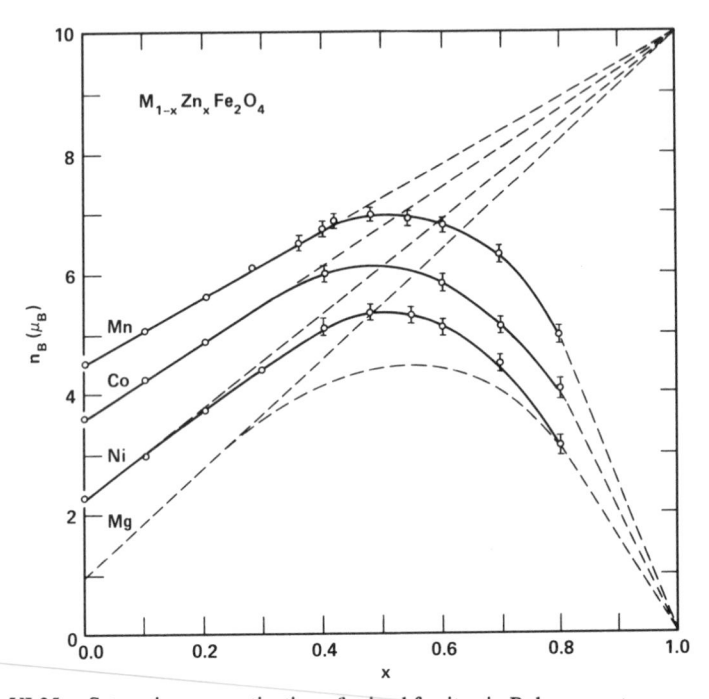

FIG. VI.25.   Saturation magnetization of mixed ferrites in Bohr magnetons per unit formula. After Guilland.[68]

b. *The Garnets:* The term garnet commonly refers to the silicate mineral used as a semiprecious stone. The fact that the garnet structure also supports ferrimagnetism was discovered[70] in 1956. The general garnet formula is $M_3Fe_5O_{12}$ which is more informatively written as $(3M_2O_3)^c$ $(2Fe_2O_3)^a$ $(3Fe_2O_3)^d$. An enormous number of chemical substitutions are possible which makes it possible to tailor the magnetic properties of garnets to specific requirements of microwave communications, lasers, magnetic bubbles, etc. The latter topic is discussed in Chapter VII. The "a" ions are arranged on a bcc lattice in octahedral sites surrounded by six oxygens as shown in Fig. VI.26. The "c" and "d" ions are on the faces surrounded by eight and four oxygens in dodecahedral and tetrahedral sites, respectively. All the above is important in understanding the crystal chemistry used to design garnets with special properties. The primary magnetic interaction is antiferromagnetic between the a- and d-sites because it has the shortest XOX distance, and the angle (126°) is closer to 180° (see Chapter IV) than the other possibilities. The rare earth ions on

[70] F. Bertaut and F. Forrat, *C. R. Hebd. Seances Acad. Sci.* **242,** 382 (1956).

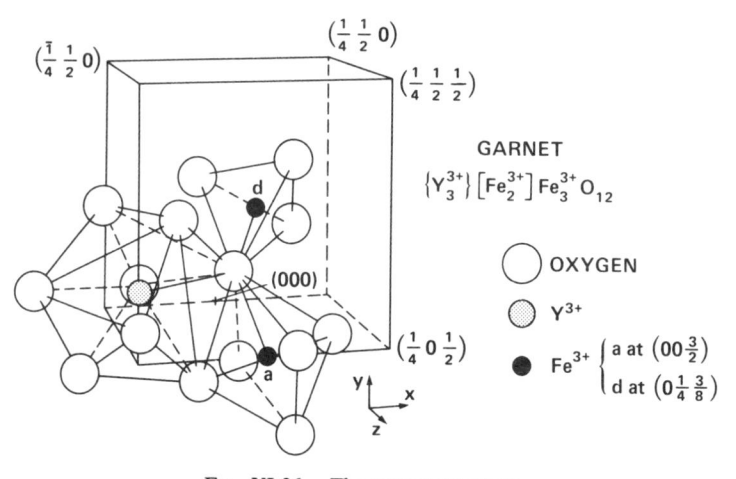

$(\frac{\bar{1}}{4}\frac{1}{2}0)$

$(\frac{\bar{1}}{4}\frac{1}{2}0)$

$(\frac{1}{4}\frac{1}{2}0)$

$(\frac{1}{4}\frac{1}{2}\frac{1}{2})$

(000)

$(\frac{1}{4}0\frac{1}{2})$

GARNET

$\{Y_3^{3+}\}[Fe_2^{3+}]Fe_3^{3+}O_{12}$

◯ OXYGEN

◉ Y^{3+}

● Fe^{3+} $\begin{cases} \text{a at } (00\frac{3}{2}) \\ \text{d at } (0\frac{1}{4}\frac{3}{8}) \end{cases}$

FIG. VI.26. The garnet structure.

the c-sites orient antiferromagnetically to the moments on the d-sites (XOX angle = 122°) giving a net moment per formula $(M_3Fe_2Fe_3O_{12})$ = $3\mu_M - \mu_{Fe}$. The magnetic properties of the rare earth garnets are listed in Table VI.5. It can be seen from this table that $\mu_M = (L + S) g\mu_B$ for > half-filled f-shell ions and $(L-S) g\mu_B$ for < half-filled f-shell ions. Compensation points, or temperatures where μ_M is equal and opposite to $\Sigma\mu_{Fe}$ are evident in Fig. VI.27.[71]

An example of the simple way in which chemical substitutions may be used to manipulate the magnitude of the moments of the ordering temperature is illustrated by ions for Fe in YIG as shown in Fig. VI.28.[72] Since the radius of Sc is large, $r_{Sc} > r_{Fe} > r_{Al}$, it substitutes on the large octahedral a-sites and Al on the smaller tetrahedral d-sites. The ferrimagnetic moment for pure YIG is given by the three irons on the d-sites opposed by the two irons on the a-sites, hence as can be seen Sc reduces the a-site moment. Thus Sc substitution increases the saturated moment while decreasing the Curie temperature because of reduction in the Fe^{3+}-O-Fe^{3+} superexchange.

c. The Perovskites: The perovskites correspond to a large family of compounds that have the same crystal structure as the mineral perovskite, $CaTiO_3$. The unit cell is illustrated by the cube in the lower right

[71] S. Geller, J. P. Remeika, R. C. Sherwood, H. J. Williams, and G. P. Espinosa, *Phys. Rev. A* **137**, 1034 (1965).

[72] M. A. Gilleo and S. Geller, *Phys. Rev.* **110**, 73 (1958).

TABLE VI.5[a]
MAGNETIC AND CRYSTALLOGRAPHIC PROPERTIES OF GARNETS

Ion M	Y	La	Pr	Nd	Sm	Eu	Gd	Td	Dy	Ho	Er	Tm	Yb	Lu
a_o(Å)	12·376	[12·767][c]	[12·646]	[12·600]	12·529	12·498	12·471	12·436	12·405	12·375	12·347	12·323	12·302	12·283
No. of 4f electrons	0	0	2	3	5	6	7	8	9	10	11	12	13	14
L	0	0	5	6	5	3	0	3	5	6	6	5	3	0
2S	0	0	2	3	5	6	7	6	5	4	3	2	1	0
$3m_c - 5$; $m_c = L \pm 2S$[d]	−5	−5	4	4	0	4	16	22	25	25	22	17	7	−5
$3m_c - 5$; $m_c = 2S$	−5	−5	1	4	10	13	16	13	10	7	4	1	−2	−5
n_B(at 0°K)[b]	5·01	[5·0]	[9·8]	[8·7]	5·43	2·78	16·0	18·2	16·9	15·2	10·2	1·2	0·0	5·07
M_s e.m.u. cm⁻³(20°C)	139				135	93	135	4	43	78	103	110	130	140
Compensation temperature (°K)							236	246	226	137	83	none	0–6	
Curie temperature (°K)	553				578	566	564	568	563	567	556	549	548	549

[a] After Geller et al.[71]

[b] n_B is the magnetic moment in Bohr magnetons per unit formula $M_3Fe_5O_{12}$ (results are sometimes given per unit formula $(3M_2O_3)(2Fe_2O_3)(3Fe_2O_3)$, i.e., $M_6Fe_{10}O_{24}$, giving twice n_B).

[c] The information in brackets [ ] is extrapolated from substituted compounds.

[d] m_c is the net moment associated with the c-ions.

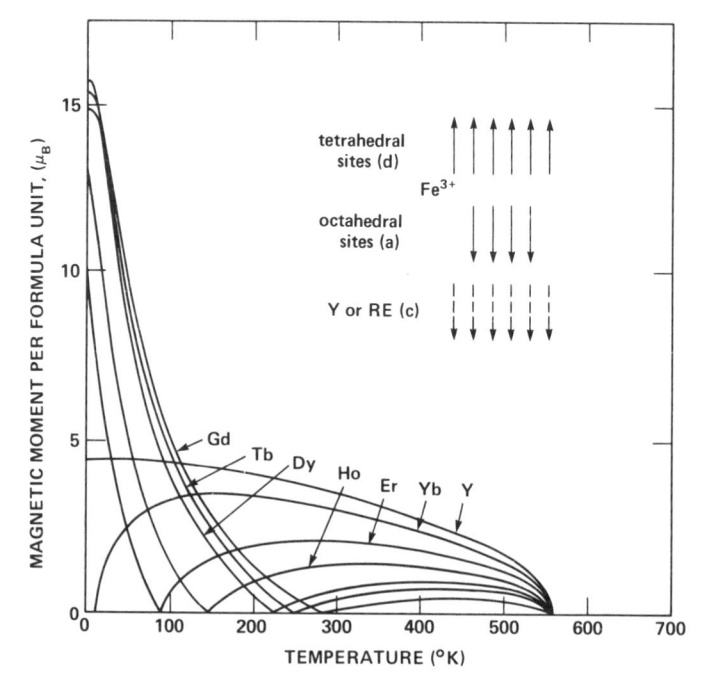

FIG. VI.27.   Saturation moment in Bohr magnetons per formula unit as a function of temperature for various rare earth iron garnets. The temperature at which the magnetization crosses zero is called the compensation temperature; here the magnetization of the rare earth sublattice is equal and opposite to the net magnetization of the ferric ion sublattices. The rare earth contribution drops more rapidly with temperature than the ferric contribution because of the small rare-earth–iron exchange interaction. After Geller et al.[71]

hand side of Fig. VI.29. We have already seen that this structure exhibits structural phase transitions ($SrTiO_3$) and ferroelectricity ($PbTiO_3$, $BaTiO_3$, $KNbO_3$) and ferromagnetism ($SrRuO_3$).

The perovskites are also interesting because they are one of the several structures which allow simultaneous occurrence of ferroelectricity *and* ferro- or antiferromagnetism. Such a *magnetoelectric* material is characterized by a linear relation between an electric field and the medium's magnetic polarization as well as between a magnetic field and the medium's electric polarization. Naively one might suspect that time reversal arguments might preclude such an effect. However, in 1960 Landau and Lifshitz[73] casually mentioned that the possibility does indeed

---

[73] L. Landau and E. Lifshitz, "Electrodynamics of Continuous Media," p. 119. Pergamon, Oxford, 1960.

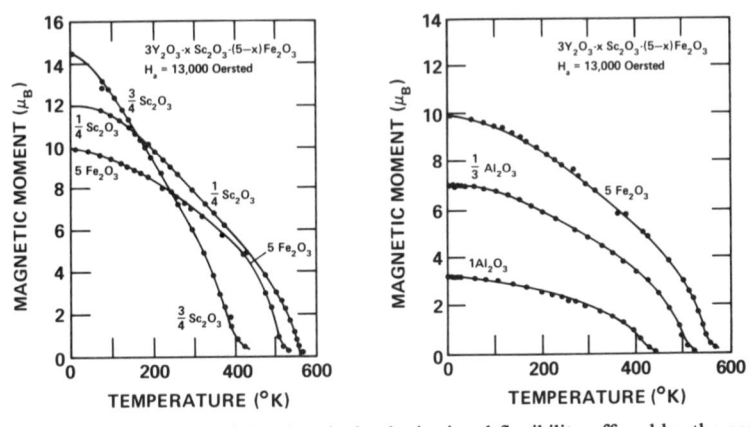

FIG. VI.28.   Illustration of the chemical substitutional flexibility offered by the garnets. On the left Sc^{3+} is introduced into YIG. Being a larger ion than Fe^{3+} it enters the larger octahedral sites increasing the net ferric moment (see relative spin orientations in the insert of Fig. VI.27). On the right the small Al^{3+} ion is introduced which goes into the tetrahedral sites reducing the net moment. After Gilleo and Geller.[72]

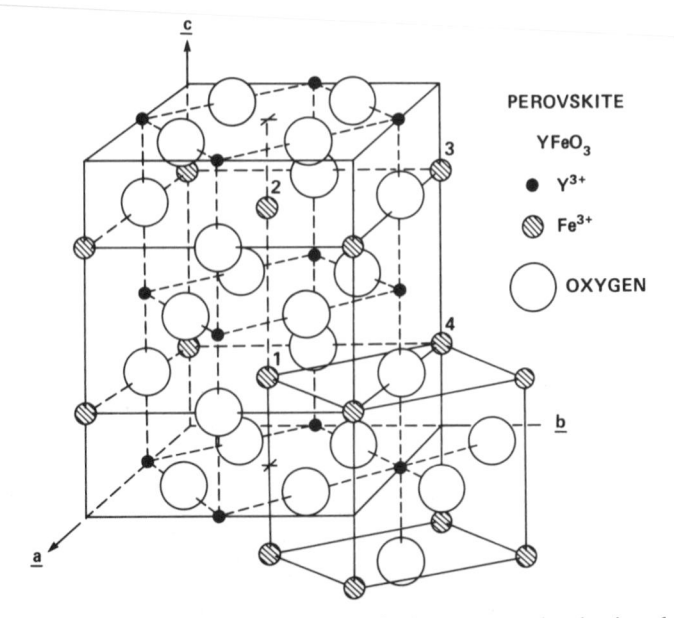

FIG. VI.29.   The orthorhombic cell of the orthoferrite structure showing its relation to the cubic perovskite unit cell. The numbers label the four inequivalent Fe^{3+} sites.

exist. Dzyaloshinski,[74] after reading the proofs of this book, predicted that Cr_2O_3 should be magnetoelectric, and this was subsequently confirmed by Astrov,[75] also in 1960. Since then magnetoelectricity has been observed in numerous materials including the perovskites $GdAlO_3$, $TbAlO_3$, and $DyAlO_3$. Unfortunately, the effect is inherently rather small and, therefore, not suitable for device applications.

Materials with the formula $RFeO_3$, where R is a rare earth or yttrium, have an orthorhombic distortion of the perovskite structure and are called *orthoferrites*. The orthorhombic cell is also shown in Fig. VI.29. The numbers 1–4 label the four inequivalent Fe^{3+} sites. There are also four inequivalent rare earth sites. The spins in this structure are coupled through the oxygen ions via superexchange. The complexity of this structure allows for the possibility of *antisymmetric* exchange of the form $D \cdot S_i \times S_u$ mentioned in Chapter IV. Symmetry considerations impose limitations on the directions of D, and, consequently, the allowed spin configurations. The configuration commonly found has the sublattice moments oriented antiferromagnetically in the ab plane with a slight canting along the c-axis. Since the canting angle is only of the order of $\frac{1}{2}$ degree, these materials are referred to as "weak" ferromagnets. The small net magnetization and the axial anisotropy were responsible for the orthoferrites originally being used as magnetic bubble[76] materials. However, since the bubble size turns out to be proportional to $1/M_s^2$ the small magnetization placed a lower limit of about 100 microns on the bubble size. Therefore, when it was discovered that an axial anisotropy could be "growth induced" in the cubic garnets, these replaced the orthoferrites as the bubble medium.

[74] I. E. Dzyaloshinski, *Sov. Phys.—JETP (Engl. Transl.)* **10**, 628 (1960).
[75] D. N. Astrov. *Sov. Phys—JETP (Engl. Transl.)* **11**, 708 (1960).
[76] For a discussion of magnetic bubbles, see Chapter VIII.

VII. Impurities and Long Range Order

We are all aware of the important role played by impurities in transport. Their effect on long range order is, for the most part, less dramatic. Ferromagnetism, for example, can tolerate large concentrations of impurities. Superconductivity can also tolerate large concentrations of *nonmagnetic* impurities. In fact, Anderson[1] has shown that if the normal Fermi surface is isotropic the superconducting transition temperature is unaffected. The pairing just becomes that between time-reversed exact eigenstates of the doped normal metal. If, however, the Fermi surface is anisotropic the presence of impurities "smears out" this anisotropy leading to mild decreases in T_c. This is illustrated by tin as shown in Fig. VII.1(a).[2] A related effect occurs in the case of charge density waves whose existence, as we have already discussed, depends critically on special features of the Fermi surface. This is illustrated by the variation of the wave vector of the charge density wave in TaS_2 [Fig. VII.1(b)].[3]

This insensitivity of superconductors to nonmagnetic impurities is basically due to the fact that a Cooper pair has zero angular momentum, i.e., s-wave pairing. If the pair carried angular momentum it would be strongly affected by nonmagnetic impurities. Therefore any hope of observing p-wave pairing would have to be done on an extremely pure material. We have already mentioned in Chapter I that in superfluid ^{3}He the helium atoms exhibit p-wave pairing. Fortunately, by the time one has reached the milliKelvin temperature range, any impurities have frozen out leaving the pure liquid.

Superconductors are, however, very sensitive to *magnetic* impurities. The theory of this effect was first developed by Abrikosov and Gor'kov in 1960. This theory assumes the existence of a paramagnetic impurity spin S which interacts with the superconducting electrons through the exchange interaction. However, it is not at all clear that any given impurity will have a well-defined moment when dissolved in any given host. The suppression of superconductivity may, in fact, be taken as an operational definition of "magnetic." This suppression does not have the same form,

[1] P. W. Anderson, *J. Phys. Chem. Solids* **11**, 26 (1959).
[2] D. Markowitz and L. P. Kadanoff, *Phys. Rev,* **131**, 563 (1963).
[3] J. A. Wilson, F. J. DiSalvo, and S. Mahajan, *Adv. Phys.* **24**, 117 (1975).

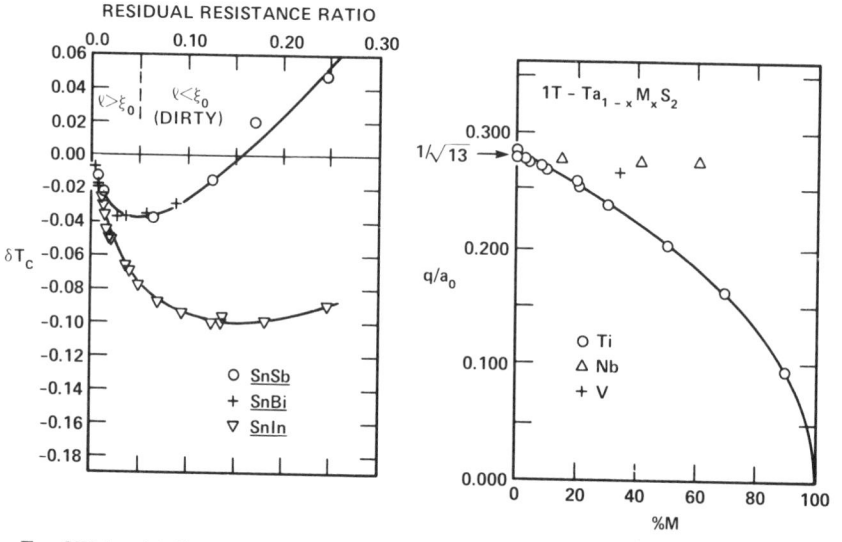

Fig. VII.1. (a) Change in the superconducting transition temperature of tin with addition of impurities. After Markowitz and Kadanoff.[2] (b) Variation of the wave vector of the charge density wave in 1T-TaS$_2$ with addition of metals M. Notice that q is not affected by metals from the same column of the periodic table as Ta. After Wilson et al.[3]

however, for different impurity-host systems attesting to the fact that there are different kinds of "magnetic" impurities.

The behavior of magnetic impurities in a metallic host is also important for our understanding of itinerant magnetism. In particular, it enables us to separate the question of moment formation from that of long range order. Thus, this chapter is complementary to Chapter IV in the sense that it focuses more on the question of moment formation.

It was not until the 1960's that magnetic impurities were studied systematically. The delay is attributable to problems in materials science —it was necessary to gain control over impurities in the parts per million range. Such control evolved with semiconductor physics in the decade of the 1950's with the development of techniques such as zone refining. The more difficult task of controlling impurities in metals rather than covalent semiconductors required high-vacuum–high-temperature procedures.

For example, it was long suspected that the low-temperature resistance minimum observed in noble metals was caused by the presence of small concentrations of "magnetic" impurities. Yet, experimentally it was found that tin (certainly not a magnetic ion) when dissolved in Cu caused the appearance of a resistance minimum. It took careful experimental work to establish that upon dissolving the tin in the copper traces of insoluble iron-oxide (present as a second phase in Cu which had previously been annealed in O_2) were reduced with the result that Fe was put back into solution. Trace concentrations of Fe dissolved in the Cu phase magnetize and are responsible for the observed minimum in the resistivity.

A.   THE VIRTUAL BOUND STATE

Friedel[3a] and his colleagues in Orsay have contributed greatly to our understanding of dilute alloys (Friedel sum rule, Friedel oscillations, etc.). Much of this work is based on a model in which the role of the impurity is to produce a change in the periodic potential of the host. The consequences of this are examined from a scattering theory approach.

Scattering theory is concerned with what the electron wave function looks like in the presence of an impurity. If the impurity is represented by a real spherical potential (i.e., we neglect inelastic scattering) then the wave function, call it $\chi_k(\mathbf{r})$, may be expanded in spherical harmonics,

$$\chi_k(\mathbf{r}) = \sum_{l=0}^{\infty} [4\pi(2l + 1)]^{1/2} i^l A_l(k, r) Y_l^0(\theta) \tag{7.1}$$

where $\mathbf{k}$ is the quantum number of the unperturbed state which is taken to be a plane wave. The coefficient $A_l(k, r)$ is called the partial wave amplitude. This is an important quantity for, when the energy variable k is regarded as a complex number, the poles of $A_l(k, r)$ along the positive imaginary axis correspond to bound states. Far from the impurity this wave function has the form

$$\lim_{r \to \infty} \chi_k(\mathbf{r}) = e^{i\mathbf{k} \cdot \mathbf{r}} + f(k, \theta)(e^{ikr}/r) \tag{7.2}$$

where $f(k, \theta)$ is called the scattering amplitude and is a function of the scattering phase shift, δ_l,

$$f(k, \theta) = (1/k) \sum_l [4\pi(2l + 1)]^{1/2} e^{i\delta_l} \sin \delta_l Y_l^0(\theta). \tag{7.3}$$

This phase shift depends upon the energy $\epsilon(\mathbf{k})$ of the scattering particle and on the details of the impurity potential. For example, if the potential is a delta function of strength V_0 then there is only s-wave scattering and

[3a] See, for example, J. Friedel, *Nuovo Cimento* (Suppl.) **2**, 287 (1958).

$$\delta_0(E) = \tan^{-1}[\pi V_0 N(E)/(1 - V_0 I(E))]$$

where $N(E)$ is the host density of states and

$$I(E) = \mathscr{P} \int dE' N(E')/(E' - E),$$

where $\mathscr{P}$ means that we take the principal part of the integral. If we define E_0 as the value of E where $V_0 I(E_0) = 1$ and expand $V_0 I(E)$ about $E = E_0$, then $\delta_0(E) = \tan^{-1}[\Delta/(E - E_0)]$ where Δ is proportional to $N(E_0)$. Since we do not actually know the details of the impurity potential, the French school treats the phase shift itself as the fundamental quantity.

Notice that when the phase shift goes through $\pi/2$ the scattering amplitude goes through a maximum. This is referred to as a *resonance* or *virtual bound state* (vbs).

In its simplest form the Friedel model assumes that the conduction electrons are independent. Suppose, for example, that we put Mn into Cu. Then we might start with core potentials associated with a Cu^+ matrix (filled d-band) and a Mn^{7+} site. The charge differential $Z = 6$ will then, through the Friedel sum rule

$$(2/\pi) \sum_l (2l + 1)\delta_l(E_F) = Z, \tag{7.4}$$

lead to a build-up of charge in the vicinity of the impurity that will resemble five d-orbitals. The remaining electron becomes s-like. If the potential is strong enough these orbitals can become bound as illustrated in Fig. VII.2, where the zero of energy is taken as the bottom of the conduction band. However, if the potential is not quite strong enough to produce a bound state it will move up into the continuum of extended states. Here it will resonate with the lth spherical components of these plane waves and broaden into a virtual bound state (Fig. VII.2). The density of states associated with this localized state is proportional to the energy derivative of the phase shift

$$N_l(E) = \frac{2l + 1}{\pi} \frac{d\delta_l(E)}{dE}. \tag{7.5}$$

Notice that for the simple example of s-wave scattering we discussed above the derivative of $\tan^{-1}$ gives a Lorentzian density of states. This is why the virtual bound state has this shape in all our sketches.

If this virtual bound state lies close to the Fermi level it can have important consequences on the resistivity and other properties. The residual resistivity (i.e., that due to the impurities), for example, is given by

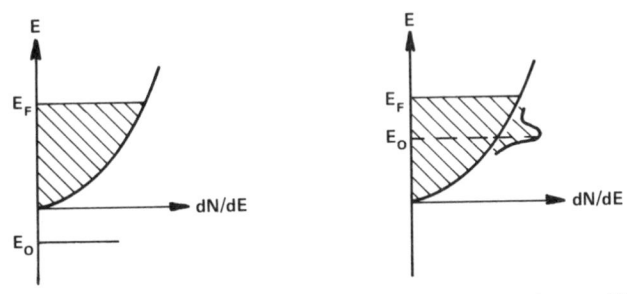

FIG. VII.2.   (a) If the energy of a localized state lies below the continuum of free electron states, we have a bound state. (b) If this energy lies within the free electron band, we have a resonance, or virtual bound state.

$$\Delta\rho = \frac{4\pi x}{ne^2 k_F} \sum_{l=0}^{\infty} (l + 1) \sin^2[\delta_l(E_F) - \delta_{l+1}(E_F)] \qquad (7.6)$$

where x is the atomic concentration of impurities and n is the number of valence electrons per atom in the host.

Problem VII.1

The differential scattering cross section $\sigma(\theta)$ is given by the square of the scattering amplitude, Eq. (7.3), i.e., $\sigma(\theta) = |f(\theta)|^2$. The effective average cross section $\bar{\sigma}$ that governs the resistivity is the angular average of $\sigma(\theta)$ weighted by $(1 - \cos\theta)$ which measures the relative change in the component of the electron's velocity along its initial direction of motion:

$$\bar{\sigma} = 2\pi \int_0^{\pi} d\theta \sin\theta \, \sigma(\theta)(1 - \cos\theta).$$

The mean free path is then $\Lambda = 1/x\bar{\sigma}$. Using these relations derive Eq. (7.6). Show that there will be a peak in $\Delta\rho$ as E_0 crosses the Fermi energy.

1. Polarization of the VBS

To determine the magnetic properties of such an impurity we must include the Coulomb interactions among the electrons on the impurity. This was first attempted quantitatively by Anderson[4] and Wolff[5] independently in 1961.

Following Friedel both Anderson and Wolff assumed that the impurity will produce a change $V(\mathbf{r})$ in the crystal potential. Since this potential is screened by the conduction electrons it is essentially confined to the impurity cell. However, they went further to include the electron–electron interaction. For simplicity they neglected the interaction

[4] P. W. Anderson, *Phys. Rev.* **124**, 41 (1961).
[5] P. A. Wolff, *Phys. Rev.* **124**, 1030 (1961).

between electrons in host cells, including it only for the impurity cell. This is a valid approximation as long as the host is not magnetic or even close to being magnetic. The Hamiltonian for an impurity at $\mathbf{R}$ then becomes

$$\mathcal{H} = \mathcal{H}_0 + \sum_i eV(\mathbf{r}_i - \mathbf{R}) + \sum_{\substack{\text{within} \\ \text{impurity} \\ \text{cell}}} v(\mathbf{r}_i - \mathbf{r}_j). \tag{7.7}$$

Here $\mathcal{H}_0$ is the Hamiltonian for the pure host.

The next step is to write this Hamiltonian in second-quantized form. The actual form will depend on the states we use to expand the field operator. For example, suppose we dissolve a transition metal such as manganese into copper. Since Mn has four fewer d-electrons than Cu, we might expect the localized state to be pushed out of the d-band of the copper host. Since the d-band is filled, it will not contribute to the properties of the dilute alloy. Therefore, we expand our field operator in terms of the s-band states and a localized nondegenerate d-like state. The one-electron terms in the Hamiltonian become

$$\sum_{\mathbf{k},\sigma} E_\mathbf{k} c_{\mathbf{k}\sigma}^\dagger c_{\mathbf{k}\sigma} + E_0 \sum_\sigma c_{0\sigma}^\dagger c_{0\sigma} + \sum_{\mathbf{k},\sigma} (V_{0\mathbf{k}} c_{\mathbf{k}\sigma}^\dagger c_{0\sigma} + V_{0\mathbf{k}}^* c_{0\sigma}^\dagger c_{\mathbf{k}\sigma}). \tag{7.8}$$

The first term is the s-band energy, the second term is the contribution from the impurity state, and the last term is the so-called s–d mixing term. Notice that this mixing is a one-electron effect. It corresponds to the hopping of an electron from the localized d orbital into the conduction band, or vice-versa. The interaction part of the Hamiltonian becomes $Un_{0\uparrow}n_{0\downarrow}$, where U is the intra-atomic Coulomb repulsion between opposite spins in the localized orbital. Equation (7.8) plus this interaction constitute what is known as the *Anderson Hamiltonian*.

If the host is characterized by a d-band that is only partially filled, then the appropriate basis in which to expand the field operator is the set of d-band states. The resulting Hamiltonian is similar to the Hubbard Hamiltonian except that the interactions occur only at one site—that of the impurity, which we label 0. If we assume a nondegenerate d-band, this Hamiltonian has the form

$$\mathcal{H} = \sum_{\mathbf{k},\sigma} E_\mathbf{k} c_{\mathbf{k}\sigma}^\dagger c_{\mathbf{k}\sigma} + E_0 \sum_\sigma c_{0\sigma}^\dagger c_{0\sigma} + \tfrac{1}{2}U \sum_\sigma n_{0\sigma}n_{0,-\sigma} \tag{7.9}$$

and is referred to as the *Wolff Hamiltonian*. Both Anderson and Wolff employed the unrestricted Hartree–Fock approximation to analyze their models. This approximation consists of replacing $Un_{0\uparrow}n_{0\downarrow}$ by $U<n_{0\uparrow}>n_{0\downarrow}$

$+ U < n_{0\downarrow} > n_{0\uparrow}$. Anderson employs a Green's function approach to calculate the density of states for the impurity, $N_{0\sigma}(E)$. The result has the Lorentzian form

$$N_{0\sigma}(E) = \frac{1}{\pi} \frac{\Delta}{(E - E_0 - U\langle n_{0,-\sigma}\rangle)^2 + \Delta^2} \tag{7.10}$$

where the width is given by

$$\Delta = \pi \sum_{\mathbf{k}} |V_{0\mathbf{k}}|^2 \delta(E - E_{\mathbf{k}}) \rightarrow \pi|V_{0\mathbf{k}}|^2 N(E_F). \tag{7.11}$$

The average occupation is then given by

$$\langle n_{0\sigma}\rangle = \int_{-\infty}^{E_F} N_{0\sigma}(E)dE = \frac{1}{\pi} \cot^{-1}\left(\frac{E_0 + U\langle n_{0,-\sigma}\rangle - E_F}{\Delta}\right) \tag{7.12}$$

with a similar equation for $\langle n_{0,-\sigma}\rangle$. These equations give both magnetic and nonmagnetic solutions depending upon the relative values of $E_F - E_0$, Δ, and U as shown in Fig. VII.3.[6] Notice that this diagram includes *negative* values of U. This would correspond to an *attractive* electron–electron interaction.

If we superimpose the density associated with the magnetic solution on top of the free electron parabolas for spin-up and spin-down we obtain the

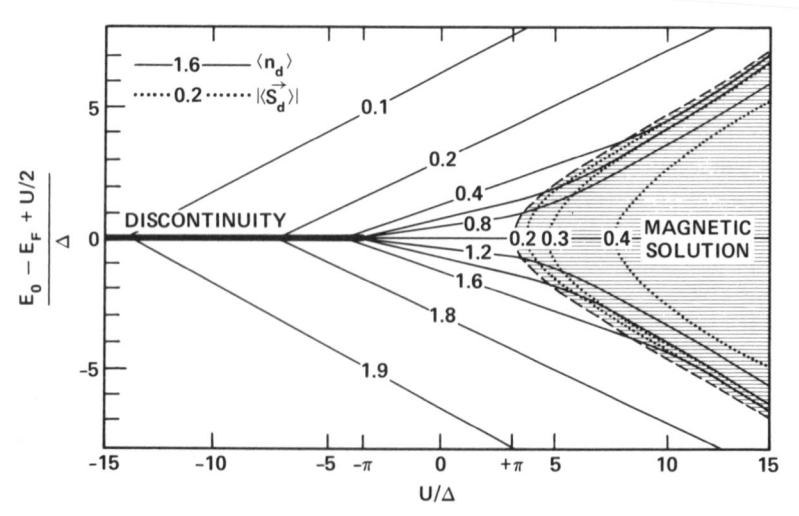

FIG. VII.3. "Phase diagram" of the Hartree–Fock state of the Anderson Hamiltonian. After Haldane.[6]

[6] F. Haldane, Ph.D. Thesis, University of Cambridge (1977).

famous Anderson figure (Fig. VII.4). In discussing experimental results we shall see that it is necessary to include spin–orbit and crystal field effects. The point is, however, that when U is large and the width of the resonance narrow we will have a localized moment. Anderson's criterion for local moment formation is

$$UN_{0\sigma}(E_F) > 1 \qquad (7.13)$$

which is analogous to the Stoner criterion for band ferromagnetism which we discussed in Chapter IV. Therefore, a low density of states for the conduction band and a small covalent admixture favor the formation of a moment. Thus, we find moments on 3d atoms dissolved in noble metals but not when dissolved in aluminum which has a relatively high density of states.

This simple model neglects the fact that d-orbitals are fivefold degenerate. The number of electrons in the localized states or d-shell is also an important factor. A half-filled d-shell is most favorable for moment formation because at this point the total spin and, therefore, the exchange energy is largest. This may be why Ni impurities show no moments in noble metals while Cr, Mn, and Fe do.

One might suspect that a magnetic impurity might polarize its free electron environment. The sd admixing does, in fact, reduce the d polarization with a corresponding polarization of the conduction electrons. However, Anderson argues that this conduction electron polarization due to mixing is compensated by an antiferromagnetic polarization due to the exchange between the local moment and the conduction electrons not unlike a "core polarization" effect (Section V,E). Thus we are left with a local moment that is slightly reduced from its "free-ion" value plus an os-

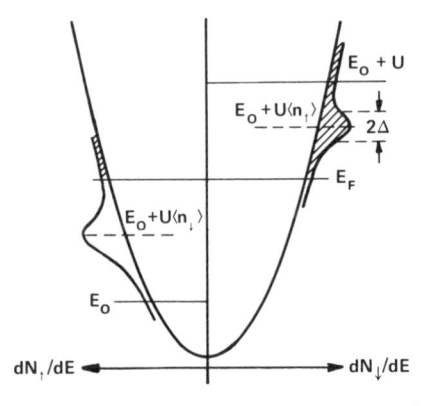

FIG. VII.4. Density of states associated with a magnetic impurity. After Anderson.[4]

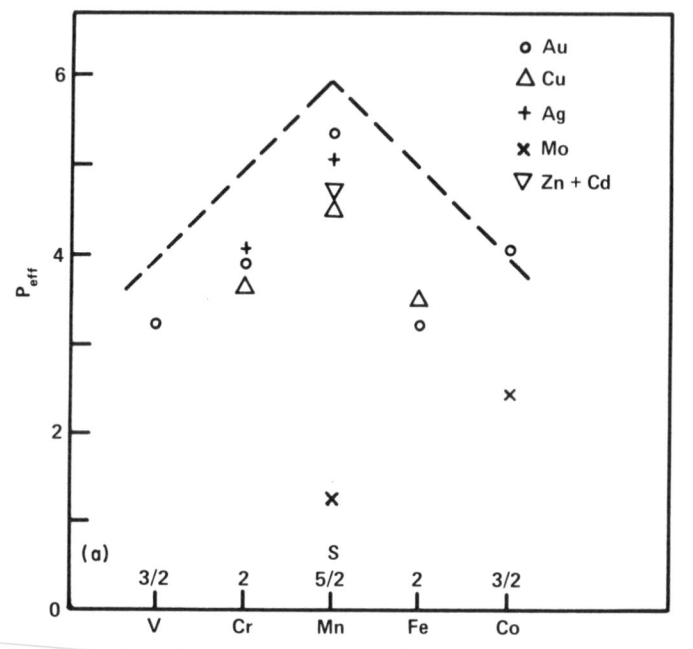

FIG. VII.5. (a) Magnetic moments associated with transition metal ions in various hosts. The dashed line is the free ion theoretical value $2\,S(S + 1)$. After Rizzuto.[7] (b) (opposite page) Variation of the Kondo temperature of first row transition metals dissolved in gold and copper. After Daybell.[18]

cillatory polarization of the conduction electrons. The reduction of the local moment is clearly illustrated in Fig. VII.5 (a) and (b).[7]

But, this is not the end of the story. For in some cases local moments only exist *above* a certain temperature. Neither the Anderson or Wolff theories can account for this behavior. This failure arises from the Hartree–Fock (H–F) approximation.

There have been numerous efforts to include correlation effects within the Anderson model. These can be divided into those which apply in the nonmagnetic region of Fig. VII.3 (small U) and those which apply in the region of the magnetic solution (large U). There are two "ways" in which the impurity may be nonmagnetic. These are illustrated in Fig. VII.6(a) and (b). In case (a) electron–electron (hole–hole) correlation is more important than electron–hole correlation and leads to a condition for a mag-

[7] C. Rizzuto, *Rep. Prog. Phys.* **37**, 147 (1974).

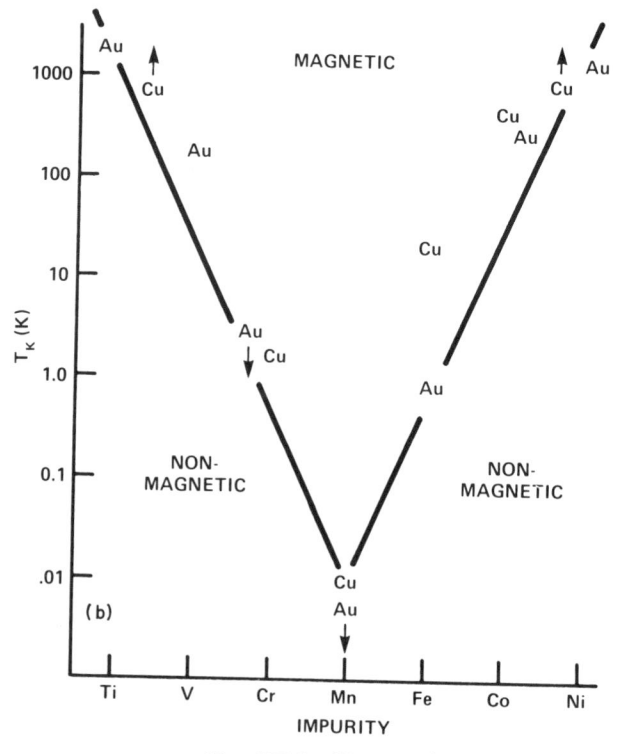

FIG. VII.5. (*Continued*)

netic solution that is less favorable than in the Hartree–Fock approximation. That is, the fluctuations suppress the transition to the magnetic state with increasing U.

In case (b) electron–hole correlation becomes important. The electron–hole interaction has the diagrammatic form:

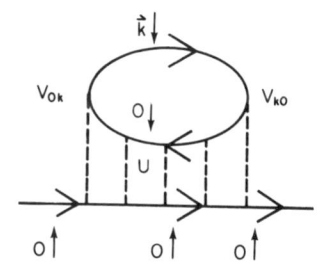

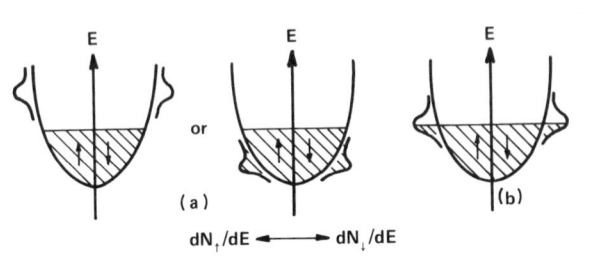

FIG. VII.6.   Two ways in which an impurity may be nonmagnetic: (a) E_0 far from E_F and (b) $E_0 = E_F - U/2$.

where the lines have the same meaning as in Chapter III. This may be twisted into:

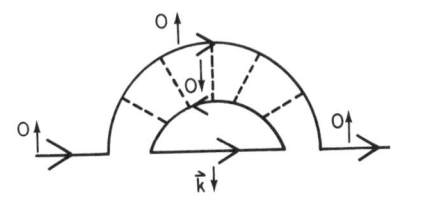

In this form it is similar to the electron–phonon interaction which we discussed in Chapter III. Therefore, one calls the repeated scattering of an electron and a hole of opposite spin a *paramagnon* or *spin fluctuation*. Mills and Lederer[8] have developed a theory of spin fluctuations for pure metals such as Pd or Pt which are "on the verge" of being magnetic. The spin fluctuation is not really a collective mode like a magnon or phonon in that it never appears as a *pole* in a response function. It does, however, lead to a peak in $\chi''(q,\omega)$ as shown in Fig. VII.7.[9] The width of this peak is a measure of the lifetime of the spin fluctuation. The spin fluctuation model has also been used to describe impurities. The lifetime associated with the localized spin fluctuation is

$$1/\tau_{sf} = \frac{\pi N_0(E_F)}{1 - UN_0(E_F)}. \tag{7.14}$$

Obviously if this spin fluctuation persists for a long time we have effectively the same situation that the Anderson model requires for an impurity to be magnetic. In fact, $\tau_{sf} \to \infty$ at the Anderson criterion $UN_0(E_F) = 1$.

[8] See, for example, D. L. Mills, M. T. Béal-Monod, and P. Lederer, *in* "Magnetism" (H. Suhl and G. T. Rado, eds.), Vol. 5, p. 89. Academic Press, New York, 1973.
[9] J. R. Schrieffer, *J. Appl. Phys.* **39**, 642 (1968).

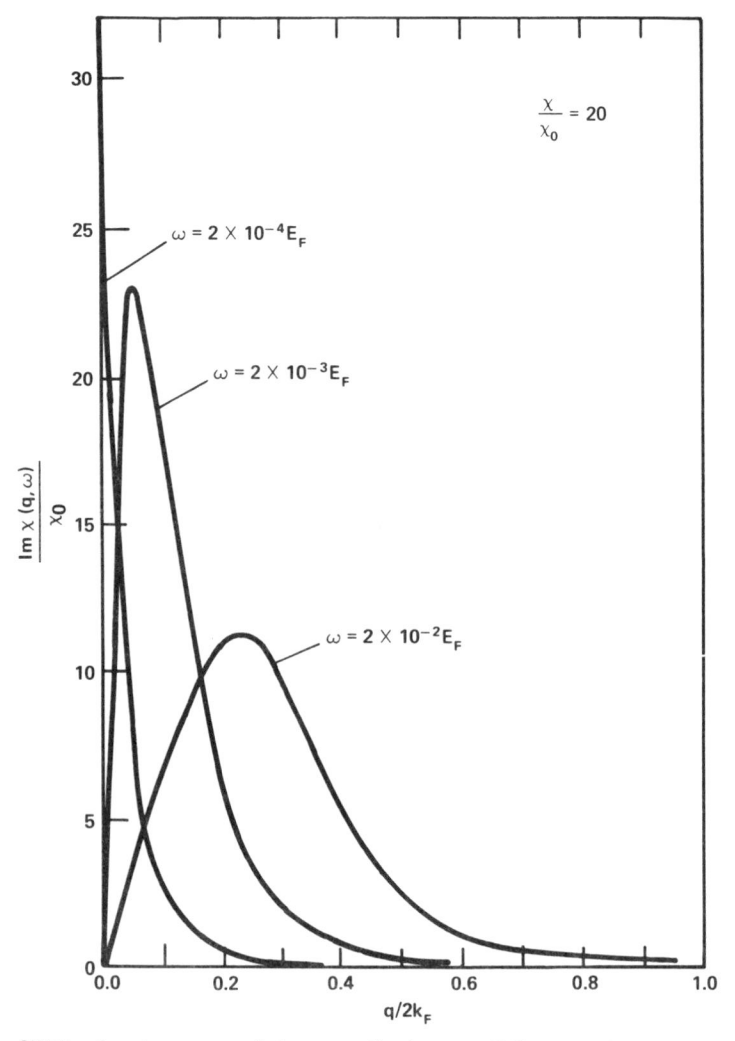

FIG. VII.7. Imaginary part of the generalized susceptibility showing the paramagnon peak. After Schrieffer.[9]

However, the point is that we do not need this very restrictive condition to achieve a magnetic response. All that is necessary is that thermal fluctuations break up this correlation, i.e., $k_B T > h/\tau_{sf}$. This will enable the localized spins to respond individually to an applied field to give a Curie law response. Thus, above the temperature

$$k_B T_K = h/\tau_{sf} \tag{7.15}$$

the impurity will behave magnetically while below T_K it will be nonmagnetic. This is called the Kondo temperature for reasons which will become clear below. Thus spin fluctuation theory appears to overcome some of the deficiencies of the Hartree–Fock approach. However, there are still problems. For example, if one attempts to include electron–electron correlations together with electron–hole correlations (by calculating what the theorists call the "parquet" diagrams) the problem becomes very complex. This approach has now largely been abandoned in favor of the renormalization techniques introduced in Chapter II. Krishnamurthy et al., [9a] have calculated the impurity susceptibility for the Anderson model over the full, physically relevant range of the parameters of that model. This is shown in Fig. VII.8. For this numerical renormalization the density of states of the conduction electrons was assumed to be constant over a width $2D$. From Fig. VII.8 we see that for $U < k_BT < D$ the empty and doubly occupied impurity states are as likely to be populated as the singly occupied state. Thus, the impurity susceptibility is that associated with a free orbital, $\chi \simeq g^2\mu_B^2/8k_BT$. As the temperature decreases below U, $k_BT\chi(T)$ increases, indicating the formation of the local moment. The local moment regime, characterized by a Curie–Weiss behavior, persists for six decades of temperatures. Finally, at very low

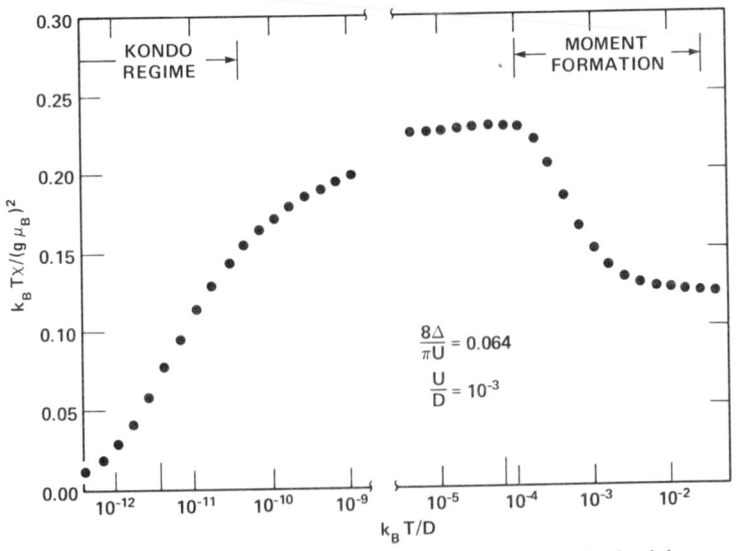

FIG. VII.8. Impurity susceptibility of the Anderson model obtained by numerical renormalization-group techniques. After Krishna-murthy et al.[9a]

[9a] H. R. Krishna-murthy, K. G. Wilson, and J. W. Wilkins, *Phys. Rev. Lett.* **35**, 1101 (1975).

temperature $k_B T \chi(T)$ drops rapidly, approaching zero at zero temperature.

In the region where the Hartree–Fock approximation gives a magnetic solution to the Anderson Hamiltonian (large U) one can transform the Anderson Hamiltonian into an effective spin Hamiltonian which offers one the hope of going beyond the Hartree–Fock approximation. This Hamiltonian has the form of the familiar s–d exchange interaction,

$$\mathcal{H} = -J\mathbf{S} \cdot \mathbf{s} \tag{7.16}$$

where $\mathbf{S}$ is the spin of the impurity and

$$J = 2|V_{0k_F}|^2 \frac{U}{E_0(E_0 + U)} \tag{7.17}$$

which is antiferromagnetic since $E_0 < 0$. In 1964 Kondo calculated the resistivity associated with the spin exchange scattering associated with this interaction to second order in the Born approximation and found it to diverge logarithmically at the temperature

$$T_K \sim e^{-1/N(E_F)J} \tag{7.18}$$

which is an alternative definition of the Kondo temperature. This opened an era of unprecedented theoretical and experimental activity in dilute alloys. One of the questions central to this "Kondo problem" is the nature of the system below the Kondo temperature. Again, the renormalization approach provided the answer. In fact, it was the s–d Hamiltonian (7.16) to which renormalization techniques were first applied.[9b] Wilson found that the susceptibility had the form shown in Fig. VII.8 *below the break*. For $T \ll T_K$ the local moment is compensated by the conduction electrons and $\chi(T)$ approaches the constant value

$$\lim_{T \to 0} \chi(T) = 0.103 g^2 \mu_B^2 / k_B T_K \tag{7.19}$$

where $k_B T_K \simeq D(N(E_F)J)^{1/2} e^{-1/N(E_F)J}$. Since the dominant scattering occurs in the vicinity of the Fermi level this "spin compensation cloud" is characterized by a length

$$\xi_K = v_F / k_B T_K. \tag{7.20}$$

The nature of the spin polarization within this cloud has been studied by NMR as we shall discuss in the next section.

[9b] K. G. Wilson, *in* "Nobel Symposia—Medicine and Natural Sciences" (B. Lundgvist and S. Lundgvist, eds.), Vol. 24, p. 68. Academic Press, New York, 1973.

B. MANIFESTATIONS OF LOCAL MOMENTS

The virtual bond state (vbs) can be "seen" by a variety of experiments. If they are done systematically for a series of either solvents or solutes, the localization and magnetization of the vbs can be followed as it approaches the Fermi level. The "variety" of experiments includes those described in Chapter V as well as the superconducting response.

1. Susceptibility

The single "magnetic" impurity interacting with the conduction electron spins gives a Curie–Weiss law at "high" temperature, and finally (below the Kondo temperature) the susceptibility saturates and becomes temperature independent. As we mentioned above, this Curie–Weiss response occurs when the interaction time with the conduction electrons is less than the thermal fluctuation time so that the impurity spin is being "buffeted" about rapidly enough to come to thermal equilibrium. Let us consider some examples.

All the 3d elements can be dissolved in PdSb, a superconducting compound. A Curie–Weiss law is found for Cr and Mn only, as shown in Fig. VII.9.[10] It is interesting to note that CrSb and MnSb, i.e., Cr and Mn with the same near neighbor configurations, are ordered (T_N = 723 K and T_c = 587 K, respectively). We shall return to the left-hand ordinate later.

As a second example consider a 3d element such as Fe dissolved in a matrix of variable composition. The case for 1% Fe in the 4d series is shown in Fig. VII.10.[11] In the Nb–Mo alloys containing 1% Fe the apparent moment per iron appears to increase from zero below 40% Mo to 2.2 μ_B above 90% Mo. Since the density of states $N(E_F)$ has a rapid variation in this region one might take this as a manifestation of the Anderson criterion mentioned above. Jaccarino and Walker,[12] however, suggest that the moment formation might be discontinuous—either zero or some maximum value. The apparent continuous change would then be associated with the probability that an individual impurity has developed a moment. This probability is determined by the near neighbor configuration. Jaccarino and Walker cite data for 1% Co in $Rh_{1-c}Pd_c$ alloys as shown in Fig. VII.11. This would suggest that the parameters in the Anderson model are local quantities. It is interesting to note that a similar dependence of moment on local environment is observed in amorphous magnetic alloys such as Re–Fe which are discussed in Chapter IX.

[10] T. Geballe, B. T. Matthias, B. Caroli, E. Corenzwit, J. Maita, and G. Hull, *Phys. Rev.* **169**, 457 (1968).

[11] A. M. Clogston, B. T. Matthias, M. Peter, H. J. Williams, E. Corenzwit, and R. C. Sherwood, *Phys. Rev.* **125**, 541 (1962).

[12] V. Jaccarino and L. R. Walker, *Phys. Rev. Lett.* **15**, 258 (1965).

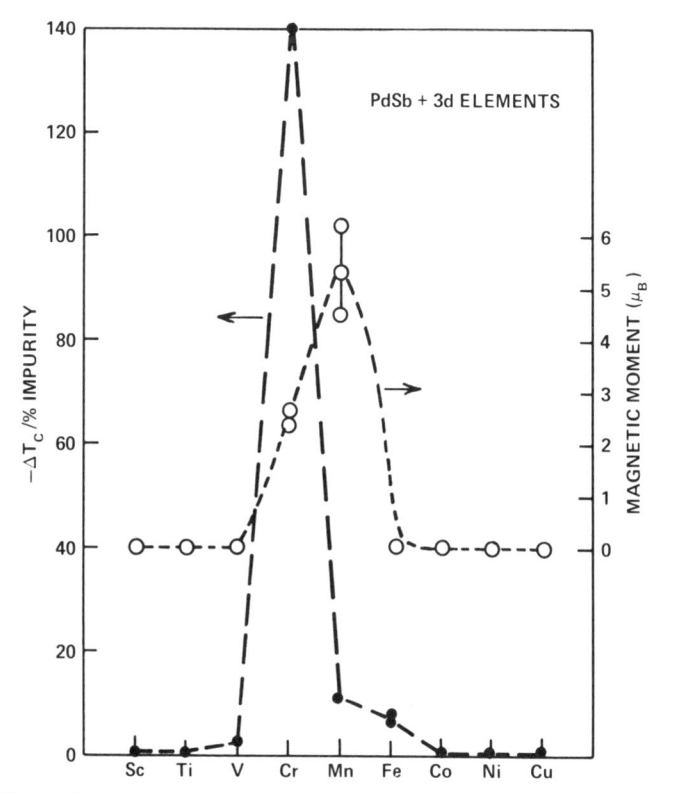

FIG. VII.9. The dashed curve gives the initial linear decrease in the superconducting transition temperature of PdSb in degrees Kelvin per percent of added 3d element. The dotted curve gives the magnetic moment of the added 3d element as deduced from susceptibility data. The fact that the superconductivity is more strongly affected by Cr and Mn impurities is connected with the fact that the virtual bound state, although split, lies closer to the Fermi level in the case of Cr. After Geballe et al.[10]

2. Resistivity

In Al, the vbs of the series of 3d transition solutes do not magnetize at room temperature presumably because E_F is too high and $N(E_F)$ is too large. There is merely a peak in the residual resistivity ($\Delta\rho$) as the vbs crosses E_F. However, at higher temperatures the moments do develop as shown in Fig. VII.12.

In the noble metals E_F is lower and the vbs "magnetizes." Operationally, this means that a Curie–Weiss law exists over a wide range of temperature and with a "reasonable" intercept. (From a practical point of view, an intercept greater than about 100° means that any analysis of the

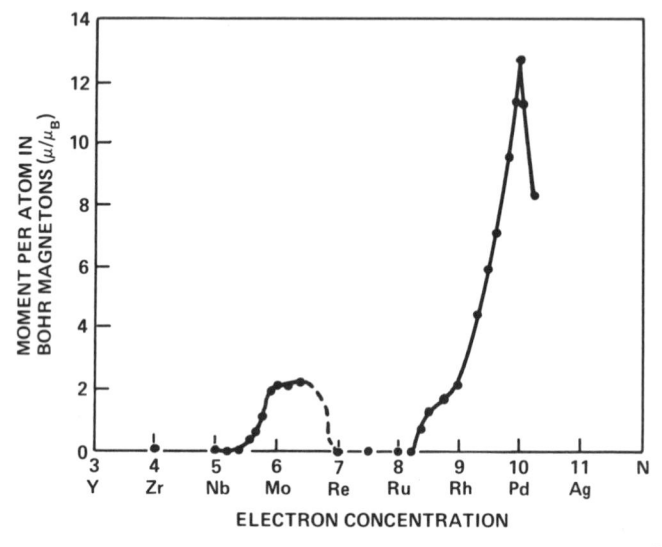

FIG. VII.10.   The average magnetic moment of Fe impurities in 4d metals and alloys as determined from the susceptibility. After Clogston *et al.*[11]

slope must involve subjectivity in subtracting out the effect of the solvent.) In Cu (Fig. VII.12), two peaks can be seen in the residual resistivities of the 3d solutes, one for the spin-up and one for the spin-down states crossing E_F. In Mn, the splitting is complete at E_F.

The magnetic vbs also can give rise to a resistance minimum or "Kondo" behavior as shown in Fig. VII.13.[11,13] The resistance minimum is the combined result of phonon scattering which decreases as the temperature is lowered and "Kondo" or spin-dependent scattering which increases with temperature, eventually saturating as shown in Fig. VII.14.

Experimentally, one must take precautions to be in the dilute range. Solute–solute interactions become important above ~10 ppm in Cu. Spin flip scattering is suppressed and one finds a *maximum* in ρ at some temperature T below T_K, as shown in Fig. VII.15.[14] The maximum eventually becomes the "spin-disorder scattering" structure which occurs in the ordering region of magnetic compounds, as discussed in Chapter V.

[13] M. Sarachik, E. Corenzwit, and L. D. Longinotti, *Phys. Rev.* **135**, A1041 (1964).
[14] F. T. Hedgcock and C. Rizzuto, *Phys. Rev.* **163**, 517 (1967).

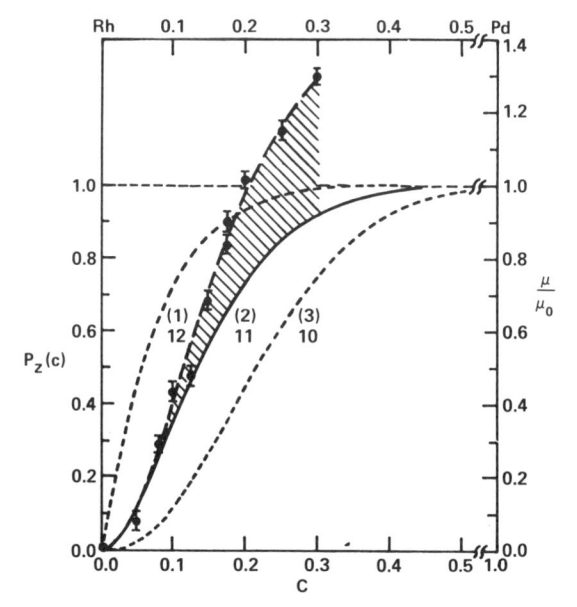

FIG. VII.11. The observed effective magnetic moment of 1% Co in the $Rh_{1-c}Pd_c$ fcc alloy system, normalized to the value $\mu_0 = 1.7$ Bohr magnetons for Co metal. Superimposed on these data are curves which give the probability $P_z(c)$ that a Co impurity will have at least 1, 2, or 3 Pd nearest neighbors. The additional moment induced in the matrix, which is proportional to Co-free susceptibility of the Rh-Pd alloys, is shown by the shaded region. After Jaccarino and Walker.[12]

3. Heat Capacity

In general there are extra "magnetic" contributions seen which give rise to an "excess" heat capacity in dilute alloys. For example, the distribution of exchange fields arising from the RKKY interaction gives rise to a contribution which starts off *linearly* with temperature and relatively independent of concentration and reaches a maximum which is concentration dependent.[15] The entropy corresponds to the ordering of the spins.

A single impurity, however, interacting with, and becoming compensated by, the electron gas also loses an entropy of the order of $R \ln(2S + 1)$ where S is consistent with the high T Curie–Weiss behavior. The heat capacity for Cr and Fe in Cu is shown in Fig. VII.16.[16] The solid line is a theoretical calculation based on the s–d or Kondo Hamiltonian.[17] The

[15] W. Marshall, *Phys. Rev.* **118**, 1519 (1960).
[16] B. Triplett, Ph.D. Thesis, University of California, Berkeley (1970).
[17] P. E. Bloomfield and D. R. Hamann, *Phys. Rev.* **164**, 856 (1967).

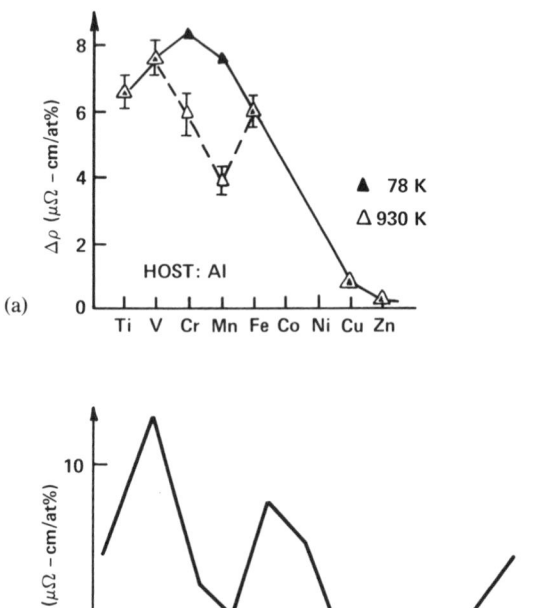

(a)

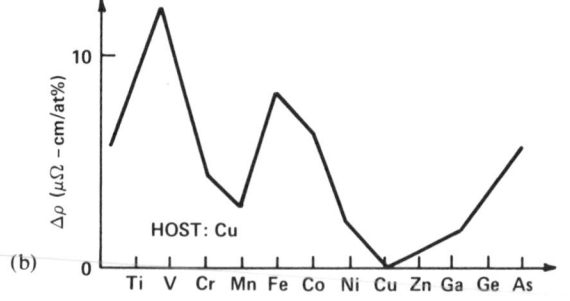

(b)

FIG. VII.12.   Residual resistivities of transition metal impurities in (a) Al and (b) Cu. The data on Al suggest that at 78 K the impurities are not magnetic, i.e., the virtual bound state is not split. Thus there is only one peak in the resistivity as this resonance moves across the Fermi level near Cr. In Al at 930 K and in Cu the impurities are magnetic. This splits the resonance and we observe two peaks as they separately cross the Fermi level.

dashed curve is proportional to T.

From experiments such as those the magnetic nature of a wide variety of dilute alloys has been determined. A summary of Kondo temperatures in Cu and Au is shown in Fig. VII.5b.[18]

4. Magnetic Hyperfine Studies

Nuclear magnetic resonance is a local probe as distinct from the susceptibility which is a spatial average of the magnetic response. Mossbauer and nuclear orientation experiments also probe the local environment, since in both cases it is the hyperfine field acting on the nucleus which is measured.

[18] M. Daybell, in "Magnetism" (G. Rado and H. Suhl, eds.), Vol. 5, p. 122. Academic Press, New York, 1973.

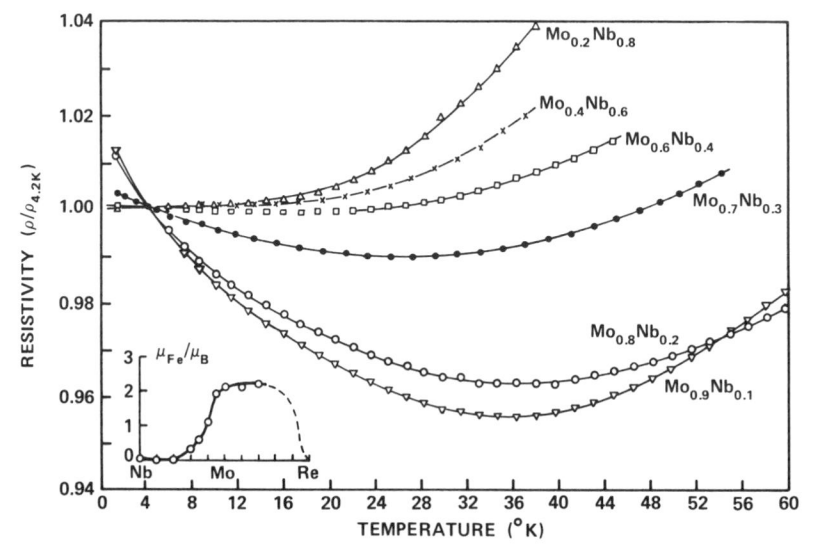

Fɪɢ. VII.13.   Characteristic Kondo resistance minimum associated with Fe dissolved in Mo-Nb alloys (after Sarachik et al.[13]). Insert shows correlation between resistance minimum with local moment formation (after Clogston, et al.[11]).

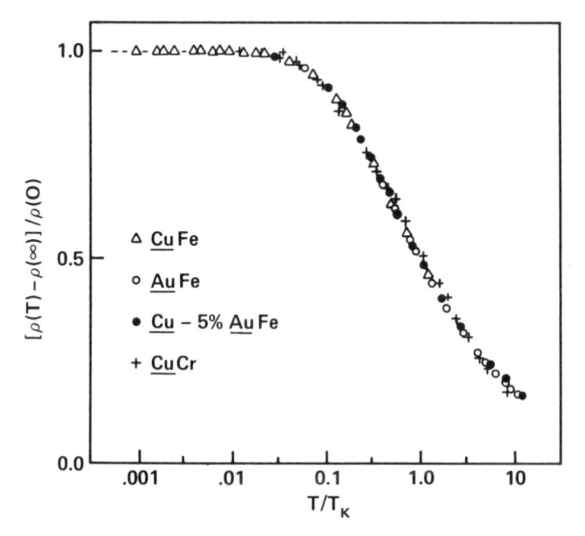

Fɪɢ. VII.14.   Excess resistivity associated with scattering from the impurity spin. At low temperatures the resistivity saturates because all the d-wave is being scattered. This is readily seen from Eq. (7.6) when $\delta_2 = \pi/2$. This is called the "unitarity" limit because it is a limit imposed by the requirement that the so-called scattering matrix which is used in deriving Eq. (7.6) be unitary.

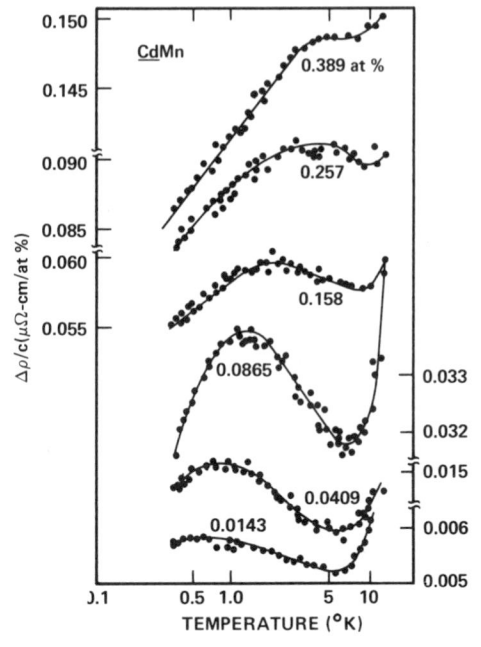

FIG. VII.15. The effect of "ordering" interactions on the low-temperature resistivity behavior of CdMn. The peak at low temperatures scales linearly with concentration, and corresponds to antiferromagnetic ordering of the Mn impurities. After Hedgcock and Rizzuto.[14]

The question that such experiments are ideal for studying is the nature of the spin polarization, $\sigma(r, T)$, of the conduction electrons at a distance r from the impurity. There have been a number of efforts, both theoretical and experimental, to determine this spin density. Above the Kondo temperature we expect $\sigma(r, T)$ to display the well-known Ruderman–Kittel–Kasuya–Yosida (RKKY) oscillations and scale with the spin susceptibility of the impurity, i.e., $\sigma(r, T)/H = \chi(T)f(r)$. The question is whether $f(r)$ retains its oscillatory shape below T_K.

To answer this question Boyce and Slichter[19] have used NMR to measure $\sigma(r, T)$ at four distinct shells of Cu neighbors surrounding isolated Fe atoms in CuFe from well below T_K to well above T_K. Since the different shells experience a different spin density, their NMR fields are shifted from that of a Cu nucleus far from the impurity. (There is also an electric quadrupole shift due to the change in electron charge density, but this is small.) This shift is related to $\sigma(r, T)$ by

[19] J. Boyce and C. Slichter, *Phys. Rev. Lett.* **32**, 61 (1974).

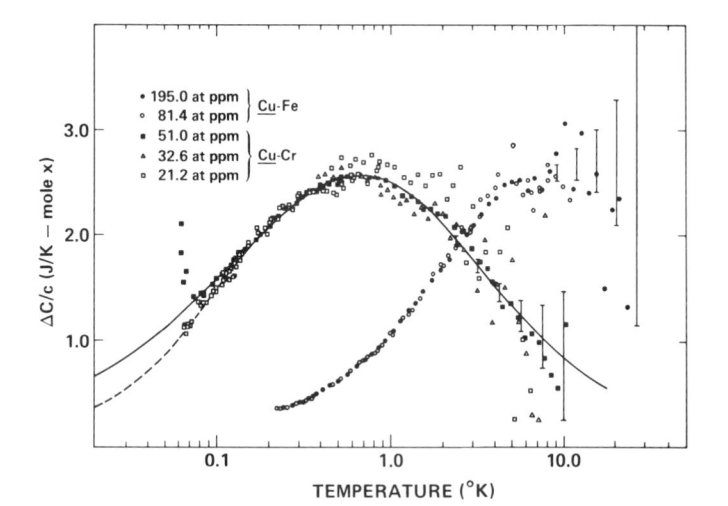

FIG. VII.16. Heat capacities of dilute solutions of Cr in Cu and Fe in Cu. The different symbols refer to measurements made in different cryostats on the same sample. After Triplett.[16]

$$\frac{\Delta K(r, T)}{K} = -\frac{\gamma e\hbar}{\chi_s} \frac{\sigma(r, T)}{\langle |\psi(0)|^2\rangle_{E_F} H} \tag{7.21}$$

where χ_s is the spin susceptibility of the conduction electron in the pure host and $\langle |\psi(0)|^2\rangle_{E_F}$ is the average of the wave function density at the nucleus for states at the Fermi level. A plot of $(\Delta K/K)^{-1}$ versus T gives a Curie–Weiss law with $\theta = 29 \pm 1$ K. This agrees with the bulk susceptibility. Figure VII.17 shows $\Delta K/K$ versus $1/(T + 29)$.

The important feature to note in Fig. VII.17 is that $\Delta K/K$ has the same temperature dependence as $\chi(T)$ both above *and below* T_K. This implies that there is no additional spin compensating electron cloud forming below T_K as some have conjectured. Boyce and Slichter identify the resonance labeled M as that due to a second nearest neighbor, B the third, and C the fourth. (The nearest neighbor resonance was too broad to be observed below 77 K.) The change in sign of C and B relative to M is a manifestation of the RKKY oscillations.

The Friedel–Anderson description of impurities in metals is analogous to a band theory of ferromagnetism. Hirst[20] has proposed an alternative description in which the mixing term in the Anderson Hamiltonian is considered as a perturbation on the many-electron ionic configuration given by

[20] See, for example, L. L. Hirst, *AIP Conf. Proc.* **24**, 11 (1975).

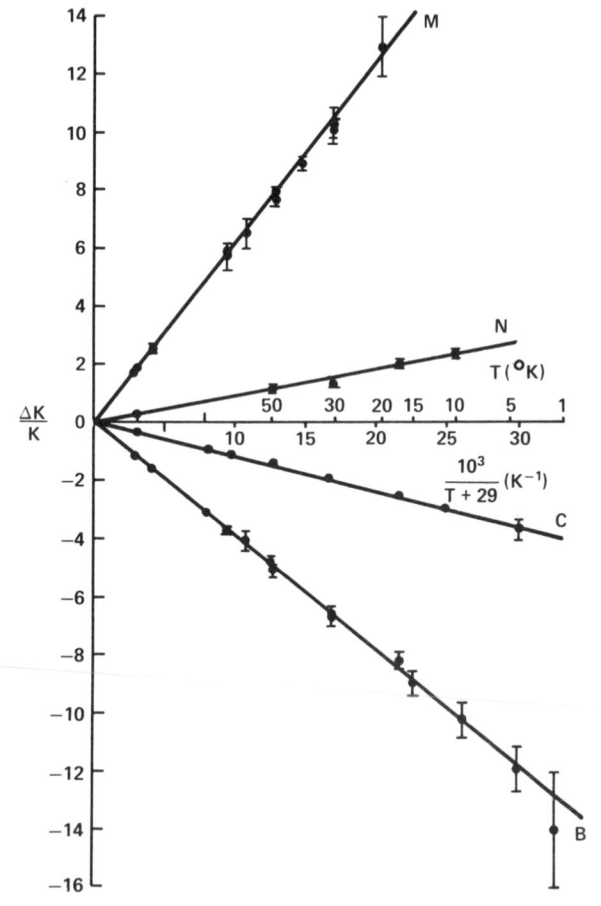

FIG. VII.17.   Knight shifts versus $1/(T + 29)$ for four ^{63}Cu resonances appearing as satellites to the pure Cu NMR in a dilute alloy of *Cu*Fe. After Boyce and Slichter.[19]

Hund's rule. The energies of such a configuration are given by $-Vn + Un(n - 1)/2$ where n is the total number of 3d electrons, V is the binding potential at the impurity site, and U is the Coulomb interaction. A comparison of these models is illustrated in Fig. VII.18. The "fingers" in the Hirst model represent splittings due to crystal field and spin–orbit effects. These splittings lead to a different impurity susceptibility than that expected for the Friedel–Anderson model. NMR measurements[20a] similar

[20a] D. C. Abbar, T. J. Aton, and C. P. Slichter, *Phys. Rev. Lett.* **41**, 719 (1978).

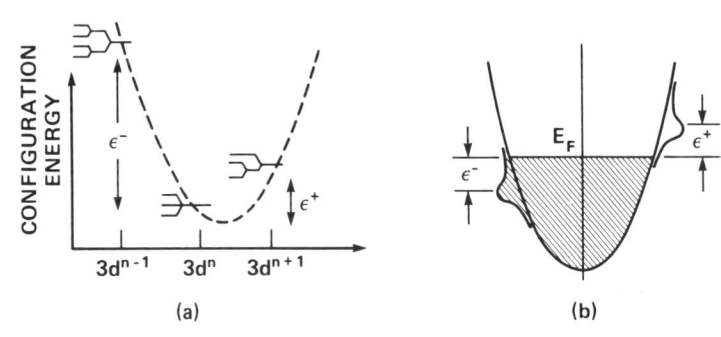

FIG. VII.18. Comparison between a configuration-based model (a) and a virtual-bound-state model (b) for a magnetic impurity.

to those above indicate that CuCr, for example, is better described by the Hirst model.

5. Superconducting Response

Since a Cooper pair consists of two electrons in time-reversed states, any mechanism which acts differently on such states tends to depair the electrons. For example, an interaction of the form $i(\mathbf{p} \times \mathbf{p}') \cdot \boldsymbol{\sigma}$, although it may flip the spin of one of the electrons, it is invariant under time reversal and therefore will not lead to pair breaking. The theoretical aspects of this problem involve the same techniques that we encountered in the strong-coupling problem treated in Chapter III. The theory of magnetic impurities in superconductors was first considered by Abrikosov and Gor'kov.[21] In Chapter III we found that the off-diagonal nature of the order parameter could be conveniently incorporated by working with two-component field operators. The Green's function for a pure superconductor is given by

$$G^{-1}(\mathbf{k}, \omega) = G_0^{-1}(\mathbf{k}, \omega) - \Sigma(\mathbf{k}, \omega). \tag{7.22}$$

From the expressions given in Chapter III we obtain

$$G^{-1}(\mathbf{k}, \omega) = Z(\mathbf{k}, \omega)\omega\sigma_0 - \epsilon(\mathbf{k})\sigma_3 - Z(\mathbf{k}, \omega)\Delta(\mathbf{k}, \omega)\sigma_1. \tag{7.23}$$

If a magnetic impurity is now introduced into the superconductor it will contribute an additional term to the self-energy, Σ_{imp}. Since the evaluation of Σ_{imp} involves the Green's function (7.23) the Green's function in the presence of the impurity will also involve the three Pauli matrices σ_0,

[21] A. A. Abrikosov and L. P. Gor'kov, *Sov. Phys. JETP* **12**, 1243 (1961).

σ_1, and σ_3. Therefore, we write

$$\mathcal{G}^{-1} = G^{-1} + \Sigma_{\text{imp}} = \tilde{\omega}\sigma_0 - \epsilon\sigma_3 - \tilde{\Delta}\sigma_1 \qquad (7.24)$$

where $\tilde{\omega}$ and $\tilde{\Delta}$ represent renormalized frequencies and gap parameters, respectively.

If we assume the impurity interacts with the electrons through an exchange interaction of the form

$$\mathcal{H}_{\text{imp}} = \sum_i J\mathbf{S}_i \cdot \mathbf{s}, \qquad (7.25)$$

then the self-energy in second order is[22]

$$\Sigma_{\text{imp}} = \frac{i\Gamma}{(\tilde{\omega}^2 - \tilde{\Delta}^2)^{1/2}} (\tilde{\omega}\sigma_0 + \tilde{\Delta}\sigma_1) \qquad (7.26)$$

where $\Gamma = 2\pi n_i N(0) J^2 S(S + 1)/4$ characterizes the pair-breaking strength. Here n_i is the number density of impurities. The radical in the denominator arises from inverting the Green's function matrix. Combining Eqs. (7.24) and (7.26), and assuming we are not in the strong-coupling regime so $Z(k, \omega) \simeq 1$, we have

$$\tilde{\omega} = \omega + i\Gamma \frac{\tilde{\omega}}{(\tilde{\omega}^2 - \tilde{\Delta}^2)^{1/2}} \qquad (7.27)$$

$$\tilde{\Delta} = \Delta - i\Gamma \frac{\tilde{\Delta}}{(\tilde{\omega}^2 - \tilde{\Delta}^2)^{1/2}}. \qquad (7.28)$$

Notice that the term proportional to Γ enters these two equations with the opposite signs. This is the result of the time-reversal noninvariance of the interaction. For nonmagnetic impurities the signs of the corresponding terms would be the same.

The order parameter Δ is now calculated as in the BCS theory except the Green's function $\mathcal{G}$ is used. The resulting integral equation for the order parameter is

$$\Delta(T, \Gamma) = N(0)V \int_0^{\omega_D} d\omega \text{Re} \frac{\tanh(\hbar\omega/2k_BT)}{[(\tilde{\omega}/\tilde{\Delta})^2 - 1]^{1/2}}. \qquad (7.29)$$

The critical temperature is obtained by solving this equation in the limit $\Delta \to 0$. Dividing both sides by Δ leaves us the task of calculating the limit of $\tilde{\omega}\Delta/\tilde{\Delta}$. From Eq. (7.28) we can see that if $\Delta \to 0$, $\tilde{\Delta}$ must also $\to 0$.

[22] S. Skalski, O. Betbeder-Matibet, and P. R. Weiss, *Phys. Rev. A* **136**, 1500 (1964).

Therefore

$$\lim_{\Delta \to 0} \frac{\tilde{\omega}\Delta}{\tilde{\Delta}} = \lim_{\Delta \to 0} \left[\omega + i2\Gamma \frac{\tilde{\omega}/\tilde{\Delta}}{[(\tilde{\omega}/\tilde{\Delta})^2 - 1]^{1/2}} \right] = \omega + i2\Gamma. \qquad (7.30)$$

For a time reversal invariant interaction the imaginary term would not be present. The critical temperature is given by

$$1 = N(0)V \int_0^{\omega_D} d\omega \, \frac{\omega}{\omega^2 + \Gamma^2} \tanh(\hbar\omega/2k_B T_c). \qquad (7.31)$$

The presence of this imaginary contribution to the energy corresponds to a relaxation mechanism. When Γ is of the order of the gap the superconductivity is suppressed. In particular, the presence of the factor Γ^2 in Eq. (7.31) reduces T_c relative to what it would be in a pure superconductor. The equation for T_c becomes

$$\ln(T_c/T_c^P) = \psi(\tfrac{1}{2}) - \psi(\tfrac{1}{2} + \rho) \qquad (7.32)$$

where $\rho \equiv \Gamma/\pi k_B T_c$ and ψ is the psi or diagamma function. The solution of this equation[22] as a function of Γ is shown in Fig. VII.19. The order parameter is also shown. Here $\Delta^P(0)$ is the BCS or "pure" value,

$$\Delta^P(0) = 2\omega_D e^{-1/N(0)V}. \qquad (7.33)$$

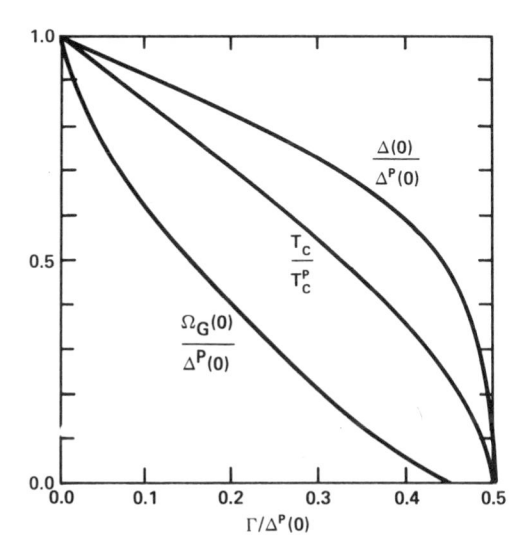

FIG. VII.19. The critical temperature T_c, the order parameter $\Delta(0)$, and the excitation energy $\Omega_G(0)$ at $T = 0$ as a function of the exchange parameter Γ. The superscript p refers to the pure material. After Skalski et al.[22]

Since Γ is linearly proportional to the impurity concentration, c, for small Γ,

$$\frac{dT_c}{dc} = -\frac{\pi^2}{8k_B} N(0)J^2S(S + 1).$$

(7.34)

This explains the left-hand ordinate in Fig. VII.9. Although, from its moment, the Cr virtual bound state is only partially split its exchange (about 1.4 eV) is an order of magnitude larger than that for Mn. This is presumably due to the proximity of the vbs to the Fermi level.

For rare earth impurities with unquenched orbital angular momentum the appropriate states for calculating thermal averages are those of the total angular momentum $J = L + S$. As we discussed in Chapter IV, the exchange interaction, which always involves the spin, then becomes

$$\mathcal{H} = \sum_i (g_J - 1)\mathcal{J}\mathbf{J}_i \cdot \mathbf{s}$$

(7.35)

where $\mathcal{J}$ is now the exchange parameter. Thus for rare earths the factor $S(S + 1)$ in Eq. (7.34) is replaced by the de Gennes factor $(g_J - 1)^2 J(J + 1)$. Figure VII.20 shows the results for adding rare earth impurities to the superconductor La. We see that, except for Ce, the data are better represented by the de Gennes factor than simply the total moment. There are, however, other substances such as $LaAl_2$ which do not follow the de Gennes curve. Such deviations indicate the importance of anisotropic contributions to the exchange given in Eq. (7.35).

Cerium is an exception because its f-level is believed to lie close to the Fermi level. *La*Ce, for example, is the only LaRE which shows a resistance minimum, hence a Kondo state. The moment indicates trivalent Ce, i.e., a 4f configuration. The corresponding Kondo temperature is well *below* the superconducting transition temperature of La. The position of the Ce f-level is very sensitive to strain. Chemically speaking, trivalent Ce can be made tetravalent under pressure which causes the f-electron to be "squeezed out" into the conduction band. The Laves phase compounds $CeCo_2$, $CeRh_2$, and $CeIr_2$ even at atmospheric pressure are tetravalent and superconducting and have been used as solutes for studying the relationship between superconductivity and magnetism as discussed in the next section. Figure VII.21[23] shows the pressure dependence of T_c for *La*Ce. Notice, in the insert of Fig. VII.21, that at high pressures the concentration dependence of T_c is much weaker than that predicted by the Abrikosov–Gor'kov theory. In fact, it is nearly exponential. In this region

[23] M. B. Maple, J. Wittig, and K. S. Kim, *Phys. Rev. Lett.* **23**, 1375 (1969).

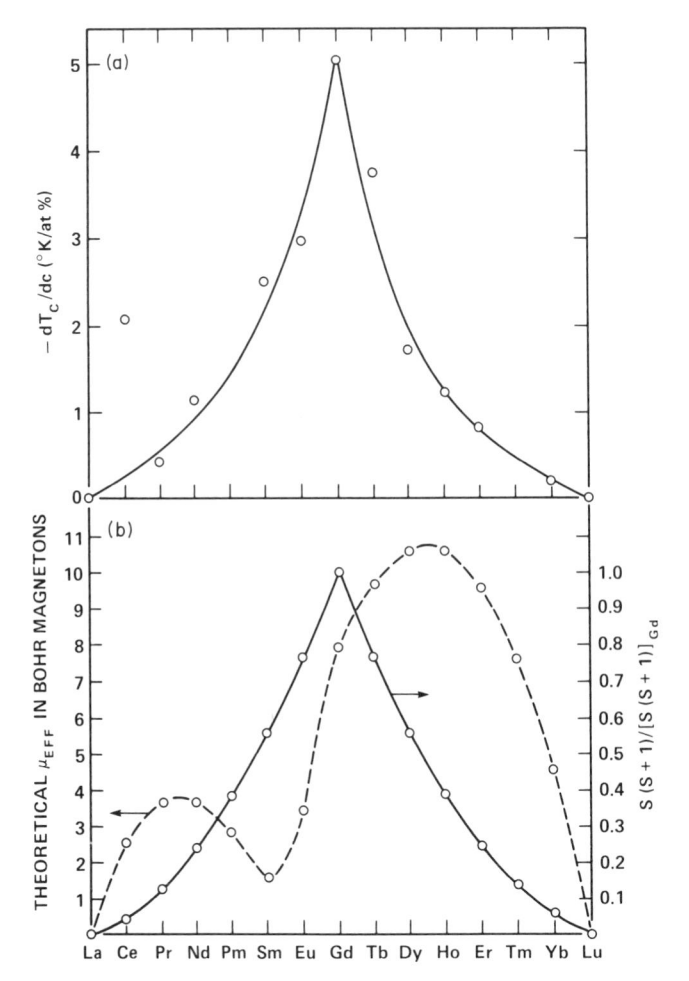

FIG. VII.20.   (a) Initial slope of the curve $T_c(c)$ for La doped with rare earth impurities. (b) The magnetic moment and the spin (normalized to Gd) of the trivalent rare earth ions.

the Ce is nonmagnetic ($T_K > T_c^P$) but there are spin fluctuations present which depress the superconductivity.

If a Kondo temperature exists which is below T_c^P there is an interesting phenomenon which can occur. Müller-Hartmann and Zittartz[24] noted that in this case the quantity Γ entering the Abrikosov–Gor'kov equation (7.31) for T_c, that is, this pair-breaking strength of one impurity, increases

[24] E. Müller-Hartmann and J. Zittartz, *Phys. Rev. Lett.* **26**, 428 (1971).

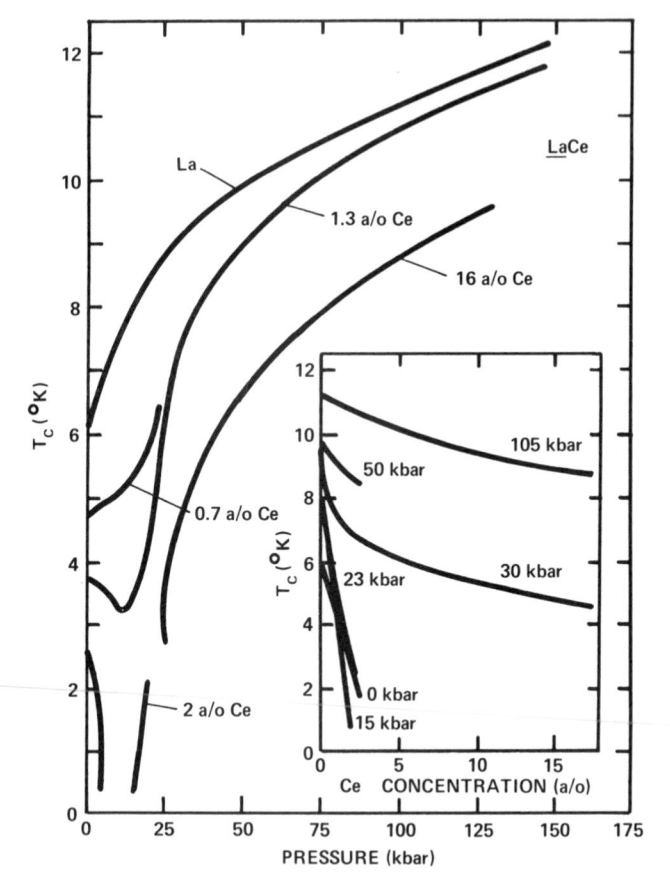

FIG. VII.21. Pressure dependence of the superconducting transition temperature of as-cast predominantly fcc La and LaCe alloys to very high pressure. Isobars of T_c versus Ce concentration are shown in the inset. After Maple et al.[23]

with decreasing temperature. This can lead to double-valued solutions. Such "re-entrant" behavior has been observed in $(La,Ce)Al_2$ as shown in Fig. VII.22.[25]

In Kondo systems the impurity loses its magnetic response because of interactions with the conduction electrons. Another way of losing this response is through the effect of a low crystal symmetry. However, in this case the pair-breaking strength of one impurity *decreases* with decreasing temperature. Consider, for example, an ion placed in a crystalline envi-

[25] B. Maple, W. Fertig, A. Mota, L. DeLong, D. Wohllenben, and R. Fitzgerald, *Solid State Commun.* **11**, 829 (1972).

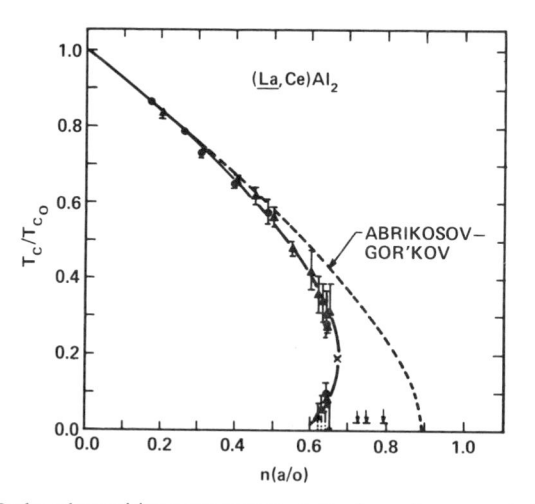

FIG. VII.22. Reduced transition temperature versus impurity concentration for the (La, Ce) Al₂ system. The symbol (x) denotes the estimated turning point of the T_c vs. n curve while the solid circles and triangles distinguish two separately prepared sets of alloys. After Maple *et al.*[25]

ronment. The electric fields from this environment will constitute a perturbing influence on the ion and, in general, lift the degeneracies of its states. If the number of electrons involved is *even,* i.e., the ion is a "non-Kramers" ion, then *all* the degeneracy could be lifted if the symmetry of the perturbing environment is low enough. In this case the ground state becomes a singlet and cannot respond to a magnetic field.

The Pr^{3+} ion is an example of such an ion. It has a $4f^2$ configuration which, according to Hund's rules, has a ground state characterized by $L = 5, S = 1, J = L - S = 4$. If this ion is placed in a hexagonal or cubic crystalline environment this $^3H_4$ ground state will split as shown in Fig. VII.23.

Let us now calculate the susceptibility associated with a system of such ions. For simplicity let us consider only two low lying singlets separated by an energy Δ. The matrix elements of the angular momentum operator J_z within either of these states is zero since they are singlets. However, the matrix element between them has some nonzero value, $\langle 0|J_z|1\rangle = \alpha$. The Zeeman interaction therefore mixes the two states. The new ground state in the presence of a field is then

$$|0\rangle = |0\rangle + \frac{g_J\mu_B H\alpha}{\Delta} |1\rangle. \tag{7.36}$$

From the moment $\langle 0|J_z|0\rangle$ we obtain the Van Vleck temperature independ-

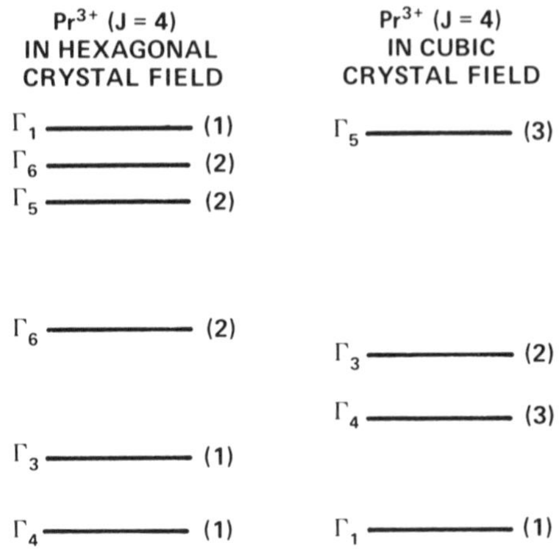

Pr³⁺ (J = 4)
IN HEXAGONAL
CRYSTAL FIELD

Pr³⁺ (J = 4)
IN CUBIC
CRYSTAL FIELD

Γ_1 ———————— (1) Γ_5 ———————— (3)
Γ_6 ———————— (2)
Γ_5 ———————— (2)

Γ_6 ———————— (2)
                            Γ_3 ———————— (2)

                            Γ_4 ———————— (3)

Γ_3 ———————— (1)

Γ_4 ———————— (1) Γ_1 ———————— (1)

FIG. VII.23.   Crystal field splitting of the Pr^{3+} ground $^3H_4$ state.

ent susceptibility

$$\chi_0 = \frac{2g_J^2\mu_B^2\alpha^2}{\Delta}. \tag{7.37}$$

Let us now add an exchange interaction between neighboring Pr ions. In the mean field approximation this becomes

$$\mathscr{H}_{ex} = -2\sum_i \sum_{j\neq i} \mathscr{J}_{ij}\langle J_z\rangle J_{iz} = -2\mathscr{J}(0)\langle J_z\rangle \sum_i J_{iz}. \tag{7.38}$$

This corresponds to an effective field

$$H_{eff} = \frac{2\mathscr{J}(0)\langle J_z\rangle}{g_J\mu_B}. \tag{7.39}$$

The average value $\langle J_z\rangle$ is therefore given by the relation

$$g_J\mu_B\langle J_z\rangle = \chi_0\left(H + \frac{2\mathscr{J}(0)\langle J_z\rangle}{g_J\mu_B}\right). \tag{7.40}$$

Thus the total susceptibility is

$$\chi = \frac{\chi_0}{1 - [2\mathscr{J}(0)/g_J^2\mu_B^2]\chi_0}. \tag{7.41}$$

The denominator contains an exchange enhancement and predicts long range order when $4\mathscr{J}(0)\alpha^2 = \Delta$. Since $\mathscr{J}(0)$ depends upon the number of neighbors it can be "tuned" by alloying. This is illustrated in Fig. VII.24[26]

[26] E. Bucher, J. P. Maita, and A. S. Cooper, *Phys. Rev. B* **6**, 2709 (1972).

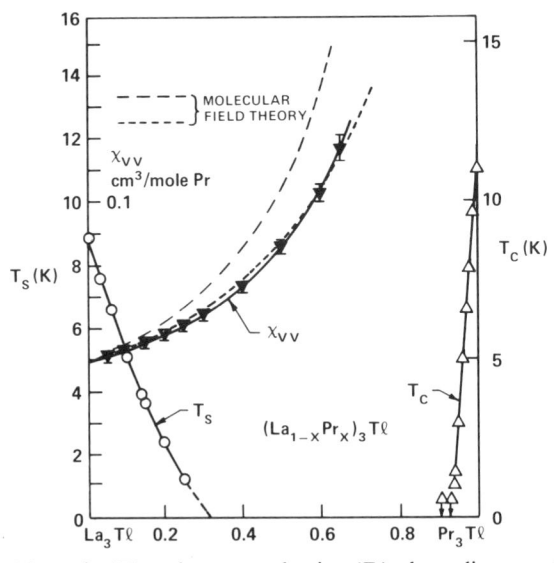

FIG. VII.24. Magnetic (T_c) and superconducting (T_S) phase diagram of $(Pr_xLa_{1-x})_3Tl$. The two dashed curves represent two different fits of Eq. (7.41) to the data. The upper one diverges at the onset of ferromagnetism at $x = 0.93$, but fails to fit the low-concentration data. The lower curve was chosen to fit these data, but it fails to diverge at $x = 0.93$. Thus molecular field theory is not adequate over the whole range of concentrations. After Bucher et al.[26]

where Pr is added to the superconductor La_3Tl. The superconducting transition temperature, denoted T_S, is driven to zero at 30% Pr. The curve for T_S would lie above the Abrikosov–Gor'kov curve in this case since the pair-breaking strength is decreasing with decreasing temperature as the excited crystal field levels are being depopulated. Then at 90% Pr $\mathcal{J}(0)$ becomes large enough to produce ferromagnetism and the Curie temperature, denoted T_c, is driven up with the addition of Pr.

Problem VII.2

Consider a triplet state $|T, (+, 0, -)\rangle$ lying an energy Δ above a ground state singlet. Let $\langle T, \pm |S_z|T, \pm\rangle \equiv \beta$. Calculate the four eigenvalues $\epsilon_i(M)$. Use these to evaluate the free energy $F = kT \ln Z$ where $Z = \Sigma_i[\exp - \beta\epsilon_i(M)]$. Discuss the possibility of obtaining a magnetization *above* some temperature, i.e., a "heat magnetization."

The presence of magnetic impurities not only affects the transition temperature of a superconductor but also the density of states. The density of states is easily obtained from the Green's function,

$$N(\omega) = \frac{N(0)}{2\pi} \int_{-\infty}^{\infty} dE \, \text{Tr}[\text{Im}\mathcal{G}(\mathbf{k}, \omega)] = N(0)\text{Re} \frac{\tilde{\omega}}{(\tilde{\omega}^2 - \tilde{\Delta}^2)^{1/2}}. \quad (7.42)$$

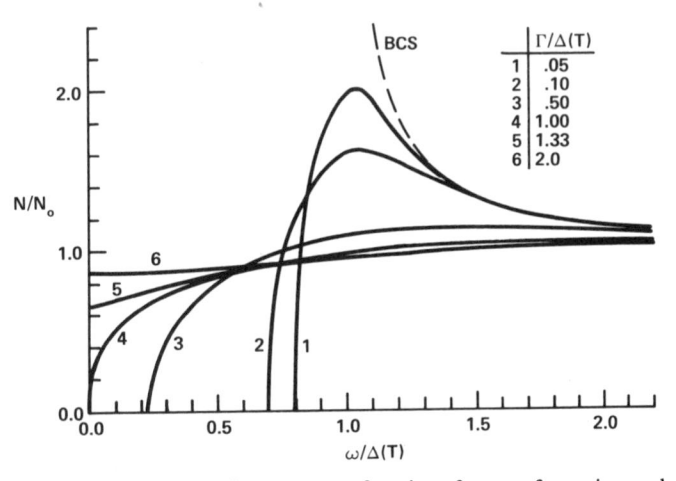

FIG. VII.25.   Density of states in energy as a function of energy for various values of the exchange parameter Γ. After Skalski *et al.*[22]

In the limit $\Gamma \to 0$ this correctly gives the BCS result

$$N(\omega) = \begin{cases} N(0) \dfrac{|\omega|}{(\omega^2 - \Delta^2)^{1/2}} & |\omega| > \Delta \\ 0 & \text{otherwise.} \end{cases} \tag{7.43}$$

The density of states for various values of Γ are shown in Fig. VII.25. The energy Ω_G at which the curve rises from a nonzero value corresponds to half the energy for an excitation (since a *pair* must be excited). Ω_G is therefore the gap parameter. This was also plotted in Fig. VII.19. We notice that the gap parameter goes to zero *before* the order parameter, $\Delta(0)$. The value of Γ at which this occurs in $\Gamma_c(\Omega_G) = 0.91 \ \Gamma_c[\Delta(0)]$. Thus there is a region of concentration in which the material is superconducting but the excitation energy is zero. This is referred to as *gapless superconductivity*. Physically, this means that although there are states in the gap, they are not effective enough to destroy the superconductivity. That is, there is still off-diagonal long range order and only this is required for obtaining infinite conductivity.

C. COEXISTENCE OF SUPERCONDUCTIVITY AND FERROMAGNETISM

Since localized moments are such an anathema to superconductivity one might think their coexistence is completely impossible. However, if we have local moments dilutely distributed in a superconducting host they are coupled through the indirect RKKY exchange. The corresponding Curie temperature, T_c, increases with impurity concentration while the superconducting transition temperature, T_s, decreases. Thus there is, in principle at least, a region in which the two long range order phenomena might coexist. Studies of solutions of Gd in superconducting La as a function of composition[27] show evidence for magnetic ordering in that there is a critical region where T_c actually increases slightly with Gd concentration over a narrow range. Similar effects can also be observed in $NbSe_2$ intercalated with 3d magnetic ions. It is also possible to find superconducting compounds which have complete solid solubility with magnetic compounds and where there appears to be an extensive overlap between the concentrations where superconductivity and magnetism coexist. In such cases there must be a weak interaction (s–d) or (s–f) between the conduction(s) electron and the magnetic d or f electrons. This occurs in the Laves phase compounds, REM_2 which follows to some extent from Laves' observation that near neighbors are always like atoms. For solutions of $CeRu_2$ ($T_c = 6.5$ K) with $GdRu_2$ ($\theta_c = 70$ K) the extrapolations of T_c from the superconducting side intersects the extrapolation of θ_c from the ferromagnetic side at 13% $GdRu_2$ with $T_c = 3.8$ K.[28] Specific heat measurements have been used to prove that the magnetic order persists in the superconducting state and that the superconductivity is a true bulk effect.[29] Small angle neutron scattering results[30] for $Ce_{0.8}Tb_{0.2}Ru_2$ give a magnetic diffraction pattern which increases smoothly through the superconducting transition as expected from the heat capacity. Analysis of these data indicate that the magnetic order is short range, about 25 correlated Tb spins, similar to the range found for the spin-glass Fe–Au.

Superconductivity and magnetic states have recently been found in the ternary compounds with the formula $RERh_4B_4$ and the Chevrel phases discussed in Chapter VI. In these systems, as distinct from the Laves phases alloys, the magnetic order is determined by a periodic structure.

The compound $HoRh_4B_4$ is ferromagnetic ($\theta_c = 6$ K). The adjacent rare earth compound, $ErRh_4B_4$, is superconducting at 8.7 K. The balance between superconductivity and magnetism in the Er compound is so delicate that when it is cooled still further it becomes magnetic and loses its

[27] R. A. Hein, R. L. Falge, Jr., B. T. Matthias, and E. Corenzwit, *Phys Rev. Lett.* **3**, 500 (1959).

[28] B. T. Matthias, H. Suhl, and E. Corenzwit, *Phys. Rev. Lett.* **1**, 449 (1958).

[29] N. E. Phillips and B. T. Matthias, *Phys. Rev.* **121**, 105 (1961).

[30] S. Roth, K. Ibel, and W. Just, *J. Phys. C* **6**, 3465 (1973).

superconductivity at 0.9 K.[31] The compound obeys a Curie–Weiss law above T_c with the full moment of the Er evident, and with a ferromagnetic intercept, $\theta_c \sim 1$ K. The ordering of the Er ions is at a low enough temperature that it may not necessarily take place via the conduction electrons (i.e., by the RKKY mechanism). It could be by superexchange or even direct dipole-dipole interaction. The internal molecular field which results is evidently sufficient to destroy the superconductivity, i.e., $H_{internal} > H_{c2}$, and thus a finite resistivity reappears below 0.9 K as seen in Fig. VII.26.[31]

Similar behavior is found in the Chevrel phases. $Ho_{1.2}Mo_6S_8$ becomes superconducting at 1.2 K and then loses its superconductivity and becomes ferromagnetic *below* 0.65 K.[32] Neutron diffraction studies have shown that in this material,[33] as well as in the $ErRh_4B_4$ discussed above,[34] the ferromagnetic long range order builds up as the superconductivity disappears. In $DyMo_6S_8$ there is evidence[35] that the compound orders antiferromagnetically at 0.4 K *without* destroying the superconducting ordering which sets in at $T_c = 1.7$ K.

Substitutions of rare earth metals for Pb and Sn in the Chevrel phases lead to interesting effects.[36] Replacement of Pb by Gd ($S = \frac{7}{2}$) or Lu ($S = 0$) leads to identical curves of T_c versus composition (and a rather universal dependence upon volume). The indifference of T_c to the spin of the Gd implies a weak exchange interaction with the conduction electrons.

The most dramatic property of the Chevrel phases is their high critical fields. The mean free path is estimated[37] to be about 30 Å, which is of the order of the diameter of the Mo_6 cluster. This suggests that these samples are in the dirty limit making them type II. As we shall discuss in Chapter VIII, when magnetic field lines enter a type II superconductor they do so in units of quantized flux. The value of the field at which the sample finally becomes completely normal is the so-called upper critical field H_{c2}. This

[31] W. A. Fertig, D. C. Johnston, L. E. DeLong, R. W. McCallum, M. B. Maple, and B. T. Matthias, *Phys. Rev. Lett.* **38**, 987 (1977).

[32] M. Ishikawa and Ø. Fischer, *Solid State Commun.* **25**, 37 (1977).

[33] J. W. Lynn, D. E. Moncton, W. Thomlinson, G. Shirane, and R. N. Shelton, *Solid State Commun.*, to be published.

[34] D. E. Moncton, D. B. McWhan, J. Ekert, G. Shirane, and W. Thomlinson, *Phys. Rev. Lett.* **39**, 1164 (1977).

[35] D. E. Moncton, G. Shirane, W. Thomlinson, M. Ishikawa, and Ø. Fischer, *Phys. Rev. Lett.* **41**, 1133 (1978).

[36] M. Sergent, R. Chevrel, C. Rossel, and O. Fischer, *J. Less Common Metals* **58**, 179 (1978).

[37] Ø. Fischer, *Appl. Phys.* **16**, 1 (1978).

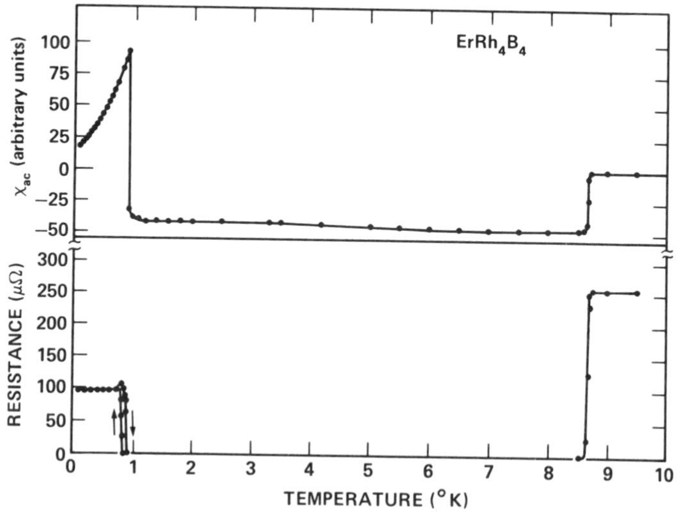

FIG. VII.26. ac magnetic susceptibility χ_{ac} and ac electrical resistivity as a function of temperature for $ErRh_4B_4$ in zero applied field. After Fertig *et al.*[31]

can range up to 600 Kg for $Gd_{0.2}PbMo_6Se_8$. The external field can break up the superconducting pairs by interacting with their orbital or spin moments. If the spin–orbit interaction is large, however, the pairing involves time-reversed states which include both spin components. This weakens the pair breaking associated with spin. The upper critical field associated with the orbital Zeeman effect is derived in Chapter VIII where it is found that $H_{c2} = \Phi_0/2\pi\xi(T)^2$. Using Eq. (2.49) the slope of the H_{c2} curve in the dirty limit is

$$\frac{dH_{c2}}{dT}\bigg|_{H=0} = 4.4 \times 10^4 \, \rho\gamma \qquad (7.44)$$

where the linear heat capacity coefficient γ is in erg/K^2-cm^3, the resistivity in Ω-cm, and the field in Oersteds. Therefore, the high values of H_{c2} are partly due to narrow bands [since $\gamma \sim N(0)$] and partly due to the short mean free paths (since $\rho \sim l^{-1}$).

Fischer *et al.*[38] have found an unusual increase in H_{c2} when some of the Sn^{2+} in $Sn_{1.2}Mo_6S_8$ is replaced with Eu^{2+} ($S = 7.2$) which has the same radius as the Sn^{2+} ion. H_{c2} (2 K) $\sim$ 400 kOe and $T_c = 10.2$ K when half the Sn is replaced by Eu, whereas H_{c2} (2 K) $\sim$ 275 kOe and $T_c = 10.4$ K for the pure Sn compound. The increase in H_{c2} with Eu is presumably due

[38] Ø. Fischer, M. Decroux, S. Roth, R. Chevrel, and M. Sergent, *J. Phys. C* **8**, L474 (1975).

to a *negative* exchange field at the Mo_6S_8 clusters due to the polarization of the conduction electrons. This negative polarization compensates the effect of the applied field thereby increasing H_{c2}. This enhancement of H_{c2}, which was actually predicted earlier,[39] could possibly occur in other compounds but is overwhelmed by the reduction in T_c due to exchange scattering from the localized moment. Mossbauer measurements on the ^{151}Eu nuclei and NMR measurements on the ^{95}Mo nuclei support this interpretation of the H_{c2} results.[40] The Mossbauer results on ^{151}Eu indicate a low s-electron density at the Eu site (from the isomer shift) and a small positive conduction electron polarization (from the hyperfine field). The low density of s-electrons is also confirmed by the small magnitude of T_1T [see Eq. (5.25)] obtained from the Eu NMR.

This magnetic behavior of the RE Chevrel phases is consistent with a simple two-band model. The superconducting properties derive from the 4d-band of the Mo ions hybridized to some extent with the chalcogenide.[41] When 4f paramagnetic ions are added, they couple weakly with the s-band conduction electrons. The resulting polarization of the s-band when the Eu^{2+} moments are aligned in an external field is parallel at the Eu site, but has a spatial dependence (as expected from the oscillating nature of the RKKY interaction) such that it is negative at the Mo site and remains so below T_c. The interaction between these s-electrons and the superconducting d-electrons is presumably too weak to produce pair breaking, and thus little dependence of T_c upon the rare earth.

D. Coexistence of Superconductivity and Charge Density Waves

Superconductivity and charge density waves can coexist as in $2H\text{-}TaS_2$ and $2H\text{-}NbSe_2$. However, since both phenomena compete for those portions of the Fermi surface with a high density of states, the appearance of a CDW tends to suppress superconductivity. If $2H\text{-}TaS_2$, for example, is intercalated with several different organic molecules the CDW is suppressed (i.e., either its amplitude or its onset temperature, or both) while T_c increases. Measurement of the magnetic susceptibility shows that the

[39] V. Jaccarino and M. Peter, *Phys. Rev. Lett.* **9**, 290 (1962).

[40] F. Y. Fradin, G. K. Shenoy, B. D. Dunlap, A. T. Aldred, and C. W. Kimball, *Phys. Rev. Lett.* **38**, 719 (1977).

[41] Isotope effect measurements [F. J. Culetto and F. Pobell, *Phys. Rev. Lett.* **40**, 1104 (1978)] on the binary Chevrel compound Mo_6Se_8 indicate that the six Mo and eight Se atoms contribute about equally to the phonon modes which govern its superconductivity.

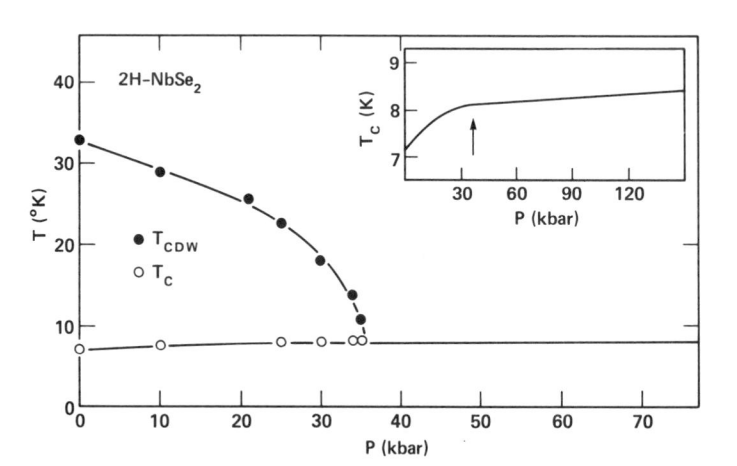

FIG. VII.27.   Phase diagram of the charge density wave state and superconductivity in 2H-NbSe$_2$. After Berthier *et al.*[23]

density of states does indeed increase with intercalation. The CDW may also be suppressed by pressure as shown in Fig. VII.27.[42] As the insert shows the superconducting transition temperature increases as the CDW disappears.

[42] C. Berthier, P. Molinié, and D. Jérome, *Solid State Commun.* **18,** 1393 (1976).

VIII. Domain Structures

Up to this point our discussions have dealt with homogeneous ground states. We have considered spatial variations in the order parameter only as fluctuations. There exist, however, equilibrium configurations in which the order parameter varies in space to give a domain structure. Magnetic domains are generally associated with different *orientations* of the order parameter as illustrated in Fig. VIII.1.[1] However, in Fig. VIII.2[2] we show domains in which the *nature* of the order parameter varies from antiferromagnetic to ferromagnetic. Superconductors also exhibit domain structure as illustrated in Fig. VIII.3.[3,4] In this case the domains correspond to regions in which the magnitude of the order parameter is finite or zero. In a magnetic system of the type illustrated in Fig. VIII.1 domains exist from zero magnetic field up to the saturation field. In a type I superconductor or an antiferromagnet the domains exist in an intermediate region of field strengths that depend upon the geometry of the sample.

The formation of domains is governed by the free energy of the system. Normally, it costs energy to create a domain wall. This leads to an optimum domain size. However, as we saw in Chapter II, a type II superconductor is characterized by a *negative* surface energy between normal and superconducting regions. This leads to a continual subdivision of domains limited only by the quantization of flux.

A. MAGNETIC DOMAINS

The first direct observation of ferromagnetic domains was made by Bitter in 1931. He applied a drop of a ferromagnetic colloidal suspension, i.e., a ferrofluid, to the surface of a ferromagnet. The resulting patterns reflect the domain structure and are known as "Bitter patterns." The first attempt to theoretically calculate a domain structure was by Landau and Lifshitz in 1935.

[1] A. H. Bobeck and E. Della Torre, "Magnetic Bubbles." North-Holland Publ., Amsterdam, 1975.

[2] A. R. King and D. Paquette, *Phys. Rev. Lett.* **30**, 662 (1973).

[3] T. Chen and G. T. McKinley, *IEEE Trans. Magn.* **mag-13**, 1580 (1977).

[4] T. Faber, *Proc. R. Soc. London, Ser. A* **248**, 460 (1958).

FIG. VIII.1. Domain structures in an YGdTm epitaxial garnet film, 6 μm thick, showing the appearance of bubble domains: (a) unmagnetized; (b) 80 Oe; (c) 100 Oe; (d) 110 Oe. After Bobeck and Della Torre.[1]

The fact that a domain structure can lower the total energy is easily seen by calculating the energy associated with various configurations. Consider first a uniformly magnetized film of thickness l as shown in Fig. VIII.4a. The energy associated with the magnetostatic fields can be expressed in terms of the surface magnetic pole density or, equivalently, the demagnetization field arising from these poles. The demagnetizing field in this case is $H_D = -4\pi M$. Therefore the magnetostatic energy per unit

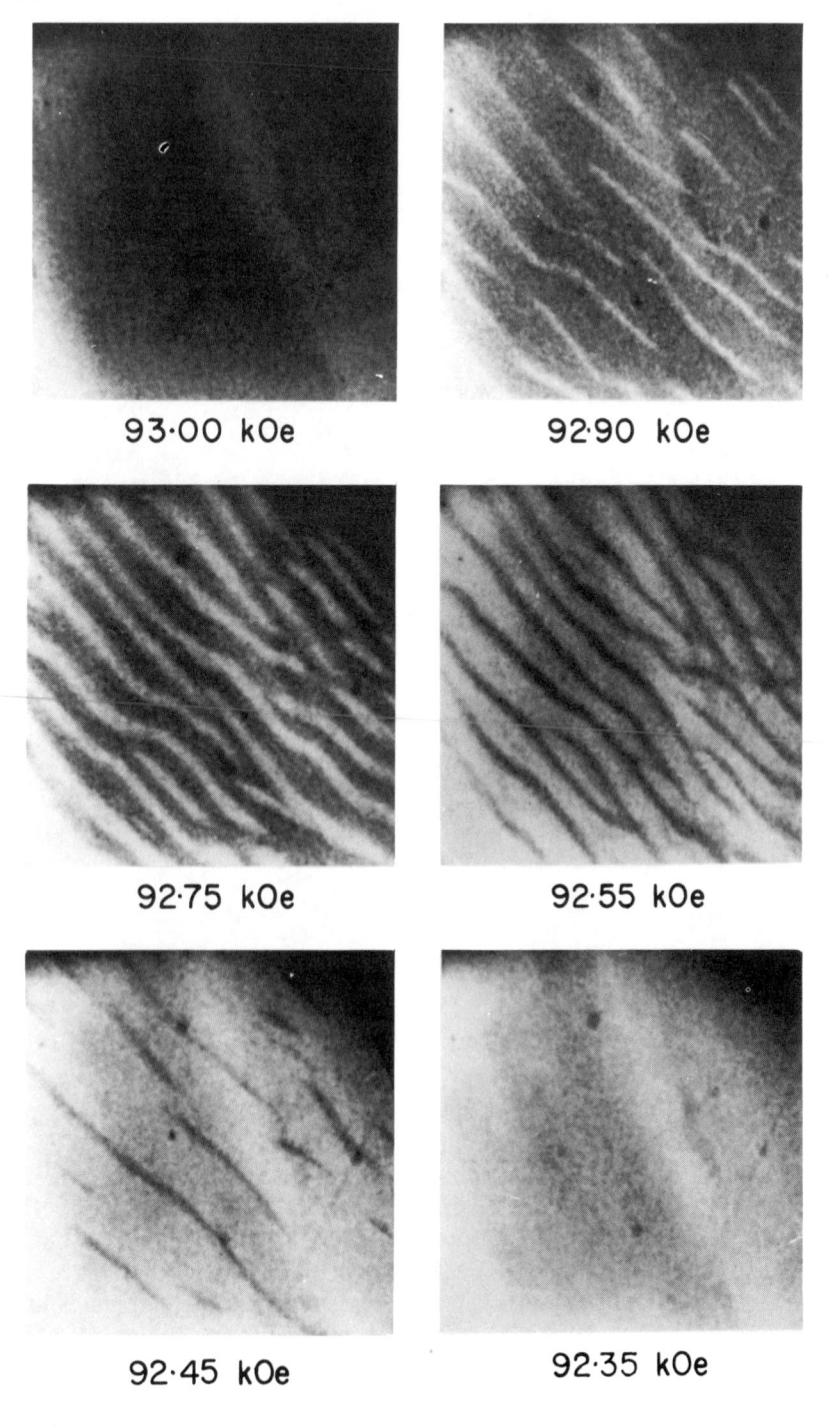

93·00 kOe

92·90 kOe

92·75 kOe

92·55 kOe

92·45 kOe

92·35 kOe

 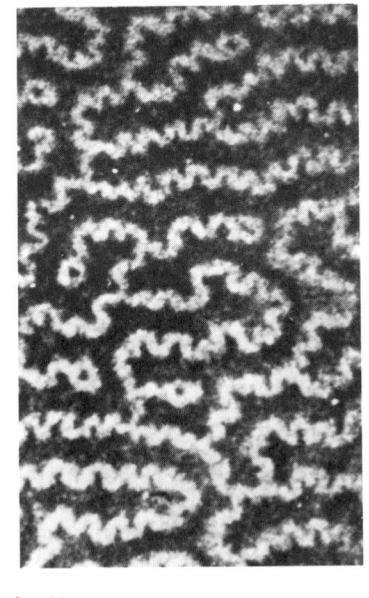

Fig. VIII.3.   (a) Magnetic domain patterns obtained by the polar Kerr effect for MnBi. The sample contains a crack which angles off into the medium towards the bottom of the photo. Thus, the "gear wheel" patterns are seen in the thick regions and evolve into "rickrack" patterns in the thin ledge. The magnification is 400× (after Chen and McKinley[3]). (b) Intermediate state normal-superconducting domains in Al at $H = 0.27\ H_c$. Magnification is about 3×. After Faber[4].

area of surface is

$$\epsilon_m = -\frac{1}{2A} \int d^3r \mathbf{M} \cdot \mathbf{H}_D = 2\pi M^2 l. \tag{8.1}$$

Kittel[5] has calculated the corresponding energies for the configurations shown in Figs. VIII.4b and c.

$$\epsilon_m(\text{stripes}) = 1.71\ M^2 d \tag{8.2}$$

$$\epsilon_m(\text{checkerboard}) = 1.06\ M^2 d. \tag{8.3}$$

[5] C. Kittel, *Rev. Mod. Phys.* **21**, 541 (1949).

Fig. VIII.2.   Photographs of domain structure in the vicinity of the antiferromagnet-to spin-flop transition in MnF_2. Contrast is achieved by illuminating the sample with monochromatic light of wavelength 3560 Å which corresponds to an optical absorption in the spin-flop state which is shifted by 4 Å from its position in the antiferromagnet state. After King and Paquette[2].

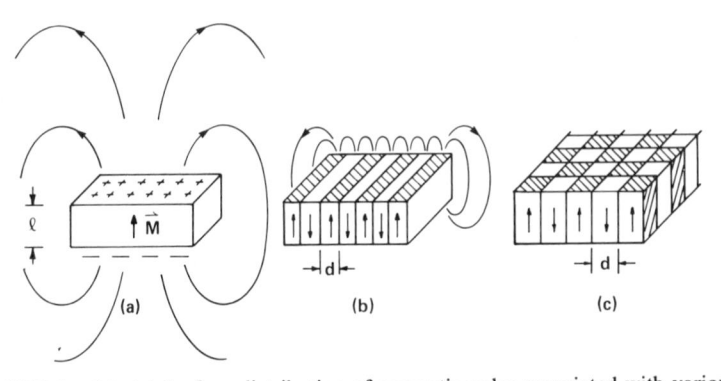

F_{IG}. VIII.4. (a)–(c) Surface distribution of magnetic poles associated with various domain configurations.

To these energies we must add the wall energies. If σ_w is the surface energy associated with a wall, then the total wall energies per unit (sample) surface area for these configurations are

$$\epsilon_w(\text{stripes}) = \sigma_w l/d \qquad (8.4)$$

$$\epsilon_w(\text{checkerboard}) = 2\sigma_w l/d. \qquad (8.5)$$

Adding this wall energy to the magnetostatic energy and minimizing with respect to the domain size d, leads to a minimum total energy of

$$\epsilon_{\min}(\text{stripes}) = 2.62\ M\sqrt{\sigma_w l} \qquad (8.6)$$

$$\epsilon_{\min}(\text{checkerboard}) = 2.91\ M\sqrt{\sigma_w l}. \qquad (8.7)$$

Thus we see that the striped domains are preferred to the checkerboard. Furthermore, the striped domain configuration will have a lower energy than the single domain if

$$\lambda \equiv \frac{\sigma_w}{4\pi M^2} < \frac{5.75}{4\pi}\ l. \qquad (8.8)$$

However, nature has an inexhaustible number of configurations available as is obvious from Fig. VIII.1. Thus, we can never hope to predict a particular configuration, merely make its appearance plausible.

We shall see later that the material length λ defined by Eq. (8.8) is an important parameter in determining certain device performance.

Domains also exist in antiferromagnets. Since there is no magnetostatic energy their origin is quite different. It is generally believed that domain boundaries occur at dislocations in the crystal and are only of the order of a few lattice spacings thick. Figure VIII.5 illustrates antiferromagnetic domains.

FIG. VIII.5. Illustration of domains in an antiferromagnet. The solid squares and circles refer to chemically inequivalent sites. In MnF_2, for example, there are two inequivalent Mn^{2+} sites related by a nonprimitive translation plus a 90° rotation.

B. MAGNETIC DOMAIN WALLS

Bloch was the first to recognize, in 1932, that a domain wall is not sharp, but must have a finite width in order to minimize the exchange energy. In this section we shall consider the origin of the wall energy, σ_w, introduced in the previous section. There are various types of walls depending upon the relative orientation of the magnetization on either side and the manner in which the magnetization rotates through the wall. Two examples are illustrated in Fig. VIII.6 along with the spherical coordinate system we shall use to specify the direction of the magnetization, or order parameter.

1. Wall Structure

The nature of the wall is governed by a Ginzburg–Landau-like free energy. We shall assume that the dominant energies in the system are a uniaxial anisotropy and the exchange. The Néel wall will also involve *vol-*

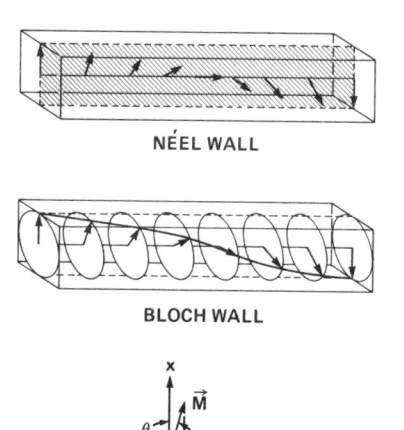

NÉEL WALL

BLOCH WALL

FIG. VIII.6. Illustration of a Néel and a Bloch wall. In the static Néel wall the magnetization rotates in the x-z plane while in the Bloch wall it rotates in the x-y plane.

ume magnetostatic energy since $\nabla \cdot \mathbf{M} \neq 0$ for this configuration. However, for the present let us neglect this aspect.

The free energy $\sigma(z)$ at some point z is then a functional of $\theta(z)$ and $\phi(z)$. For an easy axis parallel to the x-axis,

$$\sigma(z) = K \sin^2 \theta + A \left[\left(\frac{\partial \theta}{\partial z} \right)^2 + \sin^2 \theta \left(\frac{\partial \phi}{\partial z} \right)^2 \right] \qquad (8.9)$$

where K and A are the anisotropy and exchange parameters, respectively. Requiring that the total wall energy per unit area,

$$\sigma_w = \int \sigma(z) dz, \qquad (8.10)$$

be stationary leads to the Euler equations

$$\frac{\delta \sigma_w}{\delta \theta} = \frac{\partial \sigma}{\partial \theta} - \frac{d}{dz} \left(\frac{\partial \sigma}{\partial (\partial \theta / \partial z)} \right) = K \sin 2\theta + A \sin 2\theta \left(\frac{\partial \phi}{\partial z} \right)^2$$

$$- 2A \left(\frac{\partial^2 \theta}{\partial z^2} \right) = 0 \quad (8.11)$$

$$\frac{\delta \sigma_w}{\delta \phi} = \frac{\partial \sigma}{\partial \phi} - \frac{d}{dz} \left(\frac{\partial \sigma}{\partial (\partial \phi / \partial z)} \right) = - 2A \frac{d}{dz} \left(\sin^2 \theta \frac{\partial \phi}{\partial z} \right) = 0. \quad (8.12)$$

The second equation, Eq. (8.12), requires that $\sin^2 \partial(\partial \phi / \partial z)$ be a constant. Clearly one way this can be satisfied is if ϕ itself is constant. In the case of the Néel wall shown in Fig. VIII.6 this would be $\phi = \pi/2$ while for the Bloch wall $\phi = 0$. Later when we consider *moving* walls we shall see that ϕ will depend on velocity. With $\phi = $ constant, Eq. (8.11) reduces to

$$\frac{\partial^2 \theta}{\partial z^2} = \frac{K}{2A} \sin 2\theta. \qquad (8.13)$$

Multiplying both sides of $\partial \theta / \partial z$, integrating over z from $-\infty$ to z, and setting $\theta(-\infty) = \partial \theta / \partial z|_{-\infty} = 0$, gives

$$\frac{\partial \theta}{\partial z} = \sqrt{\frac{K}{A}} \sin \theta. \qquad (8.14)$$

Integrating once more gives the wall structure,

$$\theta(z) = 2 \tan^{-1}(\exp (\sqrt{K/A}z)). \qquad (8.15)$$

If we define the thickness of the wall, δ, as the length between $\theta = \pi/4$ and $3\pi/4$, then $\delta = 0.76 \sqrt{A/K}$. For anisotropy and exchange parameters equal to those of iron we find $\delta \approx 500$ Å.

2. Wall Energy

The wall energy is easily calculated by integrating Eq. (8.10) from $-\infty$ to $+\infty$ with the help of Eqs. (8.9) and (8.14):

$$\sigma_W = \int_{-\infty}^{\infty} \sigma(z)dz = \int_0^{\pi} \sigma(\theta) \sqrt{\frac{A}{K}} \frac{d\theta}{\sin\theta} = 4\sqrt{AK}. \qquad (8.16)$$

Thus the wall energy is the geometric average of the exchange and the anisotropy. A typical value would be 1 erg/cm².

Problem VIII.1
Calculate the energy of the Néel wall including its volume magnetostatic energy.

When surface magnetostatic energy is included the geometry of the sample becomes important in establishing the type of wall that will develop. This may be seen by considering a thin film of some material such as permalloy in which the magnetization prefers to lie in the plane of the film. This is the usual configuration encountered in magnetic tape or disc recording. A Bloch and a Néel wall would then appear as illustrated in Fig.VIII.7.[6] Notice that the Bloch wall will induce magnetic "charge density" on the surface of the film. A Néel wall avoids this magnetostatic energy. In 1958 Huber, Smith, and Goodenough[7] also suggested another

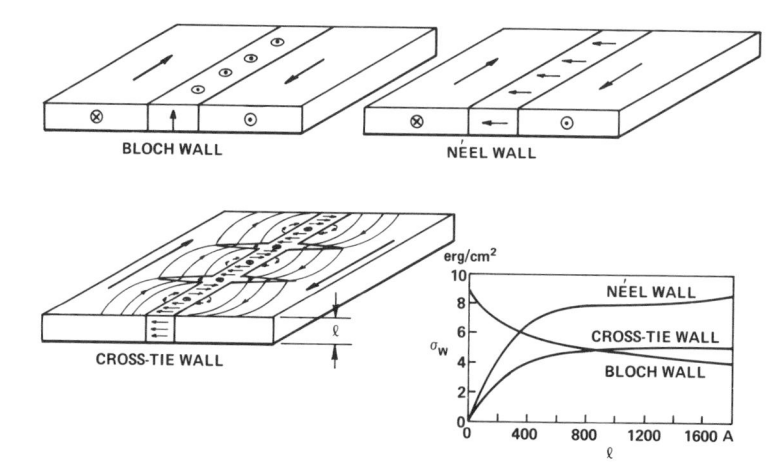

FIG. VIII.7. Surface energy of three types of walls as functions of film thickness using the parameters $A = 10^{-6}$ erg/cm, $K = 1000$ erg/cm³, $M_s = 800$ G. After Middelhoek.[6]

[6] S. Middelhoek, *J. Appl. Phys.* **34**, 1054 (1973).
[7] E. E. Huber, D. O. Smith, and J. B. Goodenough, *J. Appl. Phys.* **29**, 294 (1958).

type of wall called a *cross-tie* wall. In this wall the polarity alternates so as to reduce the magnetostatic energy of the wall. Cross-ties form to smooth the flux discontinuities associated with these polarity changes as illustrated in Fig. VIII.7.[6] The prevailing wall depends upon the thickness of the film as also shown in Fig. VIII.7. Finally, we note that it is possible for a Bloch wall to be left- or right-handed. When a left- and right-handed Bloch wall meet they generate a *Bloch line,* or Néel segment, as illustrated in Fig. VIII.8. If the Néel segment also contains two polarities one can obtain a magnetic singularity called a *Bloch point* also shown in Fig. VIII.8.

3. Domain Wall Pinning

Let us now add an external field parallel to the easy axis. The region which is favorably oriented with respect to this field will grow at the expense of the unfavorably oriented region. The ease with which the domain wall moves determines the *coercive force* of the material. The coercive force ranges over seven orders of magnitude from 0.002 Oe in supermalloy to 10,000 Oe in $SmCo_5$.

The anisotropies of these materials are 1.5×10^3 ergs/cm^3 and 1.5×10^8 ergs/cm^3, respectively, suggesting that the anisotropy plays an important role in determining the coercive force. It is also known that the coercive force may be increased by using very fine particles or by the per-

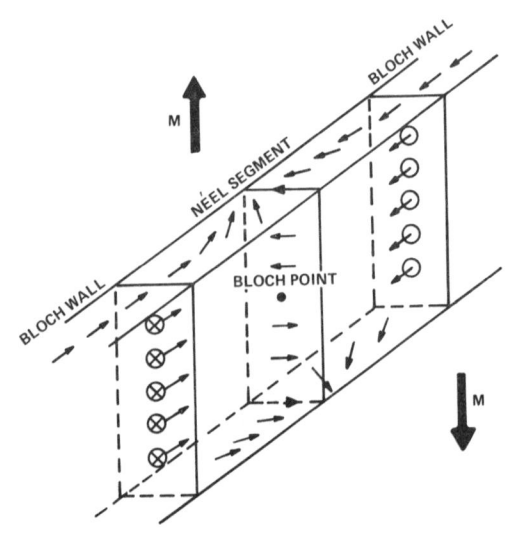

FIG. VIII.8.   Spin configuration within a Bloch line or Néel segment which contains a Bloch point.

cipitation of a second phase. Such heterogeneities introduce spatially dependent free energy minima which pin domain walls.

A model for such pinning has been suggested by Friedberg and Paul.[8] They represent the defect, such as a grain boundary, as a region of width w in which the saturation magnetization, the exchange and the anisotropy parameters have different values from those of the homogeneous material. This model is illustrated in Fig. VIII.9. Within each of the three regions indicated in Fig. VIII.9, the energy density is given by Eq. (8.9) plus a Zeeman term, $-M_i H \cos \theta$, where i specifies the region. Thus we obtain three Euler equations analogous to Eq. (8.11). Continuity of θ and $A_i \partial\theta/\partial z$ at the two interfaces gives two additional equations. The coercive force is defined as the maximum external field for which a solution to this problem exists. In the case where $h \equiv M_2 H_c / K_2$ is small, Friedberg and Paul obtain a coercive force

$$H_c = \frac{1}{\sqrt{27}} \frac{2K_1}{M_1} \frac{w}{\delta_1} \left(\frac{A_1}{A_2} - \frac{K_1}{K_2} \right). \tag{8.17}$$

In Table VIII.1 we compare the resulting calculated coercive forces with those observed. We see that this model works very well.

The large range in coercivity makes possible a variety of applications. In materials with a low coercivity, referred to as "soft", the magnetization is easily reversed. Such materials are used for transformers, inductors, microwave components, and computer memory cores. Nickel-iron alloys, with small additions of Cu and Mo called *permalloys,* have, in addition to low coercivities, very large permeabilities. Amorphous me-

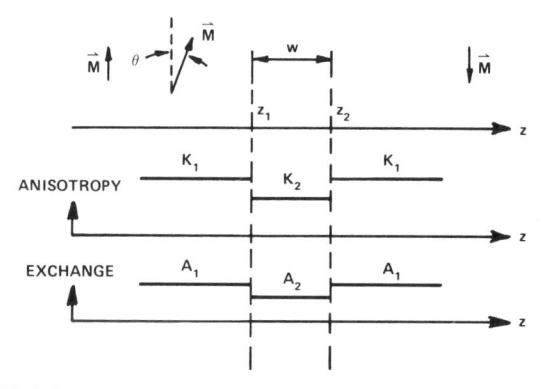

FIG. VIII.9.   Model used by Friedberg and Paul[8] to calculate the coercivity associated with a defect of width w.

[8] R. Friedberg and D. I. Paul, *Phys. Rev. Lett.* **34,** 1234 (1975).

TABLE VIII.1[a]

VALUES OF THE COERCIVE FORCE DUE TO GRAIN BOUNDARIES

Material	M_1 (magnetization)	K_1 (anisotropy) (ergs/cm³ × 10⁻⁵)	A_1 (exchange) (ergs/cm × 10⁶)	2δ (domain wall width) (cm × 10⁶)	h (expansion parameter)	H_C (coercive force) (Oersteds)	Observed coercive force
Supermalloy[b]	630.0	0.015	1.5	64.0	0.0002	0.0004	0.002
Permalloy[b]	860.0	0.02	2.0	64.0	0.0002	0.0006	0.5
Iron-Si13%[c]	1590.0	3.7	2.2	4.8	0.0009	0.2	0.1
Iron-Si14%	1570.0	3.2	2.1	5.2	0.003	0.5	0.5
Nickel	485.0	0.7	0.5	5.2	0.003	0.3	0.7
Iron	1707.0	4.8	2.4	4.4	0.003	0.7	1.0
Cobalt	1400.0	45.0	4.7	2.0	0.007	20.0	10.0
Alnico[d]	915.0	260.0	2.0	0.56	0.024	600.0	600.0
Co₅Sm	800.0	1500.0	2.0	0.24	0.056	9000.0	10000.0

[a] Values used are $w = 6 \times 10^{-8}$ cm, $A_1/A_2 = 1.1$, $K_2/K_1 = 0.85$, and $M_1/M_2 = 1$.

[b] For these materials, the coercive force may be dominated by magnetostatic effects. In particular, for permalloy, the rapid quenching of the disordered state should produce high stress fields.

[c] This material is grain oriented. Therefore, Friedberg and Paul use $w = 2 \times 10^{-8}$ cm.

[d] Modern theory suggests a spin-rotation mechanism rather than domain-wall motion for alnico. Friedberg and Paul have used an effective anisotropy taken from theoretical estimates of the intrinsic coercive force.

tallic alloys, which we shall discuss in Chapter IX, also have very low coercivities.

This, of course, is not the only origin of coercivity. The theory of the shape of a hysteresis loop, in general, is still a very active area of research. A situation of particular interest is a magnet consisting of small, single-domain particles. Stoner and Wohlfarth[8a] calculated the hysteresis curves associated with ellipsoidal particles oriented at random. They assumed that the particles were noninteracting and that the magnetization within each particle rotates coherently. The coercive force was found to be of the order of the anisotropy field which, in this case is due to the anisotropy in the demagnetization factors, i.e., $H_c \sim (N_\perp - N_\parallel)M$. More recent calculations have attempted to include the magnetic interactions among such particles.

Large coercivities are desirable for permanent magnets in order to sustain the induction in the presence of an air gap with its demagnetizing field. Since one also wants a large remanent induction, B_r, the figure of merit for a permanent magnet is the energy product, $B_r H_c$, i.e., the area under the hysteresis curve. In order to maximize H_c most permanent magnets consist of small particles. *Alnico,* for example, consists of rodlike precipitates of iron imbedded in a matrix of Ni and Al. The highest energy product is obtained from compounds of cobalt and rare earths, particularly samarium. The elements are melted together, then ground into fine particles of the order of a few microns in diameter, and finally densified by sintering.

4. Domain Wall Propagation

Let us now assume that the applied field exceeds the coercive force and consider how the wall propagates. The order parameter corresponds to the magnetization, whose equation of motion is the torque equation

$$\tau = \frac{d\mathbf{L}}{dt} = -\frac{1}{|\gamma|}\frac{d\mathbf{M}}{dt} = \mathbf{M} \times \mathbf{H} \tag{8.18}$$

where $\mathbf{H}(z) = \delta\sigma_w(z)/\delta\mathbf{M}$ which is essentially the *time-dependent Ginzburg–Landau equation* we discussed in Chapter II. Instead of adding a random noise source we add a phenomenological relaxation term proportional to $\dot{\mathbf{M}}$ (Gilbert damping),

$$\frac{d\mathbf{M}}{dt}\bigg|_{\text{relaxation}} = -\frac{\alpha}{|\mathbf{M}|}\mathbf{M} \times \frac{d\mathbf{M}}{dt}. \tag{8.19}$$

[8a] E. C. Stoner and E. P. Wohlfarth, *Phil. Trans. Roy. Soc. (London)* A **240**, 599 (1948).

The damping parameter, α, is related to the ferromagnetic resonance line-width, ΔH, by $\alpha = |\gamma| \Delta H / 2\omega_0$. This form insures that $\mathbf{M}$ relaxes to its equilibrium direction while preserving its magnitude. The equations of motion then become

$$M \frac{d\theta}{dt} = \alpha M \sin \theta \frac{d\phi}{dt} - \frac{|\gamma|}{\sin \theta} \frac{\delta \sigma_w}{\delta \phi} \tag{8.20}$$

$$M \sin \theta \frac{d\phi}{dt} = -\alpha M \frac{d\theta}{dt} + |\gamma| \frac{\delta \sigma_w}{\delta \theta}. \tag{8.21}$$

It is interesting to consider lossless propagation first. In this case Eq. (8.20) together with Eq. (8.12) shows that a time-dependent θ necessarily leads to a variation in ϕ. We shall also see that the associated volume mag-netostatic energy plays an important role. This contribution to $\sigma(z)$ is $2\pi M_z(z)^2$. Writing $2\pi M^2$ as D to distinguish this contribution and assum-ing ϕ is independent of z and remains small, Eqs. (8.20) and (8.21) become

$$\dot{\theta} \simeq -\frac{|\gamma|}{M \sin \theta} (2D\phi \sin^2 \theta) \tag{8.22}$$

$$\dot{\phi} = \frac{|\gamma|}{M \sin \theta} \left(K \sin 2\theta - 2A \frac{\partial^2 \theta}{\partial z^2} \right). \tag{8.23}$$

Combining these we obtain

$$\frac{\partial^2 \theta}{\partial z^2} - \frac{1}{c^2} \frac{\partial^2 \theta}{\partial t^2} = \frac{K}{2A} \sin 2\theta \tag{8.24}$$

where

$$c = \frac{2|\gamma|}{M} \sqrt{AD} = 2|\gamma| \sqrt{2\pi A}. \tag{8.25}$$

This is the same sine-Gordon equation we encountered in Chapter V in connection with vortex motion in Josephson junctions. The solution is a *soliton,* or localized traveling wave, of the form

$$\theta(z, t) = \sin^{-1} \left(\mathrm{sech} \sqrt{\frac{K}{A}} \frac{z - vt}{[1 - (v/c)^2]^{1/2}} \right). \tag{8.26}$$

This reduces to Eq. (8.15) when $v = 0$.

If we now add damping and a Zeeman "driving" term Eq. (8.24) be-comes

$$\frac{d^2 \theta}{dz^2} - \frac{1}{c^2} \frac{d^2 \theta}{dt^2} + \frac{\alpha M}{2|\gamma| A} \frac{d\theta}{dt} = \frac{K}{2A} \sin 2\theta + \frac{MH \sin \theta}{2A}. \tag{8.27}$$

It turns out that Eq. (8.26) is *also* a solution to this equation, but with a unique velocity,

$$v = \frac{|\gamma|}{\alpha} \sqrt{\frac{A}{K}} H. \tag{8.28}$$

The coefficient of the field defines the domain wall mobility.

Notice that we have had to make several approximations in reducing the general equations (8.20) and (8.21) to an equation with a soliton solution. Walker[9] has shown that these general equations do in fact possess a steady state propagating solution with $\phi = $ const. as long as $v < 2\pi|\gamma|M\sqrt{A/K}$ which is of the order of 5000 cm/sec for GaYIG. For velocities exceeding this value the walls develop complex internal motions. In thin films the surface demagnetizing fields lead to an even lower critical velocity. This imposes a limit on the rate at which domains can be manipulated.

Problem VIII.2

Add an "out-of-plane" anisotropy energy of the form $K\Delta \sin^2 \phi \sin^2 \theta$ to Eq. (8.9). Determine the critical velocity below which the general equations (8.20) and (8.21) will possess steady-state propagating solutions with $\phi = $ const. At what angle ϕ does the instability occur?

C. MAGNETIC BUBBLES

In Fig. VIII.1 we see that as the field increases some strip domains contract into cylindrical domains, or "bubbles." As the field increases further these bubbles collapse. The critical field for bubble stability is plotted in Fig. VIII.10a[10] as a function of the film thickness in units of the material length we defined above. The diameters of stable bubbles are plotted in Fig. VIII.10b also as a function of film thickness. We notice that there is a minimum bubble size of 3.9 λ when the thickness is 3.3 λ.

Bubbles are obviously "bits" in the digital information sense. In 1967 Bobeck[11] suggested that bubbles would be manipulated to code information. Bubbles may be moved by driving them with a field gradient which is generated by a permalloy overlay. Bubble propagation is illustrated in Fig. VIII.11.[12]

---

[9] Walker's analysis has never been published. But it is summarized by J. F. Dillon, *in* "Magnetism" (G. Rado and H. Suhl, eds.), Vol. 3, p. 450. Academic Press, New York, 1963.

[10] A. A. Thiele, *J. Appl. Phys.* **41**, 1139 (1970).

[11] A. H. Bobeck, *Bell Syst. Tech. J.* **46**, 1901 (1967).

[12] A. J. Perneski, *IEEE Trans. Magn.* **mag-5**, 554 (1969).

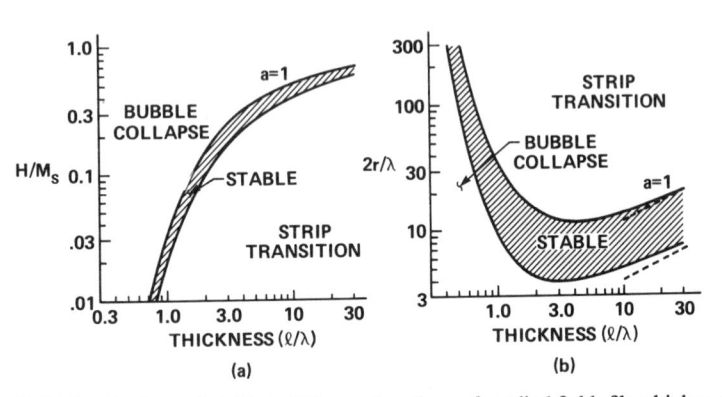

FIG. VIII.10.   Regions of stable bubbles as functions of applied field, film thickness l and the material parameter λ defined in Eq. (8.8). a is the aspect ratio defined by $a = 2r/l$ where r is the bubble radius. After Thiele.[10]

One of the limiting factors in achieving high bubble densities for memory applications lies in the photolithographic process for generating the permalloy overlays. One way to avoid this is to establish a uniform lattice of bubbles. In such a "lattice file" the information is stored in the *wall* of the bubbles through the number of Bloch lines.

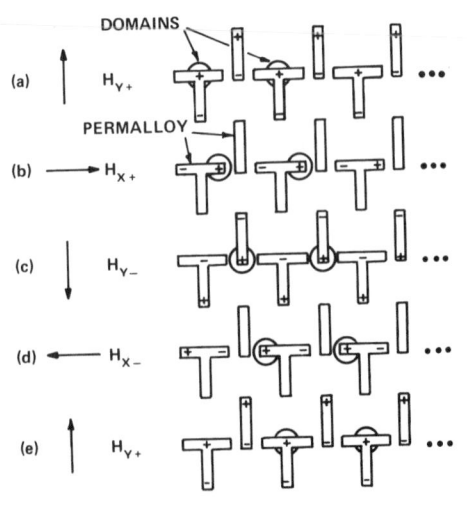

FIG. VIII.11.   Illustration of T-bar propagation. Cylindrical domains are attracted to $+$ magnetic poles that appear when the in-plane field is directed along a long dimension of a T or T bar. As the field sequences, or rotates clockwise, $+$ poles always appear immediately to the right of the domains, causing them to propagate toward the right. In this way, information may be shifted in and out of this "register." After Perneski.[12]

D. The Magneto-Optical Probe

The most practical method of studying magnetic domain structures utilizes their interaction with polarized light. In this section we shall review
magneto-optical effects. In Fig. VIII.12 we indicate the various
magneto-optical configurations. Faraday effects generally refer to nonfer-
romagnetic materials where the rotation is proportional to the applied
field. However, the rotation associated with transmission through thin
magnetic films is often referred to as Faraday rotation. Magneto-optical
Kerr effects (MOKE) refer to reflection from ferromagnetic materials.

In the transverse Kerr configuration if the electric vector of the light is
perpendicular to the plane of incidence (usually denoted E_s, where s
stands for *senkrect,* the German word for *perpendicular*) the reflectivity is
only slightly affected by the magnetization. However, since the reflection
process itself has a sense of rotation associated with it, the reflectivity of

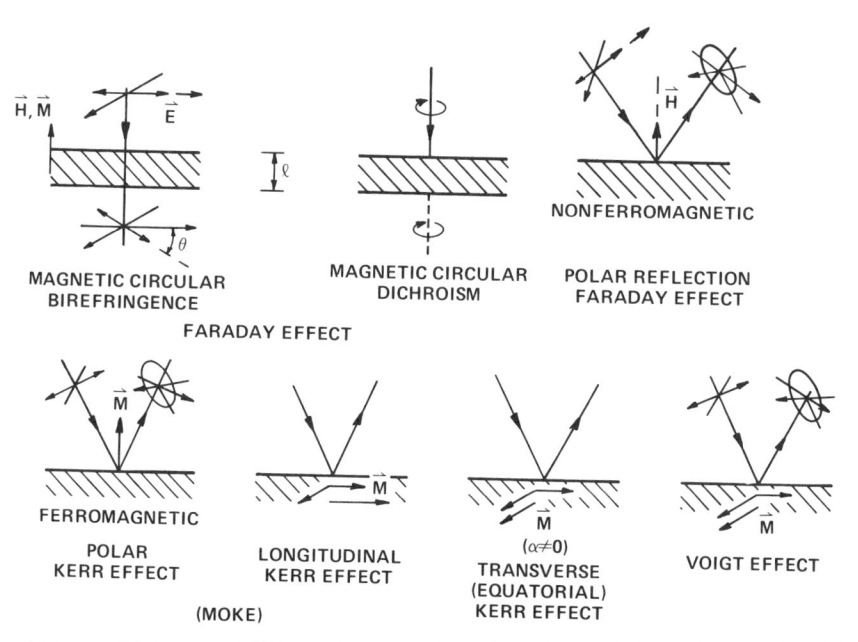

Fig. VIII.12. Summary of the various magneto-optical effects. In the *polar* Kerr effect M
is normal to the reflecting surface. If the incident light is linearly polarized and normal to the
surface the reflected light will be elliptically polarized with the axis of the ellipse rotated in a
sense determined by the direction of **M**. In the *longitudinal* Kerr effect M is parallel to the
surface and also lies in the plane of incidence. In the *transverse* Kerr effect M is parallel to
the surface but perpendicular to the plane of incidence. The configuration for the Voigt effect is the same as that for the transverse Kerr effect.

light polarized in the plane of incidence depends linearly upon M. Thus, if unpolarized light is incident on an absorbing medium with $\mathbf{M}$ perpendicular to the plane of incidence the reflected power will depend upon the direction of $\mathbf{M}$. This makes it possible to observe domain structures *without* polarizers. This effect can only be observed if the medium is absorbing. A nonabsorbing medium in the same transverse Kerr configuration will convert linearly polarized light into elliptically polarized light by an amount proportional to M^2. This magnetic birefringence is known as the Voigt effect.

1. Faraday Rotation

Let us begin by relating the Faraday rotation to the optical conductivity. Landau and Lifshitz[13] have emphasized that at wavelengths small compared with sample dimensions there is no physical meaning to the magnetization M. The point is that when we average over the microscopic Maxwell equations there are, in addition to any free currents, two "media" contributions to the average current,

$$\langle \mathbf{j} \rangle = c\nabla \times \mathbf{M} + \partial \mathbf{P}/\partial t. \tag{8.29}$$

Here $\mathbf{M}$ is defined by $(\mathbf{B} - \mathbf{H})/4\pi$ where $\mathbf{B} = \langle \mathbf{h} \rangle$ and $\mathbf{H}$ is given by $\nabla \times \mathbf{H} = (1/c)\partial \mathbf{P}/\partial t$. The magnetic moment is defined by

$$\frac{1}{2c} \int \mathbf{r} \times \langle \mathbf{j} \rangle \, dr = \int \mathbf{M} \, d^3r + \frac{1}{2c} \int \mathbf{r} \times \frac{\partial \mathbf{P}}{\partial t} \, d^3r. \tag{8.30}$$

If $\mathbf{M}$ is to be associated with the magnetization, $\partial \mathbf{P}/\partial t$ must be negligible compared with $c\nabla \times \mathbf{M}$. Since $\nabla \times \mathbf{E} = (1/c)(\partial \mathbf{B}/\partial t)$, $E/l \sim \omega B/c$ where l is the dimension of the sample. Taking $\epsilon - 1 \sim 1$ and $\mu \sim 1$, $\partial P/\partial t \sim \omega E \sim \omega^2 lH/c$ and the condition becomes $\omega^2 lH/c \ll c\chi H/l$ or $l^2 \ll \chi c^2/\omega^2$. In the optical frequency range magnetic susceptibilities are of the order of v^2/c^2 while $\omega \sim v/a$ where v is the velocity of an atomic electron and a the atomic dimension. This means l must be $\ll a$. But the macroscopic fields involve averages over at least several atomic distances. Thus, in the optical region at least, we cannot associate $\mathbf{M}$ with the magnetization of the medium, which means that we cannot define a measurable susceptibility. We therefore set $\mu = 1$ and describe the behavior of electromagnetic waves in matter by the dielectric constant $\epsilon(\omega)$, or equivalently, the conductivity $\sigma(\omega) = i\omega\epsilon(\omega)$. [In the microwave region one *can* define a susceptibility and, indeed, the quantity measured

---

[13] L. D. Landau and E. M. Lifshitz, "Electrodynamics of Continuous Media," pp. 251–253. Pergamon, Oxford, 1960.

by magnetic resonance is just $\chi''(\omega)$.] If one is in a region of the optical spectrum dominated by a magnetic dipole transition, then one can define a meaningful magnetic susceptibility. However, even in this case one generally characterizes the system by the conductivity.

The general form of the conductivity tensor for a cubic crystal magnetized along one of the cubic axes (the z-axis) is

$$\boldsymbol{\sigma} = \begin{pmatrix} \sigma_{xx} & \sigma_{xy} & 0 \\ -\sigma_{xy} & \sigma_{xx} & 0 \\ 0 & 0 & \sigma_{zz} \end{pmatrix} \tag{8.31}$$

where the components are complex, i.e., $\sigma_{\mu\nu} = \sigma'_{\mu\nu} + i\sigma''_{\mu\nu}$. The so-called Onsager relations,[14] which specify the functional dependence of response functions, requires that σ_{xx} and σ_{zz} be even functions of H or M and σ_{xy} an odd function.

Because of the axial symmetry it is convenient to work with circularly polarized quantities. The complex index of refraction for right and left circularly polarized electric fields, $n_\pm + i\kappa_\pm$, is related to the complex conductivity by

$$(n_\pm + i\kappa_\pm)^2 = 1 + i4\pi\sigma_\pm/\omega = \epsilon_\pm \tag{8.32}$$

where $\sigma_\pm = \sigma_{xx} \pm i\sigma_{xy}$. The rotation θ is one-half the phase angle change between the right and left circularly polarized waves. Thus for transmission

$$\theta_F = (\omega l/2c)(n_+ - n_-). \tag{8.33}$$

In zero field $n_+ = n_- = n$, $\kappa_+ = \kappa_- = \kappa$ and the conductivity tensor is diagonal with all the components equal to $\sigma_{xx}^{(0)}$. Thus

$$(n + i\kappa)^2 = 1 + i4\pi\sigma_{xx}^{(0)}/\omega. \tag{8.34}$$

In nonzero field if we assume $4\pi\sigma_{xy}/\omega < (n + i\kappa)^2$

$$\theta_F = -\frac{2\pi l}{c}\frac{n\sigma'_{xy} + \kappa\sigma''_{xy}}{n^2 + \kappa^2}. \tag{8.35}$$

For small κ this reduces to $\theta_F = -(2\pi l/cn)\sigma'_{xy}$. Thus the rotation is governed by the off-diagonal component of the conductivity tensor.

In a nonferromagnet material σ_{xy} is an odd function of the field H. If we write this as

$$\sigma_{xy} = \sigma_{xy}^{(1)}H, \tag{8.36}$$

[14] See, for example, L. D. Landau and E. M. Lifshitz, "Statistical Physics," Sect. 119. Addison-Westley, Reading, Massachusetts.

then the Faraday rotation becomes

$$\theta_F = VHl \tag{8.37}$$

where $V = -(2\pi/cn)\sigma_{xy}^{(1)'}$ is called the Verdet constant.
The absorption coefficient is given by

$$\alpha = 2n\kappa\omega/c. \tag{8.39}$$

Therefore the magnetic circular dichroism (MCD), which is the difference in the absorption of left and right circularly polarized waves, is

$$\alpha_+ - \alpha_- = \frac{4\pi}{c}\,\sigma_{xy}'' + \frac{2n\kappa}{n^2 + \kappa^2}\,\sigma_{xy}'. \tag{8.39}$$

Notice that when κ is small the Faraday rotation is proportional to σ_{xy}' while the MCD is proportional to σ_{xy}''. These two magneto-optical effects are therefore related thru the Kramers–Kronig relation. The specific rotations, θ_F/l, for several materials are shown in Fig. VIII.13.[15]

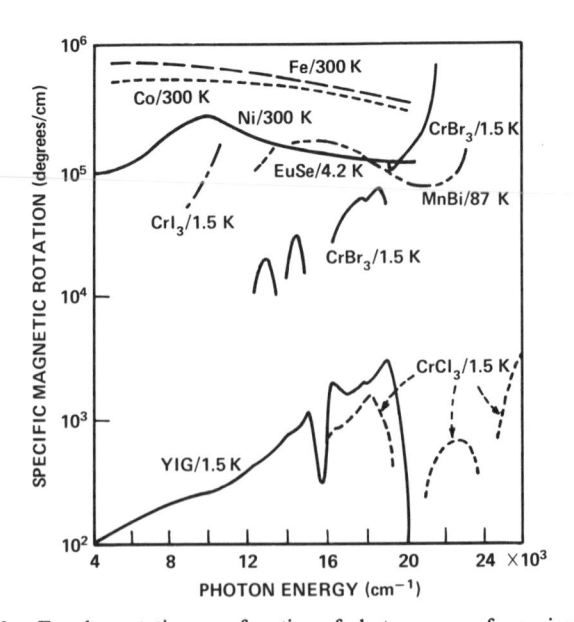

FIG. VIII.13.   Faraday rotation as a function of photon energy for various ferromagnets (compiled by Dillon[15]).

[15] J. Dillon, in "Magnetic Properties of Materials" (J. Smit, ed.), p. 149. McGraw-Hill, New York, 1971.

2. Kerr Rotation

Let us now consider the corresponding result for the *polar* Kerr effect at *normal* incidence. The ratio of the amplitude of the incident electric field, E_0, to the reflected field, E_1, is related to the complex index of refraction according to

$$\frac{E_1}{E_0} = r = -\frac{(n + i\kappa) - 1}{(n + i\kappa) + 1}. \tag{8.40}$$

If we write this complex quantity as $|r| \exp(i\phi)$, then the ratio of the reflected amplitude of right circularly polarized light to that of left circularly polarized light is

$$\frac{r_+}{r_-} = \left|\frac{r_+}{r_-}\right| e^{i(\phi_+ - \phi_-)}. \tag{8.41}$$

If the incident light is linearly polarized it may be resolved into two circularly components with equal amplitudes. After reflection these amplitudes are different, with the result that the light is now elliptically polarized with the major axis rotated relative to the incident polarization by an amount

$$\theta_K = -\tfrac{1}{2}(\phi_+ - \phi_-). \tag{8.42}$$

Approximating $\sin(\phi_+ - \phi_-)$ by $(\phi_+ - \phi_-)$ and assuming $n_+ - n_-$ is small we find

$$\theta_K = -\operatorname{Im} \frac{(n_+ + i\kappa_+) - (n_- + i\kappa_-)}{(n_+ + i\kappa_+)(n_- + i\kappa_-) - 1}. \tag{8.43}$$

Using $n_\pm + i\kappa_\pm \simeq (n + i\kappa) \mp 2\pi\sigma_{xy}/\omega(n + i\kappa)$ and again taking κ small for simplicity, this becomes

$$\theta_K = (2\lambda/cn)\sigma''_{xy}. \tag{8.44}$$

Comparing this with θ_F we see that in reflection the wavelength of the light takes the place of the sample thickness. We also notice that θ_K is proportional to the imaginary part of σ_{xy} which would be zero for a transparent medium. The ratios of reflected to incident amplitudes (Fresnel coefficients) for the different Kerr effects at oblique incidence are given by Freiser.[16]

Typical results for polar and longitudinal Kerr rotations are shown in

---

[16] M. J. Freiser, *IEEE Trans. Magn.*, **mag-4**, 152 (1968).

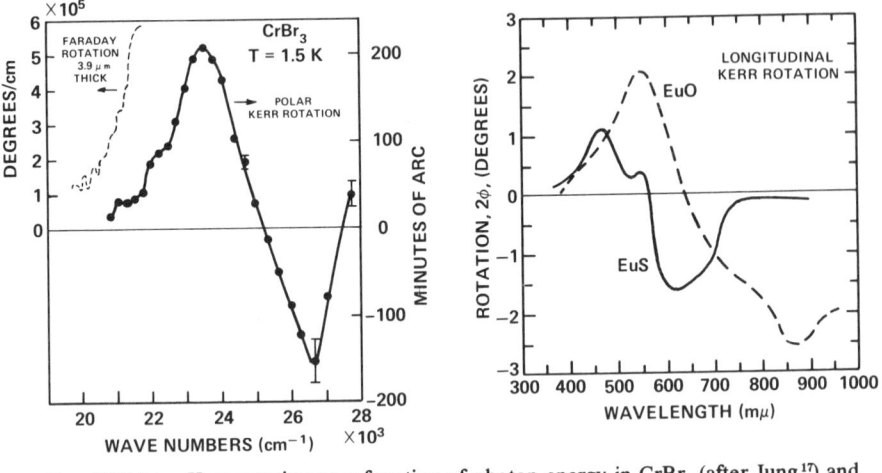

FIG. VIII.14.   Kerr rotation as a function of photon energy in CrBr₃ (after Jung[17]) and EuO and EuS (after Greiner and Fan[18]).

Fig. VIII.14.[17,18] Notice that there is well-defined structure in the Kerr rotations. In the next section we shall indicate the origin of this structure.

3. Microscopic Considerations

As we saw in the previous sections the various magneto-optical effects arise through the off-diagonal component of the conductivity tensor, σ_{xy}. Since the real and imaginary parts are related by the Kramers–Kronig relation we need only calculate one of these. The absorptive part, σ''_{xy}, is the easier to calculate since it is proportional to the transition probability per unit time given by the Golden Rule. The unperturbed states, $|\alpha\rangle$, are the eigenstates of the single particle Hamiltonian, i.e.,

$$\mathcal{H}_0|\alpha\rangle = E_\alpha|\alpha\rangle \tag{8.45}$$

where $\mathcal{H}_0 = P^2/2m + V(\mathbf{r}) + (\hbar P/2m^2c^2) \cdot [\mathbf{s} \times \nabla V(\mathbf{r})]$. Here $\mathbf{P} = \mathbf{p} + (e/c)A_0(\mathbf{r})$ where $A_0(\mathbf{r})$ is the vector potential associated with the uniform applied field and $\mathbf{s}$ is the spin of the electron.

The perturbation has the familiar form

$$\mathcal{H}_1 = (e/mc)\mathbf{\Pi} \cdot \mathbf{A}_1(\mathbf{r}) \tag{8.46}$$

where $\mathbf{A}_1(\mathbf{r})$ is the vector potential of the optical field, and $\mathbf{\Pi}$ is the kinetic

[17] W. Jung, *J. Appl. Phys.* **36**, 2422 (1965).
[18] J. H. Greiner and G. J. Fan, *Phys. Lett.* **9**, 27 (1966).

momentum operator

$$\Pi = P + (\hbar/4mc^2)s \times \nabla V(r). \tag{8.47}$$

Calculating the absorption associated with right-circularly polarized light and subtracting that due to left-circularly polarized light leads, in the electric dipole approximation, to[19]

$$\sigma''_{xy} = \frac{\pi e^2}{4\hbar\omega m^2 V} \sum_{\beta\alpha} [|\langle\beta|\Pi^-|\alpha\rangle|^2 - |\langle\beta|\Pi^+|\alpha\rangle|^2][\delta(\omega_{\beta\alpha} - \omega)$$
$$+ \delta(\omega_{\beta\alpha} + \omega)]. \tag{8.48}$$

From the Kramers–Kronig transform,

$$\sigma'_{xy} = \frac{e^2}{2\hbar m^2 V} \sum_{\beta\alpha} \left[\frac{|\langle\beta|\Pi^-|\alpha\rangle|^2}{(\omega_{\beta\alpha}{}^2 - \omega^2)} - \frac{|\langle\beta|\Pi^+|\alpha\rangle|^2}{(\omega_{\beta\alpha}{}^2 - \omega^2)}\right]. \tag{8.49}$$

All the physics of magneto-optical phenomena are contained in these expressions. Their application, however, is very complicated. Argyres[20] has considered the general case of band ferromagnets such as the transition metals and argues that the spin–orbit part of Π dominates. Cooper[21] has applied this specifically to nickel.

Problem VIII.3
Calculate the Verdet constant for a free electron metal.

In rare earth metals, however, spin–orbit induced shifts in the energy denominators dominate the magneto-optical absorption. Erskine and Stern,[22] have used the MOKE to deduce the relative positions of the spin-up and spin-down bands in Fe, Co, and Ni.

E. INHOMOGENEOUS STATES IN SUPERCONDUCTORS

Because a superconductor is a perfect diamagnet there exists a magnetic field H_c above which the normal state has a lower energy. How the flux enters the sample, however, depends upon its geometry and whether it is type I or type II. Consider, for example, a superconducting sphere of radius R in the presence of a field. Inside the sphere $B = 0$. Outside $\nabla \cdot B = \nabla \times B = 0$ with the boundary conditions $B \to H$ as $r \to \infty$ and

[19] H. S. Bennett and E. A. Stern, *Phys. Rev.* **137**, A448 (1965).
[20] P. N. Argyres, *Phys. Rev.* **97**, 334 (1955).
[21] B. R. Cooper, *Phys. Rev.* **139**, A1504 (1965).
[22] J. L. Erskine and E. A. Stern, *Phys. Rev. Lett.* **30**, 1329 (1973).

$\mathbf{B} \cdot \mathbf{n} \to 0$ as $r \to R$. The solution for $\mathbf{B}(\mathbf{r})$ is

$$\mathbf{B} = \mathbf{H} + \frac{H}{2} R^3 \nabla \left(\frac{\cos \theta}{r^2}\right) \tag{8.50}$$

which is illustrated in Fig. VIII.15a.[23] The tangential component at $r = R$ is $B_\theta(R) = (\frac{3}{2})H \sin \theta$. In particular, at the equator $B_\theta = (\frac{3}{2})H$. Therefore when $H = (\frac{2}{3})H_c$ the equatorial edges will go normal. This means that for fields $(\frac{2}{3})H_c < H < H_c$ some portions of the sphere will be superconducting while others are normal. The configuration that the system adopts depends upon the wall energy. Figure VIII.15b shows the distribution of normal regions in this *intermediate state* as deduced from a bismuth probe which utilizes its extreme magnetoresistance at low temperatures. The intermediate state domains shown in Fig. VIII.3b were obtained by placing tin powder on a superconducting indium disk. Since the tin is also superconducting it is repelled by the normal regions.

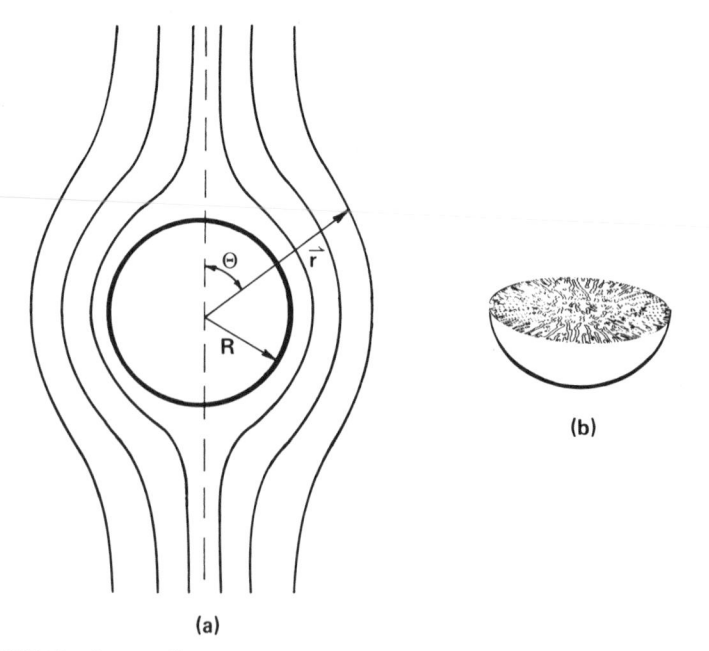

(b)

(a)

FIG. VIII.15. Intermediate state patterns constructed from field mappings made with a thin bismuth wire in a 0.2 mm gap between two tin hemispheres of diameter 4 cm.

[23] A. Meshkovsky and A. I. Shalnikov, *J. Exptl. Theoret. Phys. (U.S.S.R.)* **17**, 851 (1947).

1. Normal–Superconducting Wall Energy

The wall energy is calculated just as in the magnetic case by integrating the energy density in the presence of the wall. In this case the energy density is given by the Ginzburg–Landau free energy [Eq. 2.20] (with the Legendre transform term). The wall is only stable when the field deep inside the normal region is H_c. The energy density in the presence of the wall is

$$\sigma = \int_{-\infty}^{\infty} dz \, \mathcal{G}_S(z).$$ (8.51)

We must subtract from this the energy per unit area of the medium *without* a wall. At $z = \infty$ the normal state free energy density is $\mathcal{G}_{NO}$ plus the field energy density $H_c^2/8\pi$ plus the Legendre term, $-H_c^2/4\pi$, or $\mathcal{G}_{NO} - H_c^2/8\pi$. We see that this just equals the energy density deep inside the superconductor where the field is zero. Thus, it does not matter whether we take the medium without the wall to be normal (with a field H_c) or superconducting. Furthermore, the fact that the energies at $z = \pm\infty$ are equal also means that the wall will not move towards one side or the other. Thus, the wall energy is

$$
\begin{aligned}
\sigma_{NS} &= \int_{-\infty}^{\infty} dz \left[\mathcal{G}_S(z) - \mathcal{G}_{NO}(z) + \frac{H_c^2}{8\pi} \right] \\
&= \int_{-\infty}^{\infty} dz \left[A(T)|\psi|^2 + \frac{1}{2} C|\psi|^4 + \frac{1}{2m^*} \left| \left(-i\hbar\nabla + \frac{e^*A}{c} \right) \psi \right|^2 \right. \\
&\quad \left. + \frac{(H_c - h)^2}{8\pi} \right].
\end{aligned}
$$ (8.52)

If we minimize the free energy [Eq. (2.20)] with respect to ψ we obtain the Ginzburg–Landau equation complementary to Eq. (2.21),

$$A\psi + C|\psi|^2\psi + \frac{1}{2m^*} \left(-i\hbar\nabla + \frac{e^*A}{c} \right)^2 \psi = 0.$$ (8.53)

If we multiply this equation by ψ^* and integrate over z, we obtain

$$\int_{-\infty}^{\infty} dz \left[A|\psi|^2 + C|\psi|^4 + \frac{1}{2m^*} \left| \left(-i\hbar\nabla + \frac{e^*A}{c} \right) \psi \right|^2 \right] = 0.$$ (8.54)

Therefore

$$\sigma_{NS} = \int_{-\infty}^{\infty} dz \left[-\frac{C}{2} |\psi|^4 + \frac{(H_c - h)^2}{8\pi} \right].$$ (8.55)

It is convenient to express this surface energy in terms of a length δ,

$$\sigma_{NS} = \frac{H_c^2}{8\pi} \delta. \tag{8.56}$$

Using Eqs. (2.30), (2.31), and (2.36) this becomes

$$\delta = \int_{-\infty}^{\infty} dz \left[-\left|\frac{\psi}{\psi_\infty}\right|^4 + \left(1 + \frac{h}{H_c}\right)^2 \right]. \tag{8.57}$$

The solution of the coupled Ginzburg–Landau equations is not as simple as in the ferromagnetic case. We know that the spatial dependence of the field is characterized by λ while that of the order parameter is ξ. Therefore, let us take step functions as illustrated in Fig. VIII.16. Then $\delta = \xi - \lambda$. Thus, when $\xi > \lambda$ (type I) the wall energy is positive and a N–S domain configuration will establish itself which will depend on the geometry. This is the *intermediate state* we mentioned above. In a type II superconductor, however, $\xi < \lambda$ and the wall energy becomes negative! This will lead to a maximization of N–S wall area.

2. Flux Quantization

The creation of N–S domain walls in a type II superconductor eventually stops due to the quantization of flux. This quantization is easily derived from the assumption that the GL order parameter ψ be single valued. This requires that the phase ϕ change by multiples of 2π in going

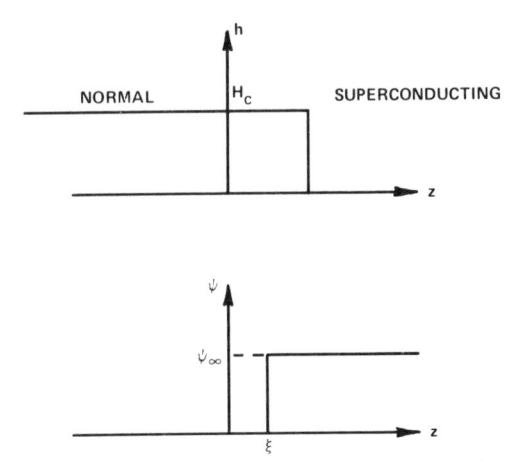

FIG. VIII.16. Simple model of the interface between normal and superconducting regions.

around a closed path, i.e., $\phi(2\pi) - \phi(0) = n2\pi$. This may be written as the integral

$$\oint \nabla\phi \cdot d\mathbf{s} = n2\pi. \tag{8.58}$$

The reason for introducing $\nabla\phi$ is that if the magnitude of the order parameter does not vary in space then $\mathbf{p}\psi = (-i\hbar\nabla)(|\psi|e^{i\phi}) = (\hbar\nabla\phi)\psi$ and

$$m^*\mathbf{v} = \mathbf{p} - \frac{e^*}{c}\mathbf{A} \rightarrow \hbar\nabla\phi - \frac{e^*\mathbf{A}}{c}. \tag{8.59}$$

Therefore

$$\oint \left(\frac{m^*}{\hbar}\mathbf{v} + \frac{e^*\mathbf{A}}{\hbar c}\right) \cdot d\mathbf{s} = n2\pi. \tag{8.60}$$

If we take our path within the superconductor where we may assume $\mathbf{v} = 0$ then the flux, $\Phi = \int d^2\mathbf{r} \cdot \mathbf{B}$ is quantized:

$$\boxed{\Phi = n(hc/e^*) = n\Phi_0.} \tag{8.61}$$

If the path cannot be chosen where $\mathbf{v} = 0$ or $\mathbf{v} \cdot d\mathbf{s} = 0$, then the fluxoid, $\Phi + (m^*c/e^*)\oint\mathbf{v} \cdot d\mathbf{s}$ is quantized. Experiments[23a] have shown that the flux is indeed quantized, and that $e^* = 2e$.

The negative surface energy plus this flux quantization therefore suggests that if flux partially penetrates a type II superconductor it does so in discrete units of Φ_0. This is known as the *mixed, or Abrikosov, state*.

3. One Vortex Line—Lower Critical Field H_{c1}

In Fig. VIII.17 we indicate the structure of a single vortex line. The energy per unit length of a vortex line is the difference between the free energy of an area containing one vortex line and that with a uniform order parameter. This net energy consists of two contributions. One is the kinetic energy of the supercurrents, $m^*v^2/2$. The other is the magnetic field energy. Thus,

$$\epsilon = \int d^2r \left[\frac{1}{2m^*}\left|\left(\frac{\hbar}{i}\nabla - \frac{e^*\mathbf{A}}{c}\right)\psi\right|^2 + \frac{h^2}{8\pi}\right]. \tag{8.62}$$

Since $\xi \ll \lambda$ we shall assume that the order parameter has the form $\psi = |\psi_\infty|e^{i\phi}$. The supercurrent is $n_s e^* v$ which, according to Eq. (8.59), becomes

$$\mathbf{j}_s = \frac{e^*}{m^*}\left(\hbar\nabla\phi - \frac{e^*}{c}\mathbf{A}\right)|\psi_\infty|^2 = \frac{c}{4\pi}\nabla \times \mathbf{h}. \tag{8.63}$$

[23a] B. S. Deaver and W. M. Fairbank, *Phys. Rev. Lett.* **7**, 43(1961); R. Doll and M. Näbauer, *ibid*, p. 51.

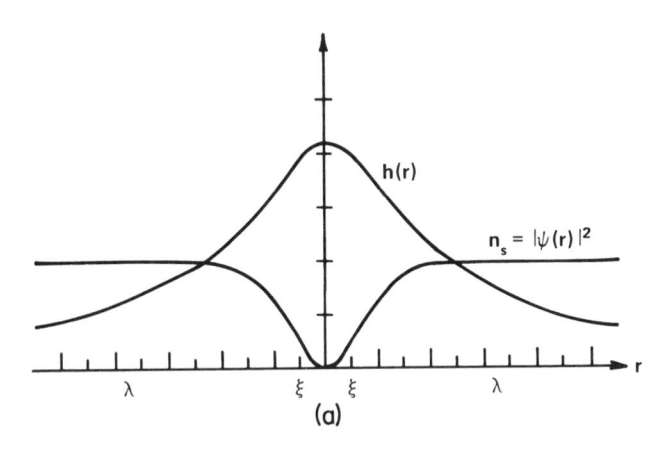

(a)

H

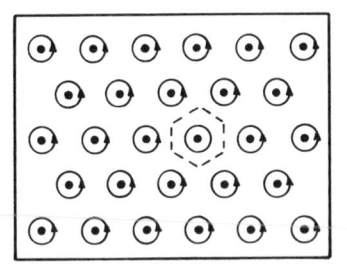

(b)

FIG. VIII.17.   (a) Structure of isolated vortex. The normal core and the decay of the magnetic field away from center are clearly evident. Curves shown for $\kappa = 10$. (b) Quantized
vortex lattice of a type II superconductor. Figure shows normal cores surrounded by circu-
lating currents. Dashed line indicates unit cells of lattice. The area of this unit cell contains
one quantum Φ_0 of magnetic flux.

The second equality is simply Maxwell's equation. This enables us to
rewrite Eq. (8.62) as

$$\epsilon = \int d^2r \left[\frac{m^* \mathbf{j}_s^2}{2(e^*)^2 |\psi_\infty|^2} + \frac{h^2}{8\pi} \right] = \frac{1}{8\pi} \int d^2r[h^2 + \lambda^2(\nabla \times \mathbf{h})^2] \quad (8.64)$$

where $\lambda^2 \equiv m^* c^2 / 4\pi |\psi_\infty|^2 e^{*2}$. The total energy is just the *volume* integral
of the integrand in Eq. (8.64). If we now minimize this integral with
respect to the field distribution h we obtain the Euler equation

$$\mathbf{h} = -\lambda^2 \nabla \times (\nabla \times \mathbf{h}) \quad (8.65)$$

which is just London's equation [Eq. (2.23)].

So far we have neglected the vortex itself. This is permissible for the order parameter, but it obviously affects h, particularly when $\lambda \gg \xi$. This is incorporated by adding a delta function source term to Eq. (8.65) whose strength is the quantum of flux, Φ_0, i.e.,

$$\lambda^2 \nabla^2 h - h = -\Phi_0 \delta(\rho)\hat{z}. \tag{8.66}$$

This equation has the solution

$$h(r) = (\Phi_0/2\pi\lambda^2)K_0(r/\lambda)\hat{z} \tag{8.67}$$

where K_0 is a Hankel function. This solution is now used to evaluate the vortex line energy given by Eq. (8.64). The result is

$$\epsilon = \left(\frac{\Phi_0}{4\pi\lambda}\right)^2 \ln(\lambda/\xi) \tag{8.68}$$

where the integral has been cut off at $r = \xi$.

We now define the critical field H_{c1} as the field at which the first flux line penetrates the superconductor. Just prior to the penetration of this flux line the superconductor is in the Meissner state ($B = 0$) and the Gibbs free energy has some value G_0. After one flux line has penetrated the free energy becomes

$$G_i = G_0 + \epsilon - \Phi_0 H_{c1}/4\pi. \tag{8.69}$$

Since $G_1 = G_0$ at this point,

$$\boxed{H_{c1} = \frac{4\pi\epsilon}{\Phi_0} = \frac{\Phi_0}{4\pi\lambda^2} \ln\left(\frac{\lambda}{\xi}\right) = \frac{\ln \kappa}{\sqrt{2}\kappa} H_c} \tag{8.70}$$

where κ is the Ginzburg–Landau parameter.

4. Vortex Lattice—Upper Critical Field H_{c2}

As H exceeds H_{c1} more and more vortices are formed. It can be shown that these vortices will arrange themselves in a *triangular lattice* as illustrated in Fig. VIII.17b. As H increases these vortices become more dense. Eventually their cores overlap and the medium becomes normal. The field at which this happens is the "upper" critical field, H_{c2}.

The spacing of the vortex lattice at H_{c2} is generally of the order of a few hundred angstroms or less. The associated magnetic field gradients at the surface are weak and have a periodicity that is too small to be easily imaged by either diamagnetic or ferromagnetic powders (as used in studying the intermediate state). Träuble and Essmann[23a] developed a

[23a] H. Träuble and U. Essmann, *J. Appl. Phys.* **39**, 4052 (1968).

clever method for imaging the flux lines by means of very small ferromagnetic cobalt particles which are nucleated in the vapor phase just above the surface. The cobalt-decorated surface is then removed from the cryostat and examined by standard transmission electron microscope procedures. In this way the triangular lattice was observed directly.

To calculate the H_{c2} we recognize that just below this point the order parameter will be very small. Therefore, we may drop the nonlinear term in the GL equation leaving us with

$$A\psi + \frac{1}{2m^*}\left(-i\hbar\nabla - \frac{e^*}{c}A\right)^2 \psi = 0. \tag{8.71}$$

Again, the reader should not confuse the scalar Landau coefficient with the vector potential. Let us now introduce the reduced quantities

$$\begin{aligned}
A' &= A/2H_c\lambda(0) \\
\psi' &= \psi/|\psi_\infty| \\
h' &= h/\sqrt{2}\,H_c \\
r' &= r/\lambda(0).
\end{aligned} \tag{8.72}$$

Equation (8.71) then takes the dimensionless form

$$\left(-\frac{i}{\kappa}\nabla' - A'\right)^2 \psi' - \psi' = 0 \tag{8.73}$$

or, dropping the primes,

$$(-i\hbar\nabla - \hbar\kappa A)^2\psi = \hbar^2\kappa^2\psi. \tag{8.74}$$

This is just the Schrödinger equation for an electron in an effective magnetic field. In the gauge $A = H \times \hat{y}$ this effective field is $\hbar c/e\kappa H$. The eigenvalues of this equation are

$$\frac{\hbar^2\kappa^2}{2m} = \left(n + \frac{1}{2}\right)\hbar\frac{eH_{\text{eff}}}{mc} + \frac{\hbar^2 k_z^2}{2m} \tag{8.75}$$

which define a discrete set of critical fields,

$$H_n = (\kappa^2 - k_z^2)/(2n + 1)\kappa. \tag{8.76}$$

The order parameter is smallest when the field is largest. Therefore H_{c2} is given by the largest field corresponding to $n = 0$, $k_z = 0$. Restoring the conventional units, H_{c2} becomes $\sqrt{2}\kappa H_c$. Using Eqs. (2.31), (2.36), (2.39), and (2.43), this may also be written

$$H_{c2} = \frac{\Phi_0}{2\pi\xi\,(T)^2}. \tag{8.77}$$

Problem VIII.4

Consider a layered superconductor consisting of superconducting sheets separated by a distance s in the z-direction. Within each layer the order parameter, ψ_l, satisfies the GL equation. Let us now assume that the layers are coupled by Josephson tunneling. This adds a term to the GL equation of the form

$$-\eta(\psi_{l+1}e^{-2ieA_zs/\hbar c} - 2\psi_l + \psi_{l-1}e^{2ieA_zs/\hbar c}).$$

Here $\mathbf{A}$ is the vector potential and η is a coupling parameter. Near T_c the order parameter in neighboring layers may be expanded about ψ_l. Also assume that the field is small enough to allow an expansion of the exponentials. These expansions transform the problem into an anisotropic GL problem. Determine the coherence lengths $\xi_\parallel(T)$ and $\xi_\perp(T)$ and the corresponding upper critical fields. [In large fields the exponentials may not be expanded and the GL equation becomes a Mathieu equation and H_{c2} diverges at a temperature T^* given by $\xi_\perp(T^*) = s/2$. This is interpreted as the point at which the normal cores of the vortices fit between the layers. When magnetic effects within the layers are included this divergence is suppressed.[24]]

5. Spin-Dependent Effects

In the case of very high-κ superconductors for which H_{c2} becomes quite high, one must take into consideration the Pauli paramagnetism of the metal. In the simple BCS picture the electrons in a Cooper pair have opposite spins. Thus, as H increases, eventually the Zeeman energy μH of the electron with its spin aligned against the field exceeds the binding energy of the pair and the precise pairing required for superconductivity becomes energetically unfavorable. This leads to a Pauli paramagnetic limiting field H_p, sometimes called the Clogston–Chandrasekhar limit.[25] Equating the condensation energy, $N(0)\Delta^2/2$, to the magnetic energy, $\chi H_p^2/2$, where $\chi = 2\mu_B^2 N(0)$, gives

$$H_p(0) = \frac{\sqrt{2}\,\Delta(0)}{\mu_B} = 18.4 \text{ kOe} \cdot T_c. \tag{8.78}$$

An important modification of this result is usually required in practice, however, due to spin–orbit scattering from impurities. When such spin–orbit scattering is present, spin is no longer a good quantum number and Cooper pairing no longer involves strictly spin-up and spin-down pairs. When the spin–orbit scattering is strong, it is found that

$$H_p^{(s-o)}(0) = \left(\frac{\hbar}{6\tau_{so}\Delta(0)}\right)^{1/2} H_p(0) \tag{8.79}$$

which can be much larger than $H_p(0)$. Here τ_{so} is the spin–orbit scattering

[24] R. A. Klemm, A. Luther, and M. R. Beasley, *Phys. Rev.* **12**, 877 (1975).
[25] A. M. Clogston, *Phys. Rev. Lett.* **9**, 266 (1962); B. S. Chandrasekhar, *Appl. Phys. Lett.* **1**, 7 (1962).

relaxation time. While Pauli paramagnetic limiting has been observed to reduce the observed critical fields in some special cases, it has not been so far a serious impediment to high-field superconductivity in compounds or alloys containing niobium or higher atomic number elements for which one expects strong spin–orbit scattering.

The various free energies associated with these different possibilities are illustrated in Fig. VIII.18.[26]

6. Vortex Pinning

Superconductors only exhibit truly zero resistance when in the Meissner state ($B = 0$). In the vortex state the presence of a net current (either a transport current or a magnetization current) produces a Lorentz force $\mathbf{J} \times \mathbf{\Phi_0}/c$ (per unit length) on the vortices (analogous to the macroscopic Lorentz force density $\mathbf{J} \times \mathbf{B}/c$). where $\mathbf{J}$ is the current density evaluated

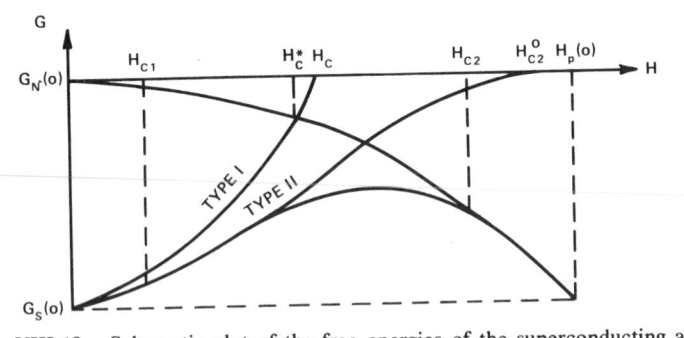

FIG. VIII.18. Schematic plot of the free energies of the superconducting and normal states vs. magnetic field, and of the associated transitions. The horizontal line at $G_N(0)$ represents the normal state ignoring the Pauli spin susceptibility, while the parabolic curve emanating from $G_N(0)$ represents the normal state *with* spin paramagnetism. The horizontal line dahsed at $G_S(0)$ represents the superconducting state, assuming it to have no response to the magnetic field. The intersection point of this line with the parabola at $H_p(0)$ defines the Clogston–Chandrasekhar paramagnetically limited first-order critical field (Eq. 8.78). A type I superconductor with a Meissner effect is represented by the parabolic curve labeled type I and its intersection at H_c gives the thermodynamic critical field. Inclusion of the spin paramagnetism of the normal state lowers the first-order transition to the point H_c^*. A type II superconductor has field penetration above a point which defines a critical field H_{c1}. If spin effects are neglected in both the normal state and the mixed state, the free energy follows the curve labeled type II, up to the point H_{c2}^0, where it becomes tangent to the no-spin normal-state line. If the spin susceptibility of the normal state and the mixed state is included, the type II curve bends over to a point at which a second-order transition occurs to the normal state at the field H_{c2}. After Werthamer et al.[26]

[26] N. R. Werthamer, E. Helfand, and P. C. Hohenberg, *Phys. Rev.* **147**, 295 (1966).

at the center of the vortex. The subsequent motion of the vortices, usually referred to as flux flow, is a dissipative process and results in a net electrical resistance in the superconductor. A complete understanding of this dissipation is not yet available, and the precise origins of flux flow resistance remain one of the outstanding incompletely solved problems of basic superconductivity, although clearly a good part of it arises from Joule heating of the normal electrons in the cores of the vortices. By pinning the vortices on physical or metallurgical defects, this vortex motion can be prevented up to some critical current density J_c which depends on the strength of the pinning. The electrical resistance is zero only at dc, however. For ac, hysteretic losses arise which can only be minimized with considerable difficulty.

It is not difficult to see how flux pinning can arise. For example, local variations of T_c (or equivalently the condensation energy $H_c^2/8\pi$) throughout the superconductor will lead to local free energy minima for the vortices, since clearly it is energetically favorable to have the "normal" cores of the vortices located where the superconducting condensation energy is minimal. Changes in electron mean free path can produce pinning through the resulting changes in the GL coherence length ξ and consequently in the diameter of the vortex core. NbTi, which is the most widely used material at present for large-scale superconducting magnets, is a ductile bcc alloy (see Chapter VI). During the drawing of the wire dislocations are formed which coalesce in cell walls which then serve as pinning sites. In the brittle A-15 superconductors grain boundaries serve as the primary pinning barriers; J_c is inversely proportional to the grain size over a wide range.

While it is easy to imagine how pinning can arise, and in fact to estimate the pinning strength of a particular defect, it is a much more subtle proposition to calculate the resulting critical current. The difficulty arises because when the vortices overlap, as is the case in high fields, they tend to act collectively and move in so-called flux bundles rather than the individual vortex lines.[27]

Fortunately, these complicated details can be avoided in practice by using the critical state model first introduced by Bean.[28] In this model, it is assumed that beginning at the surface, flux penetrates (or leaves) the superconductor in such a way that at every point inside the material to which the flux has penetrated, the gradient of the field (or the magnitude of the current, since $J \propto dB/dx$) is at its critical value, i.e., is in the "criti-

[27] See, for example, A. M. Cambell and J. E. Evetts, "Critical Currents in Superconductors." Taylor & Francis, London, 1972.

[28] C. P. Bean, *Phys. Rev. Lett.* **8**, 250 (1962).

cal state.'' This picture follows immediately by noting that if the gradient were less than the critical value, the flux would be pinned firmly and therefore would not move at all. If the gradient were larger, flux would flow until the gradient relaxed to that value which could just be sustained by the pinning barriers. For simplicity, Bean assumed J_c to be a constant.

Some typical contours of B as a function of position for an increasing (solid lines) and then decreasing (dashed lines) applied field are shown in Fig. VIII.19 for the case of the Bean model and the field applied to a slab of thickness d. The field $H_s = 2\pi J_c d/c$ is of particular importance since it represents the applied field at which in the Bean model flux first completely penetrates the superconductor. For simplicity H_{c1} and any possible surface barrier have been neglected in plotting Fig. VIII.19.

Subsequent workers have generalized the critical state model to cases where J_c is a function of the local induction B. The most prominent of these is the Kim–Anderson model[29] in which J_c has the form

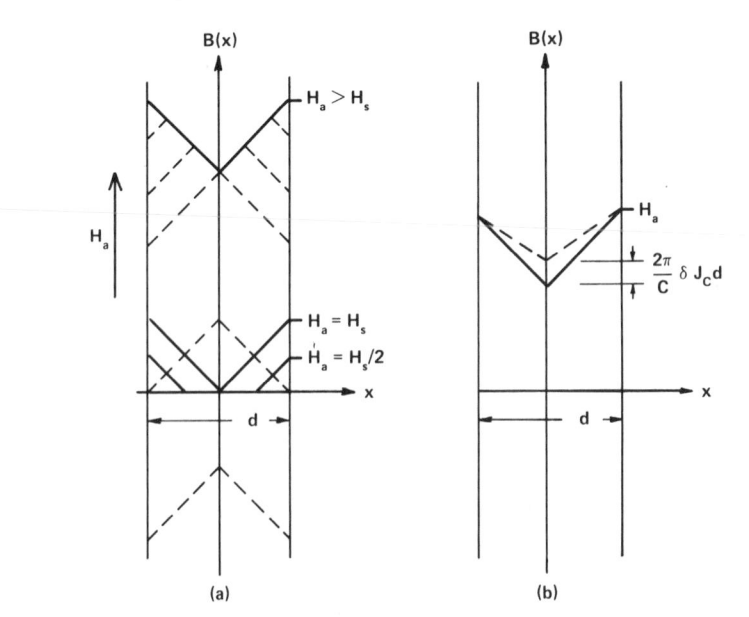

(a)   (b)

FIG. VIII.19.   (a) Field contours $B(x)$ in a flat superconducting slab for the Bean model. For increasing applied fields flux penetrates from the surface (solid lines) eventually reaching the center of the slab at field $H_s = 2\pi J_c d/c$. When the applied field is reduced, flux leaves near the surface (dashed lines). Note that when the applied field returns to zero, flux is trapped in the interior of the superconductor. (b) Change in field contour (solid-to-dashed line) after small decrease in J_c associated with a flux jump. After M. R. Beasley.

[29] P. W. Anderson and Y. B. Kim, *Rev. Mod. Phys.* **36**, 39 (1964).

$J_c = \alpha/(B + B_0)$, which correctly reflects the property that J_c is a decreasing function of B. Neither model properly shows that $J_c \to 0$ as $B \to H_{c2}$, however.

7. Analogies

Throughout this text we have stressed the analogies between various long range order phenomena. The behavior just described for type II superconductors has interesting analogs in liquid crystals and superfluid ^{4}He.

a. Smectics A: In Chapter I we mentioned that a smectic A liquid crystal many undergo a second-order transition to the nematic state. Huberman *et al.*,[30] have suggested that this occurs through the thermal appearance of "dislocations." Such a dislocation is illustrated in the top of Fig. VIII.20.

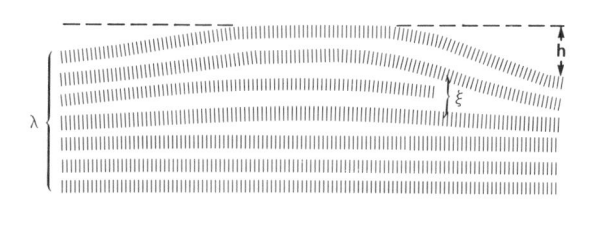

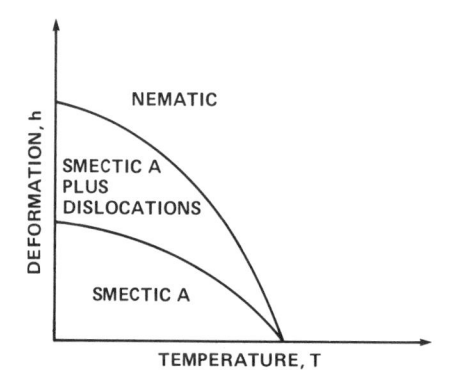

FIG. VIII.20.   Top: illustration of a dislocation in a smectic A liquid crystal. On the left a small deformation is shown penetrating to a depth λ. On the right a large deformation has induced a dislocation whose "core" size is ξ. Bottom: phase diagram associated with the smectic A to nematic transition.

[30] B. A. Huberman, D. M. Lubin, and S. Doniach, *Solid State Commun.* **17,** 485 (1975).

This figure also illustrates another aspect of the smectic A. If we deform the surface, this deformation penetrates into the sample, defining a penetration depth λ which depends upon the elastic constant of the material. If the amplitude of the deformation, h, is large a dislocation may be introduced by the stress. de Gennes[31] pointed out that the appearance of dislocations in a smectic A under an imposed deformation is completely analogous to the appearance of vortices in a type II superconductor. This leads to the phase diagram shown at the bottom of Fig. VIII.20.

b. *Superfluid* ^4He: As we mentioned in Chapter I, below 2.17K ^{4}He exhibits superfluidity. Since the superfluid velocity, v_s, is proportional to the gradient of the phase of the order parameter, this means that the superfluid flow is irrotational, i.e., $\nabla \times v_s = 0$. Treating the superfluid as also incompressible ($\nabla \cdot v_s = 0$) leads to the prediction that the superfluid does not contribute to the shape of the meniscus of the helium in a rotating "bucket." Thus, at lower temperatures where the fraction of superfluid component becomes larger, the meniscus should become flatter. Experimentally, however, the surface was found to always correspond to a classical fluid. This led Feynman to propose the existence of vortex lines where $\nabla \times v_s$ becomes singular. Onsager and Feynman further suggested that the circulation associated with such a line defined as $\oint v_s \cdot dl$, is quantized in units of h/M where M is the mass of the helium atom. As a result of this quantization, there is a critical angular velocity, ω_{c1}, below which the rotating helium is vortex-free. This is referred to as the Landau state in analogy with the Meissner state. To study this state Rudnick and his colleagues[32] measured the superfluid velocity in a rotating torus of square cross section. The torus was packed with very fine particles (diameters ~ 100Å) of Al_2O_3. This results in a "superleak," that is, a situation in which the normal fluid is unable to flow. As the powder is rotated at some angular velocity ω_n "through the superfluid" it imparts a certain angular velocity ω_s to the superfluid. Thus, the superfluid has angular momentum but is still irrotational! Assuming the powder consists of spherical grains the relation between ω_s and ω_n is simply dependent on the ratio of their volumes. Eventually, a "critical slip" velocity is reached at which point vortices are created. Beyond this point, the relative angular velocity $\omega_n - \omega_s$ tends to saturate as illustrated in Fig. VIII.21. If the angular velocity of the torus is now reduced to zero the vortices, which are pinned by the powder grains persist, giving a nonzero value of

[31] P. G. de Gennes, *Solid State Commun.* **10**, 753 (1972).

[32] H. Kojima, W. Keith, S. J. Putterman, E. Guyon, and I. Rudnick, *Phys. Rev. Lett.* **27**, 714 (1971).

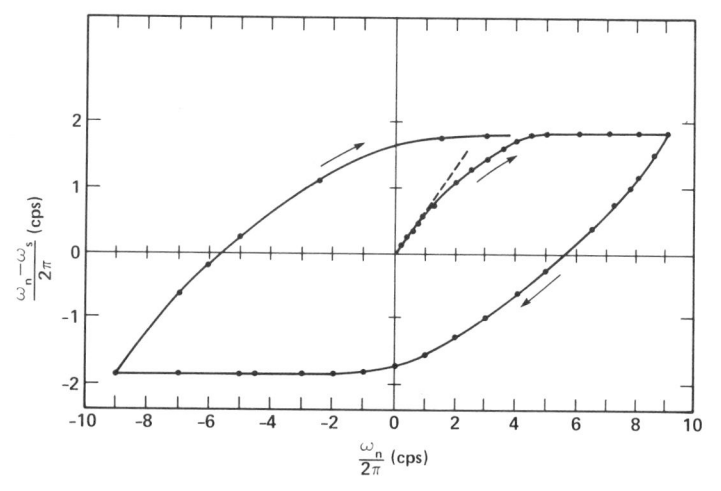

FIG. VIII.21. Frequency of rotation of the superfluid relative to that of the normal fluid for complete cycling of the normal fluid angular velocity from rest. After Kojima *et al.*[32]

$\omega_n - \omega_s$. These experiments suggest that the critical angular velocity for vortex formation is $(h/M)/Rd$ where R is the radius of the torus and d the size of the channels in the superleak. The experimental data for $\omega_n - \omega_s$ show no sign of diminishing, which suggests that the upper critical velocity ω_{c2} must be very high.

Problem VIII.5
Discuss what you would expect to happen if the torus were set into rotation at $T > T_\lambda$ and then cooled below T_λ.

F. SURFACES

Within the past few years the study of surfaces has grown enormously. This growth is the result of several factors. First of all, there have been economic incentives relating to catalysis and integrated electronics. And secondly, there have been significant advances in a wide variety of techniques for characterizing surfaces. For example, high vacuum technology now makes it possible to prepare a surface and have it remain uncontaminated long enough to study its structure by low energy electron diffraction (LEED), to study its electronic states by photoemission, and its chemical composition by Auger spectroscopy.

This activity on surfaces has not particularly focused on long range order. Because of its infinite conductivity, "surface" superconductivity does not require surface probes. Thus surface superconductivity was readily confirmed shortly after it was predicted by Saint-James and de Gennes in 1963.

In the case of magnetic materials a long-standing question has been whether the magnetization vanishes at the surface, i.e., whether there are "dead" layers. Liebermann et al.,[33] for example, monitor the saturation flux of Fe, Co, and Ni films as their thickness is continuously increased by electroplating from solution. They find that several layers remain dead as the film thickness continues to grow. Shinjo et al.,[34] on the other hand, find that the hyperfine field seen by [57]Co atoms deposited on the surface of cobalt metal by electrolytic means is the *same* as in the bulk. Spin-polarized field-emission[34a] also shows no evidence for a dead layer in Ni.

1. Magnetism

Most of the theoretical work on surface magnetism has been concerned with two questions: How does the magnetization vary with distance from the surface, and how does the magnetization at the surface vary with temperature? The answers to these questions will depend upon how the exchange varies near the surface, which is difficult to predict. Electronic configurations will be different, certain superexchange paths will be disrupted, etc. One approach has been to assume that only those spins on the surface are affected, and that their exchange, J_s, is related to that of the bulk spins, J, by $J_s = J(1 - \Delta)$, where Δ may be positive (weakened surface exchange) or negative (enhanced surface exchange).

The Ginzburg–Landau formalism we have been using is also convenient for these surface problems. The equation governing the order parameter is the same we employed in studying the microbridge, Eq. (2.52):

$$\xi^2(d^2f/fz^2) + f - f^3 = 0 \qquad (8.80)$$

where $f(z) = m(z)/m_\infty$ and ξ is the magnetic coherence length, Eq. (2.61). Mills[35] has argued that this equation may be applied throughout the crystal, including the surface, provided the following boundary condition is imposed on the solution:

$$\left.\frac{df}{dz}\right|_{z=0} = f(z = 0)/\lambda \qquad (8.81)$$

[33] L. Liebermann and J. Clinton, *AIP Conf. Proc.* **10**, 1531 (1973).

[34] T. Shinjo, T. Matsuzawa, T. Takada, S. Nasu, and Y. Murakami, *Phys. Lett. A* **36**, 489 (1971).

[34a] M. Landott and M. Campagna, *Phys. Rev. Lett.* **38**, 663 (1977).

[35] D. L. Mills, *Phys. Rev. B* **3**, 3887 (1971).

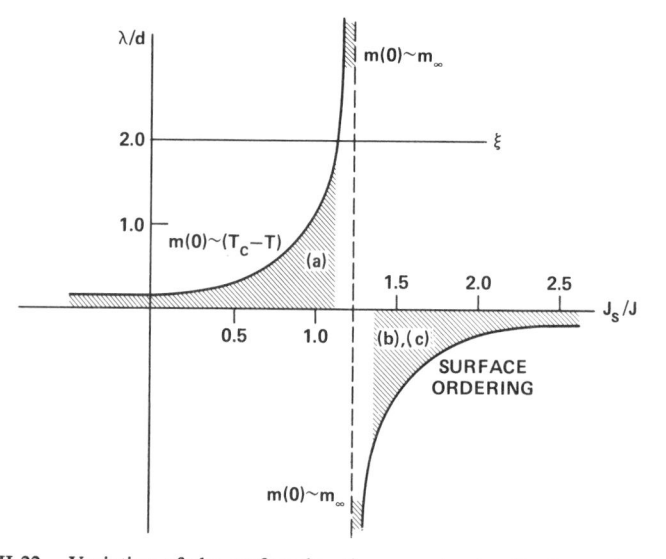

Fig. VIII.22. Variation of the surface length parameter λ with the surface exchange $J_s = J(1 - \Delta)$. The spatial variation of the magnetization in the regions (a), (b), and (c) is indicated in Fig. VIII.24.

where $\lambda = d/(1 + 4\Delta)$ and d is the lattice spacing. Notice that λ corresponds to a second characteristic length. The condition (8.81) is the same as if the medium were extended (or contracted) a distance λ beyond (within) the true surface to an extrapolated boundary at which the condition $f = 0$ is imposed. The length can become negative when the surface exchange exceeds the critical value $J_{sc} = 1.25\ J$.

The solutions of Eq. (8.80) are summarized in Fig. VIII.22. When $0 < \lambda \ll \xi$, $f(0) = 2\lambda/\xi$ which implies $m(0) \sim (T_c - T)$. That is, the critical exponent of the surface layer $\beta_1 = 1$. This is to be contrasted with the bulk value of $\frac{1}{2}$ we derived in Chapter II. Thus one effect of the surface is to modify the critical behavior. There is always the possibility that such a result might be an artifact of the mean field nature of the Ginzburg–Landau approach. Binder and Hohenberg[36] have carried out Monte Carlo calculations for the spatial and temperature dependence of the magnetization of an Ising system with a free surface. These results are shown in Fig. VIII.23. When the data of Fig. VIII.23 for $\Delta = 0$ are plotted on a log–log scale, it is found that the critical exponent $\beta_1 \sim \frac{2}{3}$. Recalling that theories beyond Ginzburg–Landau give a bulk critical exponent near $\frac{1}{3}$ we see that these numerical calculations confirm the GL result that the surface mag-

[36] K. Binder and P. C. Hohenberg, *Phys. Rev. B* **9**, 2194 (1974).

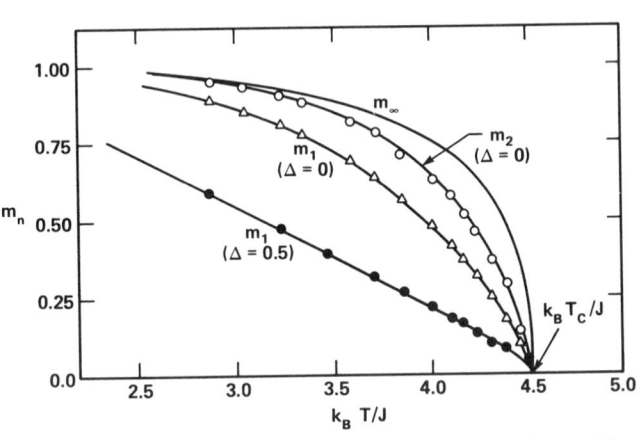

FIG. VIII.23.   Monte Carlo results for the magnetization of the Ising model as a function of temperature for different layers. The curve labeled $\Delta = 0.5$ corresponds to a weakened surface exchange. After Binder and Hohenberg[36].

netization decreases more rapidly with temperature than that in the bulk. This is sketched in Fig. VIII.24a and explains the anomalous behavior of the spin-dependent photoemission we presented in Fig. II.5.

When $-\xi \ll \lambda < 0$, Fig. VIII.22 shows we are in the region of enhanced surface exchange. In this region the surface orders at a temperature *higher* than the bulk. The spatial variation of the magnetization for two cases is shown schematically in Fig. VIII.24b and c.

A possible example of enhanced surface magnetism has been reported by Akoh and Tasaki.[37] They found that the magnetic susceptibility of very small particles (90–300Å) of vanadium contained a Curie–Weiss contribution that was inversely proportional to the average particle diameter as shown in Fig. VIII.25.

The Ginzburg–Landau formalism is, of course, not accurate as the temperature moves away from T_c. Maradudin and Mills[38] have used spin wave theory to calculate the spin deviation

$$\Delta(l_z) = S - \langle S(l_z) \rangle$$

as a function of the plane index l_z. They find that the thermal disorder is greater at the surface. In particular, the spin deviation increases from the bulk value $\Delta_\infty(T)$ to *twice* this value at the surface over a distance $\lambda_T = d[3T_c/z(S + 1)T]^{1/2}$. This corresponds to the wavelength of a spin wave with energy $k_B T$. It is interesting to note that λ_T is much smaller than

[37] H. Akoh and A. Tasaki *J. Appl. Phys.* **49**, 1410 (1978).

[38] A. A. Maradudin and D. L. Mills, *J. Phys. Chem. Solids* **28**, 1855 (1967).

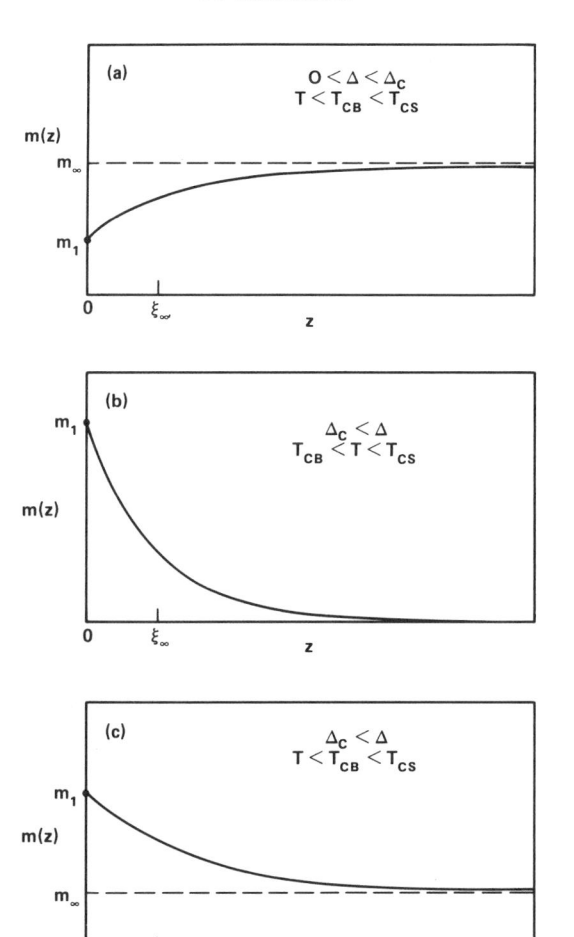

FIG. VIII.24. Schematic variation of the magnetization as a function of distance from the surface for values of surface exchange corresponding to the points (a), (b), and (c) in Fig. VIII.22.

the depth to which a *surface* spin wave penetrates, which is of the order of $\lambda_{\parallel}^2/d$ where $\lambda_{\parallel} = 2\pi/k_{\parallel}$ is the wavelength of the wave parallel to the surface. The reason for this is that the contribution to $\Delta(l_z)$ from the surface modes is precisely canceled by a contribution arising from surface modification of the bulk modes.

Surface spin wave modes have been observed[39] directly by magnetic

[39] J. J. Yu, R. A. Turk, and P. E. Wigen, *Phys. Rev. B* **11**, 420 (1975).

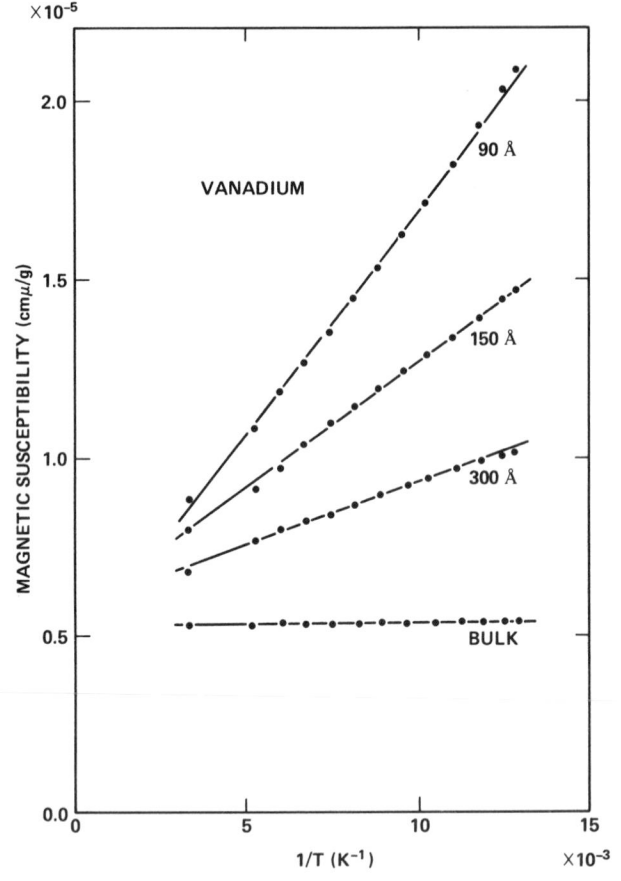

FIG. VIII.25.   Magnetic susceptibility of small particles of vanadium as a function of temperature. After Akoh and Tasaki.[37]

resonance. The geometry is indicated in the insert of Fig. VIII.26. The surface resonance, which occurs at a higher field (lower frequency) than the bulk resonance, is identified by its insensitivity to the sample thickness as Fig. VIII.26 shows. This resonance is associated with the interface between the YIG and the substrate. A second resonance associated with a mode at the YIG–air interface is sometimes observed depending upon how the sample is annealed. As the applied field is rotated towards the perpendicular orientation the high-field surface mode increases in intensity while the bulk modes all decrease in intensity. At a critical angle θ_c which depends upon the film the surface mode becomes the only mode observed and corresponds to the uniform precession.

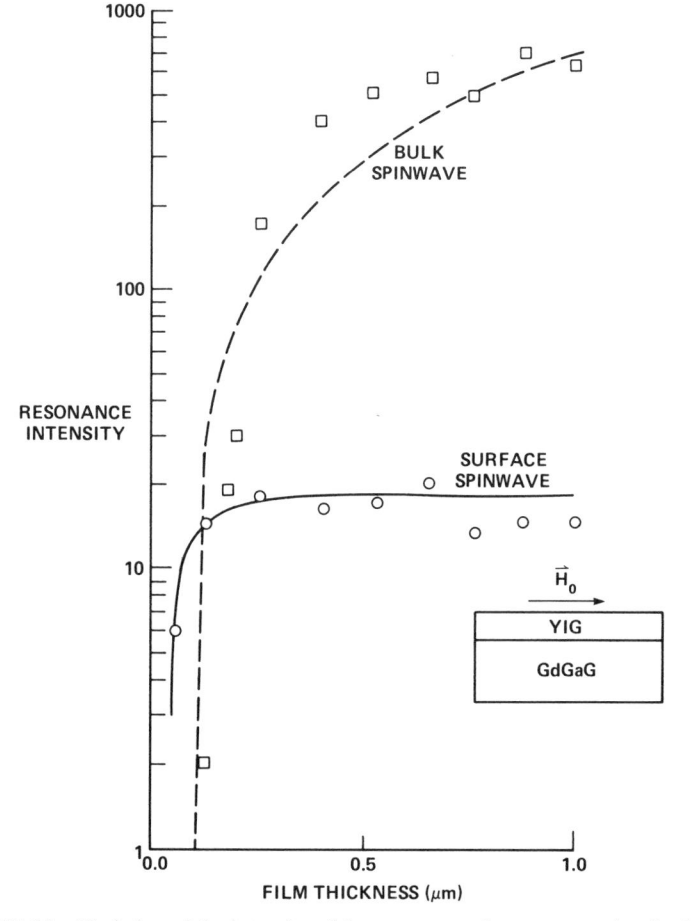

FIG. VIII.26. Variation of the intensity of the two magnetic resonance signals observed from a YIG film as a function of the thickness of the film. After Yu *et al.*[39]

The nature of these surface modes has been related not to a variation in exchange at the surface, but to a variation in anisotropy energy. That is, it is assumed that the layer of spins at the surface experiences an anisotropy field K_s which is, in general, different from the bulk anisotropy. This enables one to define a surface parameter $A = 1 - (g\mu_B/2zSJ)K_s \cdot \hat{\alpha}$ where $2zSJ/g\mu_B$ is the exchange field acting on a spin due to the interactions with its z neighbors in the adjacent plane, and $\hat{\alpha}$ is a unit vector in the direction of the magnetization. The amplitude of the transverse component of the magnetization as a function of this surface parameter is shown

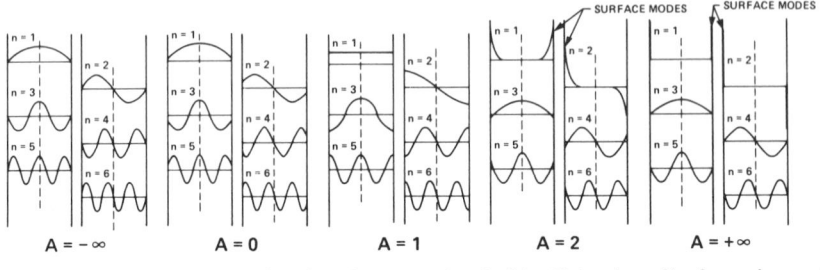

FIG. VIII.27.   Spin wave eigenfunctions associated with a Heisenberg film for various values of the surface parameter A, which characterizes the surface anisotropy.

in Fig. VIII.27 for a film consisting of eleven layers. Notice that when $A = -\infty$ the magnetization at the surface is zero. We speak of the spins as being "pinned" in this case. It is only when $A > 1$ that surface modes appear. In fact, Harada et al.[40] have shown that the spin wave results of Wigen and his coworkers can be explained by assuming that the surface anisotropy has an easy axis normal to the surface, i.e., $\mathbf{K}_s = K_s \hat{\mathbf{z}}$, and the bulk anisotropy has an easy plane parallel to the surface. The surface anisotropy energy has a magnitude 100 times larger than that of the bulk. This large anisotropy is believed to be due to the presence of Fe^{2+} ions on the surface as a result of the migration of oxygen vacancies.

2. Superconductivity—H_{c3}

In Chapter II we noted that the boundary condition on the superconducting order parameter at a superconductor–vacuum interface is $\nabla \psi \cdot \hat{\mathbf{n}} = 0$. In the presence of a magnetic field this takes the gauge-invariant form.

$$\hat{\mathbf{n}} \cdot (-i\hbar \nabla - (2e/c)\,\mathbf{A})\psi = 0. \qquad (8.82)$$

In 1963 Saint-James and de Gennes[41] showed that as a consequence of this boundary condition superconductivity persists above H_{c2} in the form of a surface sheath of thickness $\sim \xi(T)$. This surface superconductivity is eventually suppressed at a field

$$H_{c3} = 1.695(\sqrt{2}\ \kappa H_c). \qquad (8.83)$$

The derivation of this result follows the Ginzburg–Landau derivation of H_{c2}. In particular, if we choose the gauge $\mathbf{A} = (0, Hx, 0)$, Eq. (8.71) becomes

[40] I. Harada, O. Nagai, and T. Nagamiya, *Phys. Rev. B* **16**, 4882 (1977).
[41] D. Saint-James and P. G. de Gennes, *Phys. Lett.* **7**, 306 (1973).

$$\left[-\nabla^2 + \frac{4\pi i}{\Phi_0} Hx \frac{\partial}{\partial y} + \left(\frac{2\pi H}{\Phi_0} \right)^2 x^2 \right] \psi = \frac{1}{\xi^2} \psi. \tag{8.84}$$

Assuming a solution of the form

$$\psi = f(x) e^{ik_y y} e^{ik_z z}, \tag{8.85}$$

the equation for $f(x)$ is

$$-\frac{\partial^2 f}{\partial x^2} + \left(\frac{2\pi H}{\Phi_0} \right)^2 (x - x_0)^2 f = \left(\frac{1}{\xi^2} - k_z^2 \right) f. \tag{8.86}$$

This is the equation for a harmonic oscillator centered at $x_0 = \Phi_0 k_y / 2\pi H$. In the case of an infinite medium we were free to choose $x_0 = 0$ since the lowest eigenvalue is degenerate with respect to x_0. However, in the presence of a surface the Ginzburg–Landau free energy turns out to have a lower minimum when $x_0 = \xi/\pi$. This gives the larger field, H_{c3}, which is plotted in Fig. V.5.

Notice that for a type I superconductor with $0.42 < \kappa < 0.707$ H_{c3} exceeds H_c. Such a superconductor will therefore not show supercooling. For values of κ smaller than 0.42 H_{c3} is less than H_c and limits the supercooling.

Problem VIII.6

Discuss qualitatively what you would expect if the superconducting surface were plated with a normal metal.

G. Inhomogeneities in One and Two Dimensions

Before concluding this chapter, we consider two theoretical developments in lower dimensional systems that relate to domain structure.

1. Structural Phase Transitions

A standard model for structural phase transitions is that illustrated in Fig. VIII.28 in which massive A atoms produce a two-well potential for the smaller B atoms. This Hamiltonian, in a continuum representation, in one dimension is

$$\mathcal{H} = \frac{1}{a} \int dx \left[\frac{m}{2} \dot{u}(x)^2 - \frac{|A|}{2} u(x)^2 + \frac{B}{4} u(x)^4 + \frac{mc_0}{2} \left(\frac{du}{dx} \right)^2 \right] \tag{8.87}$$

where a is the lattice spacing and c_0 the velocity of the low amplitude modes (i.e., when A and B are small). If the depth of the wells is larger than the intersite interaction, then we have a collection of weakly coupled

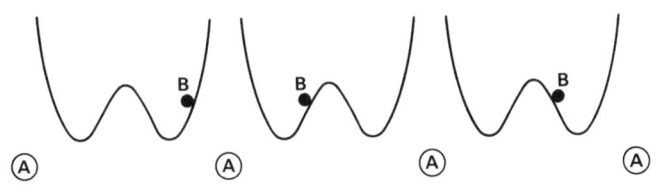

FIG. VIII.28.   Two-well model used to describe structural phase transitions. The curves represent the potential seen by the atom B due to atoms A.

anharmonic oscillators. The phase transition in this case is of the *order–disorder* type. On the other hand, if the intersite interaction is strong, the transition is governed by extended lattice modes and is of the *displacive* type.

Minimizing the energy (8.87) gives the equation of motion

$$m\ddot{u} - |A|u + Bu^3 - mc_0^2u'' = 0. \qquad (8.88)$$

This is satisfied by a function of the form $u = f(x - vt)$. Introducing the dimensionless variables

$$\eta = f/u_0 \qquad \text{where} \quad u_0 = \sqrt{|A\sqrt{/B}} \qquad (8.89)$$

$$s = (x - vt)/\xi \qquad \text{where} \quad \xi = \xi_0\sqrt{1 - v^2/c_0^2} \quad \text{and}$$

$$\xi_0 = a\sqrt{C/|A|}, \qquad (8.90)$$

then η must satisfy

$$d^2\eta/ds^2 + \eta - \eta^3 = 0. \qquad (8.91)$$

Notice that this is the zero-current limit of the microbridge equation, (2.55). Krumhansl and Schrieffer[42] have investigated the solutions of Eq. (8.91). These are illustrated in Fig. VIII.29. The top solution is simply the ordered state in which all the particles are at rest at the bottom of one of the potential wells. The second solution (b) corresponds to oscillations within these wells, i.e., phonons. But solution (c) represents an intrinsically nonlinear excitation. Analytically this has the form

$$u = u_0 \tanh[(x - vt)/2\xi] \qquad (8.92)$$

and corresponds to a *domain wall*. Notice that it has the same form as $\cos\theta$ has in the magnetic case [Eq. (8.26)]. In fact, this wall is another example of a *soliton*. This wall is characterized by a potential energy as well as a kinetic energy. Thus at any finite temperature such domain-like excitations will be present. This has been confirmed by computer simu-

[42] J. A. Krumhansl and J. R. Schrieffer, *Phys. Rev. B* **11**, 3535 (1975).

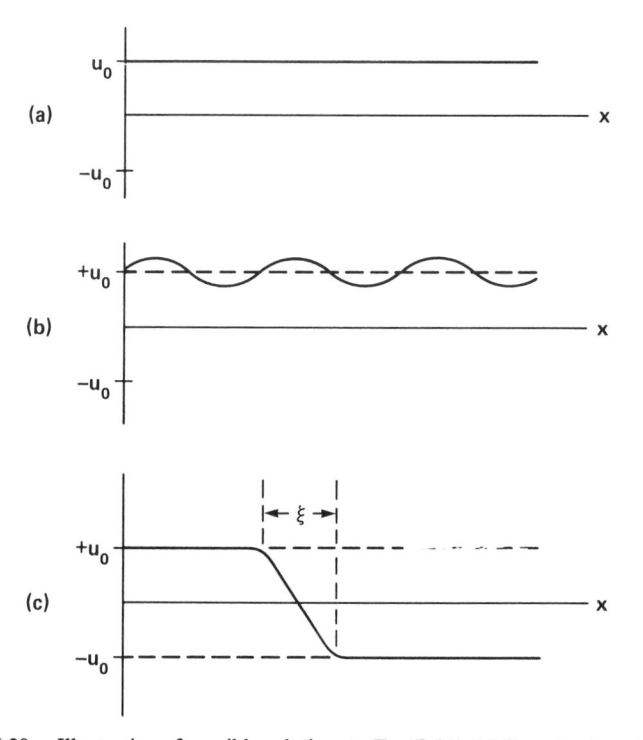

FIG. VIII.29. Illustration of possible solutions to Eq. (8.91). (a) Completely ordered state, (b) an elementary excitation, and (c) a domain wall.

lation studies[43] which show domain structure similar to that illustrated in Fig. VIII.29c. These results also show that the "phonon dressing" of the domain walls is very important. In particular, the walls appear to undergo Brownian-like motion, driven by the gas of phonons.

An interesting aspect of these domain-wall-type excitations is that they contribute a "central peak" to the spectral function given by the Fourier transform of $\langle u(o, o) u(x, t) \rangle$. How this relates to the central peak observed in three-dimensional systems (see Fig. I.22) is not clear at this time.

2. Spin Vortices

In Chapter I we mentioned that a two-dimensional spin system with continuous symmetry does not exhibit long range order. However,

[43] T. R. Koehler, A. R. Bishop, J. A. Krumhansl, and J. R. Schrieffer, *Solid State Commun.* **17**, 1515 (1975).

Stanley and Kaplan[44] found that high-temperature series expansions indicated a phase transition at which the susceptibility becomes infinite. The explanation of this paradoxical situation was given by Kosterlitz and Thouless[45] who suggested that there exist metastable states corresponding to vortices which are bound in pairs below some critical temperature, above which they become free. The model considered by Kosterlitz and Thouless was the XY model we mentioned in Chapter I. The Hamiltonian corresponds to what is "left over" after subtracting the Ising part from the isotropic Heisenberg Hamiltonian,

$$\mathcal{H}_{XY} = -2J \sum_{i>j} (S_i^x S_j^x + S_i^y S_j^y) = -2J \sum_{i>j} \cos(\phi_i - \phi_j)$$

where ϕ_i is the angle the ith spin makes with some arbitrary axis. Expanding the cosine,

$$E = E_0 \approx J \sum_{\mathbf{r}} (\Delta\phi(\mathbf{r}))^2$$

where Δ denotes the first difference operator. For a vortex as shown in Fig. VIII.30, $\Delta\phi(\mathbf{r}) = 2\pi/2\pi r$. Therefore

$$E - E_0 \cong 2\pi J \ln(R/a)$$

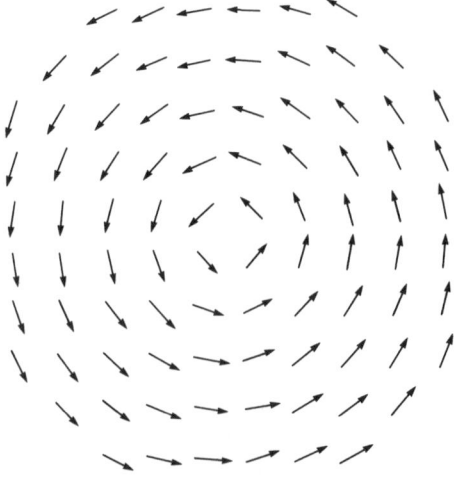

FIG. VIII.30.   Spin vortex.

[44] H. E. Stanley and T. A. Kaplan, *Phys. Rev. Lett.* **17**, 913 (1966).
[45] J. M. Kosterlitz and D. J. Thouless, *J. Phys. C* **6**, 1181 (1973).

where R is the radius of the system and a the nearest neighbor distance. Since this vortex could be centered on any one of the $\pi R^2/a^2$ sites the entropy is

$$S \cong 2k_B \ln(R/a).$$

Therefore the free energy, $E - E_0 - TS$, changes sign at

$$k_B T_c \cong \pi J.$$

In addition to these vortices there are the usual spin waves. These are responsible for destroying the long range order as we described in Chapter I. However, if a magnetic field is applied in the plane of the system below T_c, the vortices will move at right angles to the field giving the infinite susceptibility found by Stanley and Kaplan. The existence of vortices has been "confirmed" through computer simulation studies.[46]

[46] C. Kawabata and K. Binder, *Solid State Commun.* **22,** 705 (1977).

IX. Order Amidst Disorder

What happens to long range order when the periodicity associated with a crystal lattice is not present? In this chapter we shall discuss what sorts of noncrystalline solids can be obtained and, in the case of metals, the kinds of magnetic or superconducting order which are achieved.

A. INTRODUCTION

As far back as 1923 Gibson and Giauque[1] studied the properties of a super-cooled glycerine melt in the narrow vicinity of a temperature at which configurational equilibrium ceases to be maintained, now known as T_g, where g stands for glass. The heat capacity decreases in magnitude from its liquid value to very close to its crystalline value in this narrow region. The crystalline magnitude is then tracked all the way to low temperatures due to the fact that the glass has been frozen into a single or limited number of configurations described by Zachariasen[2] as "a solid with an infinite unit cell." If these other configurations are degenerate in energy, and the system can tunnel among them, the Third Law of Thermodynamics is violated and, as a consequence, the heat capacity will show a linear dependence on temperature as $T \rightarrow 0$.

Chaudhari and Turnbull[3] operationally take T_g as the temperature for which the shear viscosity, η, $\simeq 10^{13}$ poise. The time constants for molecular rearrangements scale roughly as η and are already several minutes for $\eta = 10^{13}$ poise. Changes in η by as much as 10 orders of magnitude occur over the well-defined temperature range in which the heat capacity goes from its supercooled liquid to near-crystalline value so that in practice the defined T_g is only slightly dependent upon kinetic factors such as rate of cooling. An idealized representation of the thermal and rheological manifestations of the glass-melt transition is shown in Fig. IX.1.

Once we have agreed that configurational equilibrium is not maintained an infinite number of frozen-in amorphous states are accessible in princi-

[1] G. E. Gibson and W. F. Giauque, *J. Am. Chem. Soc.* **45**, 93 (1923).

[2] W. H. Zachariasen, *J. Am. Chem. Soc.* **58**, 3841 (1932).

[3] P. Chaudhari and D. Turnbull, *Science* **199**, 11 (1978).

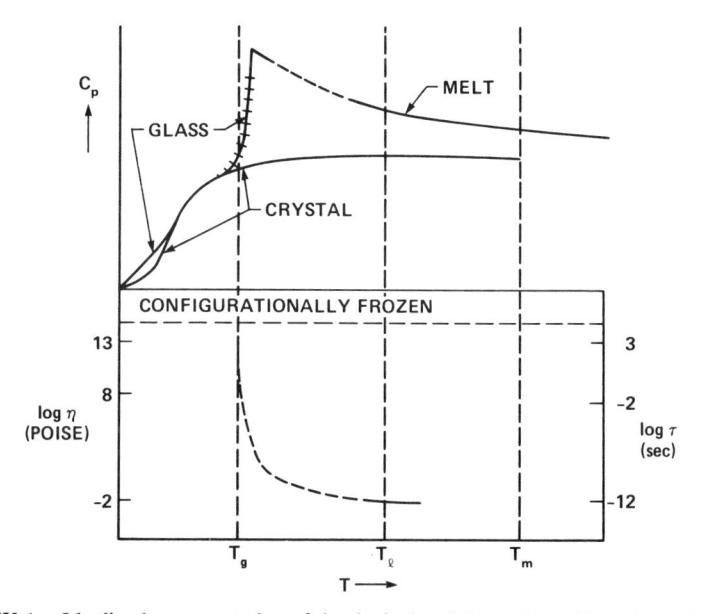

FIG. IX.1. Idealized representation of rheological and thermal manifestations of melt ↔ glass transition in metal alloys. The upper figure shows the temperature (T) dependence of the heat capacity, C_p, of the alloy in its liquid and crystalline forms near the thermodynamic melting temperature, T_m; T_l is the liquidus temperature. The dashed section is inaccessible because of the rapid crystallization at those temperatures. The low temperature region is exaggerated to indicate the fact that a glass may not satisfy the Third Law of Thermodynamics as discussed in the text. The lower figure shows the dependence of the shear viscosity, η, and the time constant for flow, τ, of a glass forming molten alloy on T.

ple, but fortunately, in practice, only two idealized limiting cases seem to be approachable.[4]

The first, already introduced, is the quenched super-cooled liquid. The appropriate structure is the continuous random network of Zachariasen.[2] This model was originally concerned with covalently-bonded glasses with

---

[4] Freezing-in of configurational states also occurs in crystalline solids although manifestations are not as dramatic as in glasses. Thus binary alloys which, under equilibrium conditions, would "phase-separate" are frequently quenched. In some cases, where the temperature at which the phase separation would take place is well below the activation energy for diffusion, it is impossible to do other than to "quench-in" the high temperture state. Stable low-temperature polymorphs may never be obtainable for kinetic reasons. An example of a borderline system is Nb_3Au. If a melt of Nb_3Au is moderately quenched in a stream of argon gas it remains in the high temperature bcc phase (and undergoes a superconducting transition temperature $T_c = 1.2$ K). If allowed to cool moderately slowly by its own radiation it transforms to the A-15 structure (with a $T_c = 11.2$ K). A related bcc phase, V_3Al, conceivably could be transformed by very slow cooling over an appropriate temperature range into the A-15 phase, but this has never been accomplished.

tetrahedral coordination. It has been extended by Bernal[5] and associates to include noncovalently bonded dense random-packed networks such as monatomic metals.

A second limiting case, which in principle can be topologically quite distinct from the first, is reached from the microcrystalline solid. As the size of the crystallites is reduced to $\leqslant 15$ Å the number of surface atoms becomes approximately equal to the number of bulk atoms. The idealized limiting amorphous state is reached by allowing the crystallite size to shrink and develop a random orientation with respect to its neighbors.

A modification of the microcrystalline case, the connected-cluster model, is suggested by calculations which show that aggregates of 50 atoms form stable, molecular clusters, which are noncrystalline in the sense they are not required to have space-filling symmetry elements (i.e., they can have fivefold axes). Chaudhari and Turnbull[3] argue that such clusters, if optimally connected (insofar as space-filling is concerned) during the growth of a film, could form a structure hardly distinguishable from the dense random-packed (DRP) model.

Of the two general methods of preparation to be discussed in the next section, it would seem that the liquid-quench methods lead naturally to the random-network models while the condensation methods might be expected to lead to microcrystalline or cluster models.

B. Preparation

1. From the Liquid

It has been known since antiquity that it is possible to supercool melted sand and obtain glass. In the early 1960's Duwez[6] and his colleagues explored what would happen to metallic melts when the rate of cooling (quench rate) was made as fast as possible. Their approach has evolved into a variety of techniques all of which have in common the rapid contact of the molten metal layer with a metal substrate. The original preparations used a shock tube for splattering molten drops tangentially onto a copper substrate, each drop forming a large interface as thin as 1000 Å. The "splat" technique, as it has come to be known, provides the highest liquid quenching rate, estimated to be up to 10^7 degrees per second for the very thin regions.

To be put into the glassy state a liquid metal must be supercooled to

[5] J. D. Bernal, *Nature (London)* **185**, 68 (1960); J. L. Finney, *Proc. R. Soc. London, Ser. A* **319**, 497 (1970); *Nature (London)* **266**, 309 (1977).

[6] W. Klement, R. H. Willens, and P. Duwez, *Nature (London)* **187**, 869 (1960); also see the review by P. Duwez, *Ann. Rev. Mat. Sci.* **6**, 83 (1976).

temperatures less than about 0.6 of its melting temperature, T_m. Homogeneous nucleation of the crystalline state by configuration fluctuations is negligible in pure melts down to ~0.8 T_m. If this were not so a typical element with $T_m \sim 2000$ K would have to be cooled through a range the order of 400° during which spontaneous nucleation is likely, and even the fastest splat-cooling would not be sufficient to retain the liquid-like configuration.

It has been found however that binary systems particularly those with deep eutectics are likely to form glasses. Typical examples are Pd–Si, Mg–Zn, Fe–B, Zr–Cu, Tc–Be, and Nb–Ni. It is likely that many more will be found. In these systems extreme rapid-quenching is not necessarily required. Simpler methods have been developed in order to make commercial samples with the potentially useful characteristics of metallic glasses which include mechanical properties (high strength and ductility), chemical properties (inert and corrosion resistant) and magnetic properties (low coercive fields). A relatively straightforward method is illustrated in Fig. IX.2 which results in uniform ribbons.

2. Preparation by Atomic Deposition

Much higher effective quench rates can be obtained by physical vapor deposition, i.e., either sputtering, or vacuum evaporation and co-

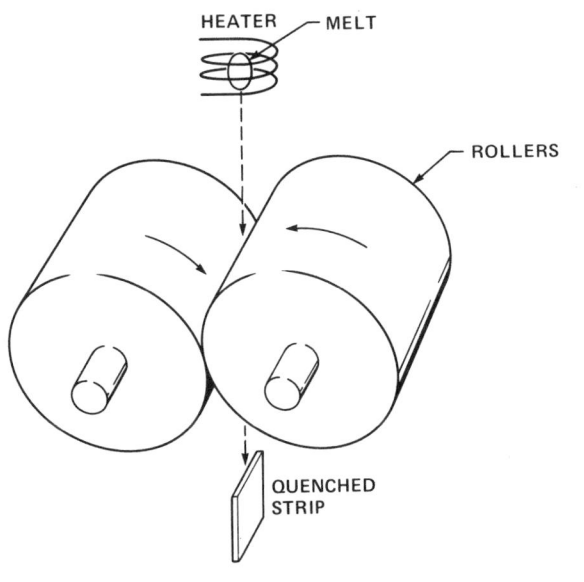

FIG. IX.2. Technique for producing amorphous ribbons.

deposition from two (or more) sources. The important parameters are the substrate temperature, rate of deposition, and presence of trace impurities which tend to restrict the mobility of the atoms once they stick on the surface. Under some conditions an atom may move only a few atomic distances after striking the surface, thus, effectively cooling the order of 10^3 degrees in 10^{-12} seconds. A variety of amorphous films have been prepared by sputtering rare earths with transition metals, such as Gd–Co, which have interesting magnetic properties as discussed later in this chapter.

Buckel and Hilsch[7] showed many years ago that amorphous phases such as Bi could be obtained by vapor condensation onto He cooled substrates. The crystallization temperatures for high purity amorphous elements is generally below liquid nitrogen temperatures. It is even possible to produce nominally pure elemental metals such as Cr and Co[8] as amorphous films.

Electrodeposition and chemical deposition from solution can also be used to produce amorphous metals. These methods bear some resemblance to the vaporization methods except now the source-material is transported through a liquid rather than a vacuum. The substrate temperature can be varied only within narrow limits. Nevertheless, in favorable cases liquid phase deposition can be a very practical method. In fact the "chrome-finished" parts used at one time by the automotive industry owed their desirable characteristics to their (unrecognized) amorphous state. The first apparent experimental observation of amorphous ferromagnetism was the work of Brenner et al.[9] in 1950 in an amorphous film of Co–P prepared by electrodeposition.

C. CHARACTERIZATION

Having produced what we hope is an amorphous state, how can we verify this? The standard approach is to analyze the diffraction patterns obtained from X-ray or electron scattering. However, as we shall see, the availability of the continuous X-ray spectrum associated with synchrotron sources is leading to the development of X-ray absorption as a structural probe.

[7] W. Buchel and R. Hilsch, Z. Phys. 131, 420 (1952).
[8] P. K. Leung and J. G. Wright, Phil. Mag. 30, 995 (1974).
[9] A. Brenner, D. E. Couch, and E. K. Williams, J. Res. Natl. Bur. Stand. 44, 109 (1950).

1. X-Ray Diffraction

When an X ray, for example, with incident wave vector $\mathbf{k}_i$ and wavelength λ strikes an atom i there is a probability, governed by the so-called scattering amplitude $f_i(k)$, that the X ray will be elastically scattered with a wave vector $\mathbf{k}_s = \mathbf{k}_i - \mathbf{k}$. If we have a system of N atoms composed of n different species the scattered intensity in the direction 2θ, where $\sin\theta = k\lambda/4\pi$, is[10]

$$I_N(k) = \sum_{i=1}^{n} N_i |f_i|^2$$

$$+ \sum_{i=1}^{n} \sum_{j=1}^{n} N_i f_i f_j^* \frac{1}{k} \int_0^\infty 4\pi r (\rho_{ij}(r) - \rho_{o,j}) \sin kr \, dr \quad (9.1)$$

where $4\pi r^2 \rho_{ij}(r)dr$ is the number of j-type atoms in a spherical shell at a radius r about an i-type atom, and $\rho_{o,j} = c_j \rho_0$ where $c_j = N_j/N$ and ρ_0 is the density. Let us consider only that contribution to the scattering that depends upon the interparticle distances. This normalized contribution defines the *interference function,*

$$I(k) = \frac{I_N(k) - N[\langle |f|^2 \rangle - |\langle f \rangle|^2]}{N|\langle f \rangle|^2} \quad (9.2)$$

where $N\langle f \rangle = \Sigma_i N_i f_i$. This becomes

$$I(k) = 1 + \frac{1}{k} \sum_i \sum_j W_{ij}(k) I_{ij}(k) \quad (9.3)$$

where

$$W_{ij}(k) = c_i c_j \frac{f_i(k) f_j^*(k)}{|\langle f(k) \rangle|^2} \quad (9.4)$$

and

$$I_{ij}(k) = \int_0^\infty 4\pi r \left(\frac{\rho_{ij}(r)}{c_j} - \rho_o \right) \sin kr \, dr. \quad (9.5)$$

Defining the Fourier transform of $F(k) \equiv k[I(k) - 1]$ as $G(r)$, we have

$$G(r) = \sum_i \sum_j \int_0^\infty W_{ij}(r - r') I_{ij}(r') \, dr'. \quad (9.6)$$

[10] See, for example, G. S. Cargill, *Solid State Phys.* **30**, 227 (1975).

The *radial distribution function*, RDF(r), is a sum of all the angular-averaged pair correlation functions weighted by their scattering amplitudes. It is related to $G(r)$ by

$$\text{RDF}(r) \equiv 4\pi r^2 \rho(r) = rG(r) + 4\pi r^2 \rho_0. \qquad (9.7)$$

In the limit $r \to 0$ the RDR(r) must go to zero. Therefore, in this limit $G(r) \to -4\pi r \rho_0$. Thus, in the range $0 < r < 2$ Å, $G(r)$ should be a straight line with slope $-4\pi\rho_0$. Experimentally, however, one often observes ripples on this line. These ripples arise primarily from the Fourier transform of an inelastic Compton scattering background that inevitably accompanies $I_N(k)$. A method for removing this background, called the "Kaplow correction," is to measure the density, ρ_0, independently, and subtract $4\pi r \rho_0$ from $G(r)$. This difference, when Fourier transformed back to $F(k)$, provides the Compton background. From the scattering amplitudes as functions of k we can construct $I(k)$ using Eq. (9.2), compute the Fourier transform, $G(r)$, and obtain the radial distribution function. It turns out that the scattering amplitudes are not strongly k-dependent. Therefore in practice one assumes they are constant. This simplifies matters, for then $W_{ij}(r - r') = W_{ij}\delta(r - r')$. The radial distribution function then separates into a weighted sum of partial radial distribution functions associated with the different "pairings." For example, suppose we have only two types of atoms, A and B. Then

$$\rho(r) = \frac{c_A f_A^2}{\langle f \rangle^2} \rho_{AA}(r) + 2c_A \frac{f_A f_B}{\langle f \rangle^2} \rho_{AB}(r) + c_B \frac{f_B^2}{\langle f \rangle^2} \rho_{BB}(r). \qquad (9.8)$$

There are basically two ways in which the RDF(r) can be used. The first is to analyze the nearest neighbor environment. Figure IX.3,[11] for example, shows the functions $F(k)$, $G(r)$, and the RDF(r) for amorphous $GdFe_2$. The RDF has been decomposed into three Gaussian components as suggested by Eq. (9.8). The first peak occurs at 2.54 Å. This is very close to the Fe–Fe distance in crystalline $GdFe_2$. Similarly, the second peak at 3.04 Å is close to the crystalline Fe–Gd distance. The third peak, however, occurs at a much greater distance (3.47 Å) than that associated with the Gd–Gd crystalline spacing (3.18 Å).

Notice that the RDF(r) in a multicomponent system is always a weighted average of the various pairs. Since the scattering amplitude, f_i, is proportional to the number of atomic electrons, Z, the heavier components will dominate the RDR(r). This presents a particular problem when we have light glass formers present in small concentrations. In amorphous $Pd_{84}Si_{16}$, for example, the RDF(r) does not enable us to determine

[11] G. S. Cargill, *AIP Conf. Proc.* **18**, 631 (1974).

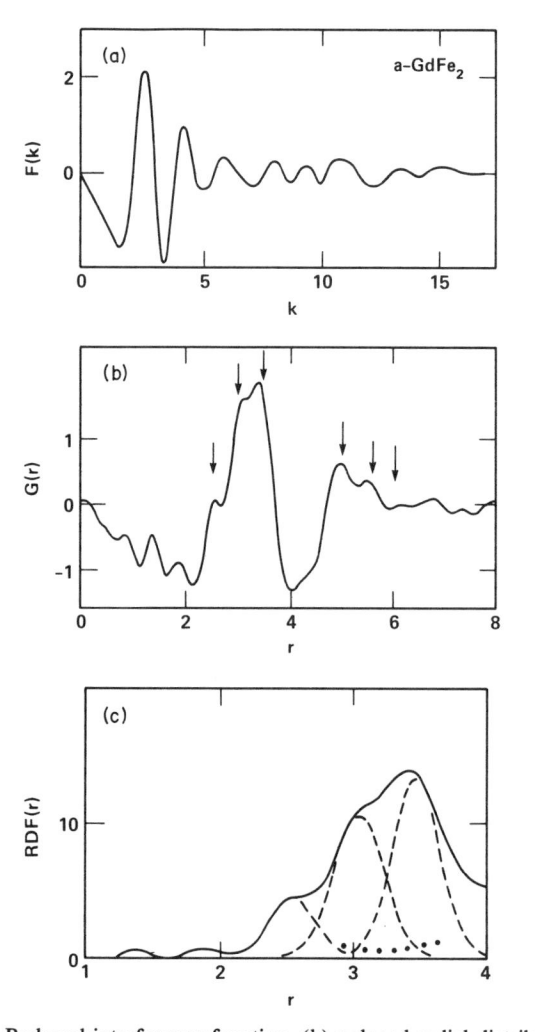

Fɪɢ. IX.3. (a) Reduced interference function, (b) reduced radial distribution function, and (c) RDF for amorphous GdFe$_2$. The RDF has been synthesized by three Gaussian components plus an estimated Fe–Fe almost-nearest-neighbor contribution (dots). After Cargill.[11]

whether or not Si–Si nearest neighbors occur with less frequency than expected from a random arrangement, i.e., of whether or not chemical ordering exists. In the next section we shall see that X-ray absorption is particularly useful in this respect.

The other way of employing the RDF(r) is to compare it with specific models. This often enables us to say what the structure is *not*. Because of the large energy associated with the covalent bond the group IV semicon-

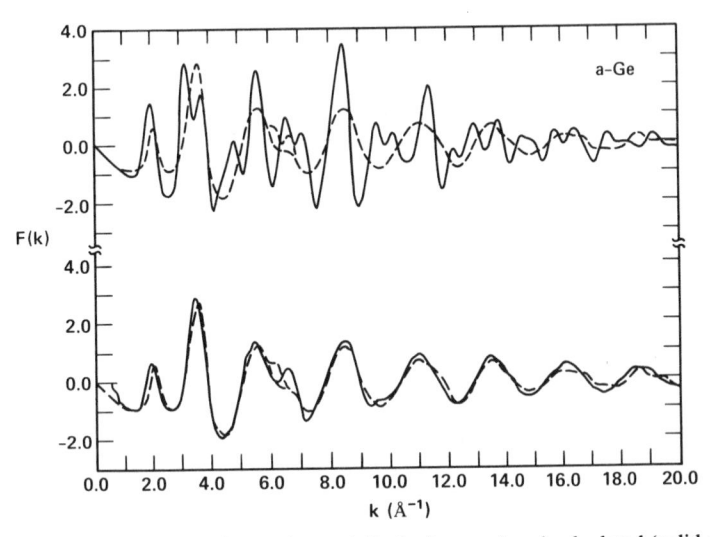

Fig. IX.4. A comparison of experimental (dashed curves) and calculated (solid curves) reduced intensity function for amorphous germanium assuming a microcrystallite model (top) and a random network model (bottom). After Graczyk and Chaudhari.[13]

ductors such as Si and Ge prefer to maintain their fourfold coordination. A random network can, in fact, be constructed[12] in which each atom is fourfold coordinated. The "randomness" appears in the variations of the bond angles. In Fig. IX.4[13] we compare the experimental reduced interference function, $F(k)$, (dashed line) with that given by a random network model and a microcrystallite model. We see that the data are better described by the random network model.

In the case of metallic glasses attempts have been made to describe the amorphous structure by the dense, random packing (DRP) of hard spheres. Models have been built by hand using ballbearings[14] and also generated by computer.[15] The packing algorithm used in most computer simulations is to add atoms in such a way that each new atom exactly touches three existing atoms. The pair correlation function can then be computed and compared with that obtained from X-ray data. Usually the hard sphere radius is taken as an adjustable parameter. This assures that the position of the nearest neighbor peaks will be correct. Nevertheless there is often considerable deviation at larger r. Another problem that

[12] D. E. Polk, *J. Non-Cryst. Solids* **5**, 365 (1971).

[13] J. Graczyk and P. Chaudhari, *Phys. Status Solidi B* **58**, 163 (1973).

[14] J. L. Finney, *Proc. R. Soc. London, Ser. A* **319**, 479 (1970).

[15] C. H. Bennett, *J. Appl. Phys.* **43**, 2727 (1972).

Boudreaux and Gregor[16] have identified with such computer simulations is an anisotropy of the pair correlations which is physically unacceptable. Therefore, at present, computer-generated pair correlation functions are not accurate enough to say whether metallic glasses correspond to a dense random packing of hard spheres.

2. EXAFS

Another technique for obtaining structural information involves the *absorption* of X rays. Figure IX.5[17] shows, for example, the absorption of X rays in crystalline Ge in the vicinity of 11.1 KeV. From similar studies on atomic (gaseous) germanium, it is known that this absorption is due to transitions from the germanium K-shell into the conduction band as illustrated in Fig. IX.6. The structure from about 30 to 1000 eV above the absorption edge is called the extended X-ray absorption fine structure (EXAFS). The origin of this structure lies in the fact that as the excited electron propagates away from the germanium atom part of its amplitude is scattered back from the surrounding atoms. Thus, the final state f in the expression for the transition rate per unit photon flux,

$$W = \frac{2\pi}{\hbar} \sum |\langle f|\hat{\boldsymbol{\epsilon}} \cdot e\mathbf{r}|i\rangle|^2 \delta(E_i + \hbar\omega - E_f) \qquad (9.9)$$

contains both an outgoing wave plus ingoing waves that can lead to constructive or destructive interference as a function of energy. $\hat{\boldsymbol{\epsilon}}$ is the X-ray polarization. This interference causes oscillations in the matrix element in Eq. (9.9). Stern[18] has calculated the total absorption from (9.9) and finds

$$\sigma_K(k) = \sigma_K^0[1 + \Delta(k)] \qquad (9.10)$$

where σ_K^0 is a smooth, free-atom-like K-shell absorption coefficient, and $\Delta(k)$ is the interference part,

$$\Delta(k) = \frac{m}{4\pi h^2 k} \sum_j \frac{N_j}{R_j^2} t_j(2k)e^{-2R_j/\lambda} \sin[2kR_j + 2\delta_j(k)]e^{-2k^2\sigma_j^2}. \qquad (9.11)$$

Here $t_j(2k)$ is the *magnitude* of the amplitude for backscattering from the jth atom and λ is the mean free path of the electron. The sum is over the different shells of atoms whose average distance from the absorbing atom is R_j and which contains N_j atoms. The phase shift $\delta_j(k)$ is the sum of the p-wave phase shift in the absorbing potential plus the phase shift in the

[16] D. S. Boudreaux and J. M. Gregor, *J. Appl. Phys.* **48**, 152 (1977).
[17] T. M. Hayes and S. H. Hunter, unpublished.
[18] E. A. Stern, *Phys. Rev. B* **10**, 3027 (1974).

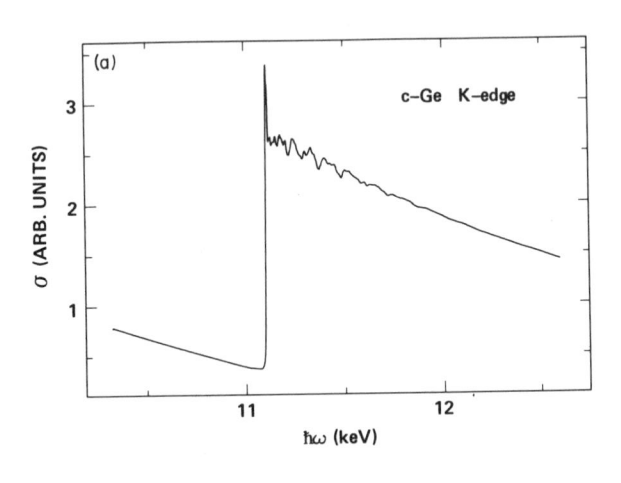

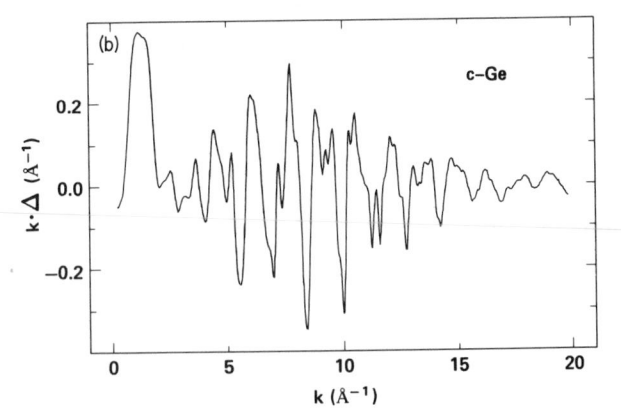

FIG. IX.5.   (a) X-Ray absorption edge of crystalline germanium in vicinity of 11 keV. (b) Extended X-ray absorption fine structure (EXAFS) extracted as described in the text. After Hayes and Hunter.[17]

backscattering amplitude from the jth atom. The exponential containing σ_j^2 is a Debye–Waller-type term where σ_j is the rms fluctuation of the atom about R_j.

Equation (9.11) neglects multiple-scattering effects, i.e., the possibility that the electron scatters from more than one atom before "returning." Such effects have been investigated[19] and do not appear to be important except, perhaps, in Cu. This is fortunate, for if multiple-scattering were

---

[19] C. A. Ashley and S. Doniach, *Phys. Rev. B* **11**, 1279 (1975); P. A. Lee and J. B. Pendry, *ibid.* p. 2795.

HIGH-ENERGY PHOTOEXCITATION

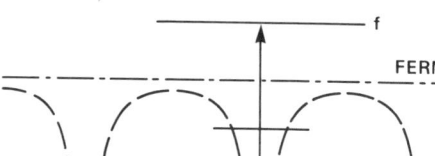
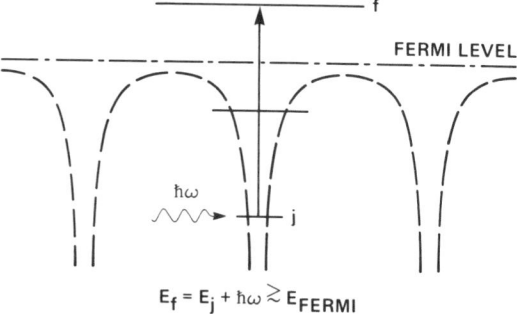

$E_f = E_j + \hbar\omega \gtrsim E_{FERMI}$

FIG. IX.6.   Schematic illustration of the X-ray absorption process.

important, as it is in low energy electron diffraction (LEED), EXAFS would be much less useful.

$\Delta(k)$ is determined experimentally by using the relation

$$\Delta(k) = \frac{\sigma_K(k) - \sigma_K{}^0}{\sigma_K{}^0}. \tag{9.12}$$

The contribution $\sigma_K{}^0$ is obtained by subtracting the extrapolated absorption due to shells other than K-shell and then Fourier filtering to extract the EXAFS. The EXAFS absorption so obtained for the c-Ge example is shown in Fig. IX.5b.

One difficulty underlying EXAFS measurements is the determination of the electron momentum, k. If the photon energy is $\hbar\omega$ then the electron momentum is given by $\hbar^2 k^2/2m = \hbar\omega - E_0$. The reference energy E_0 is difficult to determine. The edge is often complicated by the presence of exciton-like structure as well as many-body effects. The EXAFS shown in Fig. IX.5 was obtained by choosing E_0 at the center of the edge of $\sigma_K{}^0$. The first peak (~ 2 Å^{-1}) is believed to be excitonic. In a metal E_0 lies *below* the X-ray edge by E_F plus an amount due to many-body corrections which may be as large as several electron volts.

If we knew the k-dependence of the total phase shift entering $\Delta(k)$, then one could invert this expression to obtain the structural information. An attempt[20] has been made to calculate this phase shift including correlation effects. However, these results are also plagued by uncertainty in the choice of E_0.

One can turn the problem around and use EXAFS to obtain the phase

[20] P. A. Lee and G. Beni, *Phys. Rev. B* **15**, 2862 (1977).

shift. Citrin *et al.*,[21] for example, extract the phase by fitting the data in *k*-space to a functional form resulting from parameterizations of the scattering amplitude, the final state electron lifetime, and the phase shift. Hayes *et al.*,[22] on the other hand, have emphasized the advantages of transforming the EXAFS data from *k*-space to *r*-space. In Fig. IX.7 we show the Fourier transforms of the EXAFS of crystalline and amorphous Ge. The exciton is excluded by starting the *k*-integration at the first zero-crossing above this peak. The structure in the vicinity of 0.3 Å is an artifact of the data reduction. These results show that the nearest neighbor distance is not changed by the disorder while next-nearest neighbor correlations are completely destroyed. This is consistent with the random network model. This is to be contrasted with the analogous results for selenium. Selenium prefers a twofold coordination which leads to chain structures. The EXAFS is shown in Fig. IX.8. The structure in the vicinity of 3 Å in the crystalline selenium is due to interchain neighbors and the second intrachain neighbors. In the amorphous material the intrachain peak persists, indicating that the structure is consistent with a model in which the amorphous phase consists of twisted chains.

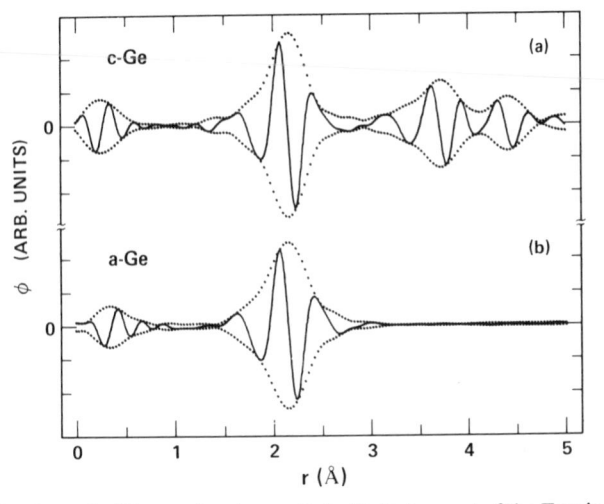

FIG. IX.7.   Real part (solid curve) and magnitude (dotted curve) of the Fourier transform of the EXAFS amorphous germanium. After Hayes and Hunter.[17]

[21] P. H. Citrin, P. Eisenberger, and B. M. Kincaid, *Phys. Rev. Lett.* **36,** 1346 (1976).
[22] T. M. Hayes, P. N. Sen, and S. H. Hunter, *J. Phys. C* **9,** 4357 (1976).

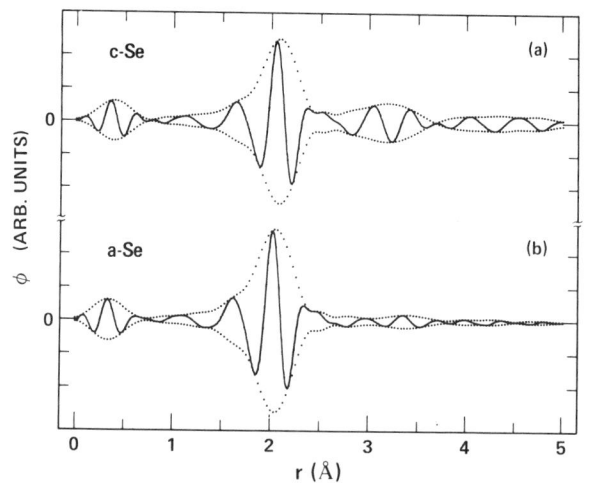

FIG. IX.8.   Real part (solid curve) and magnitude (dotted curve) of the Fourier transform of the EXAFS for crystalline and amorphous selenium. After Hayes and Hunter.[17]

The Fourier transform of the EXAFS may be written[22]

$$\phi(r) = \sum_{\alpha} \int_0^{\infty} \frac{dr'}{r'^2} \, p_{\alpha}(r') \xi_{\alpha}(r - r') \tag{9.13}$$

where p_{α} is the radial distribution of the αth atom species relative to the excited atom species and ξ is essentially the Fourier transform of the product of the scattering matrix element, the final state electron lifetime factor, the initial phase shift due to the excited atom potential, and a window function, which represents the limited range of the Fourier transform. The advantage of the EXAFS technique lies in the fact that one can select a particular species in a multicomponent system with respect to which $p_{\alpha}(r)$ is measured. Conventional X-ray diffraction, as we noted above, involves an average over *all* possible pairs.

In Fig. IX.9 we show the Fourier transform of the EXAFS on the Ge K-shell absorption in crystalline PdGe and amorphous $Pd_{78}Ge_{22}$. By analyzing the crystalline data, for which $p_{\alpha}(r)$ is known, the unknown functions $\xi_{\alpha}(r)$ are obtained. These may then be used to analyze the amorphous data. In this way, Hayes *et al.*[23] were able to show that there are fewer than 5% nearest neighbor Ge about Ge in the metallic glass. This is

[23] T. M. Hayes, J. W. Allen, J. Tauc, B. C. Giessen, and J. J. Hauser, *Phys. Rev. Lett.* **40**, 1282 (1978).

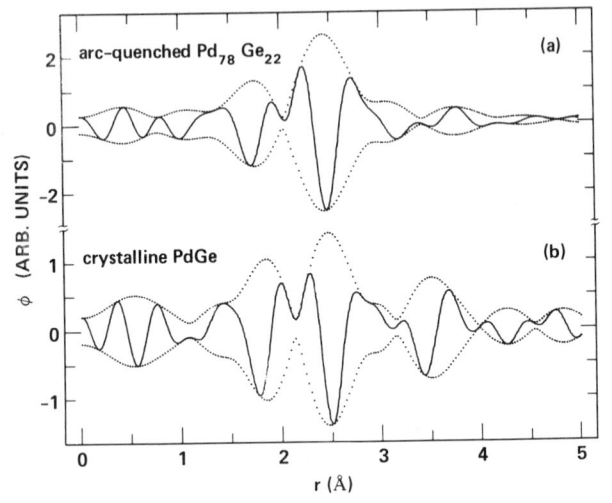

FIG. IX.9.   Real part (solid curve) and magnitude (dotted curve) of the Fourier transform of the EXAFS on the Ge edge for amorphous and crystalline PdGe. After Hayes *et al.*[23]

much less than the random packing model would predict. This suggests that these amorphous structures contain some degree of chemical ordering.

D. SUPERCONDUCTIVITY IN AMORPHOUS METALS

Superconductivity has been found in a wide variety of amorphous metals, starting with the pioneering work of Shalnikov[24] and Buckel and Hilsch.[25] The latter found Bi films, condensed from the vapor onto substrates cooled by liquid helium were metallic, and behaved more like supercooled liquid Bi than semimetallic crystalline Bi. The quenched films were superconducting near 6 K and crystallized to semimetallic Bi upon warming above liquid hydrogen temperatures.

Transition metals condensed onto cold substrates are also amorphous. Thus it is possible to study superconductivity systematically in the amorphous state and gain further insight by being able to compare with the systematics in the crystalline state.

It is not always possible to obtain metallic coordination by quench condensing. Thus Ge and Si remain semiconducting and 4-coordinated in the films

[24] A. I. Shalnikov, *Nature (London)* **142**, 74, 1938.
[25] W. Buckel and R. Hilsch, *Z. Phys.* **138**, 109 (1954).

amorphous quench-condensed state. However, their deep eutectics, with Au and Ag for example, can be supercooled from the liquid state and obtained as superconductors, as well as being quenched from the vapor. Te and Se form molecular species in the vapor phases and hence cannot be obtained as amorphous metals by quenching.

1. Nontransition Metals

It is useful to compare the superconducting properties of amorphous superconductors with their crystalline counterparts. For the polyvalent semimetals such as Bi the increase in local coordination upon becoming noncrystalline results in a large increase in $N(E_F)$. Collapsed crystalline phases can also be obtained under pressure in which there is also an increase in coordination resulting in semimetallic → metallic behavior. Superconductivity is then found with T_c's comparable to the amorphous forms.

Presumably the periodic lattice of the high pressure phase has little effect on the superconductivity. Be, on the other hand, is an exception. It is questionable whether pure quench-condensed Be($T_c = 9$ K) is amorphous,[26] but the amorphous phase can be stabilized by alloying with a number of elements and has a $T_c = 9$ K versus 0.026 for crystalline Be. Crystalline Be has an extremely high Debye temperature and is undoubtedly in the very weakly coupled regime.

Superconducting tunneling experiments show that amorphous s–p superconductors can fall in the strong coupling regime. $Pb_{0.75}Bi_{0.25}(T_c = 6.9$ K) has $2\Delta/kT_c \sim 5.0$ and $\lambda = 2.7$ for the amorphous film (see discussion of this system in Chapter VI). This is in contrast to $2\Delta/kT_c = 3.5$ and $\lambda < 1$ for a weak-coupled BCS superconductor. When the density of the states determined from a tunnel junction is analyzed to give $\alpha^2(\Omega)F(\Omega)$ (Fig. IX.10)[27] it is found that the sharp peaks observed for the transverse and longitudinal branches in crystalline films become washed out, and a somewhat increased weight is found at low frequencies.

An interesting question is whether the large increase in the λ of amorphous films over that of their crystalline counterparts is to be attributed primarily to phonon softening [i.e., the denominator of Eq. (3.26), $\lambda = N(0)\langle I^2 \rangle/M\langle \Omega^2 \rangle$], or an increase in the electronic part. Although neutron diffraction measurements have not been made on amorphous superconductors, information is available from heat capacity measurements and Mossbauer-effect results.[28] In at least one amorphous metal, Pd–Cu–Si

[26] J. Peterson, *Z. Phys.* **24,** 274 (1977).
[27] K. Knorr and N. Barth, *J. Low Temp. Phys.* **4,** 469 (1971).
[28] G. Bergmann, *Phys. Rep.* **27c,** 259 (1976).

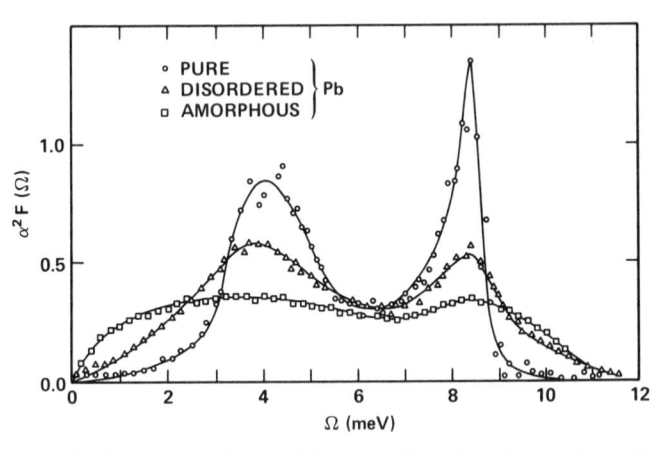

FIG. IX.10. $\alpha^2 F(\Omega)$ for crystalline and fine crystalline Pb and amorphous Pb (with 10% Cu) (after Knorr and Barth[27]). Notice the enhancement of the amorphous curve at low frequencies.

the sizeable decrease in Debye temperature has been shown to be due to a decrease in the sound velocity of transverse phonons.[29] The Debye temperature is also decreased in amorphous superconducting metals relative to their crystalline counterparts, but not sufficiently to explain the observed increase in λ, at least in the case of Sn.[30] In Table IX.1, a comparison of the tunneling results (left-hand column) with Mossbauer results for Sn (right-hand column) show that the coupling constant, α, must increase substantially and that phonon softening, i.e., an increase in $F(\Omega)$ at low Ω, is not responsible for the observed 30% increase in T_c.

There is no way for estimating α in the amorphous case as there is in the corresponding crystalline case (Chapter III) because of incomplete knowledge of phonon and electron eigenfunctions. Crystal momentum is not conserved in the amorphous state so that there is an increase in the final density of states in electron–phonon scattering processes with a re-

TABLE IX.1

Sn	$\int \alpha^2 F(\omega)\,d\omega/\omega$	$\int F(\omega)\,d\omega/\omega$
Amorphous	0.9	0.139 meV^{-1}
Crystalline	0.35	0.128 meV^{-1}

[29] B. Golding, B. C. Bagley, and F. S. L. Hsu, *Phys. Rev. Lett.* **29**, 68 (1972).
[30] J. Bolz and F. Bobell, *Z. Phys. B* **20**, 95 (1975).

sultant increase in the transition probability. Bergmann[28] shows for a simple disordered free-electron model that as a result of this increase in the low energy region $\alpha^2(\Omega)F(\Omega)$ is proportional to Ω rather than Ω^2 as it is in the crystalline case where α is constant. A quantitative expression for $\alpha^2(\Omega)F(\Omega)$ has been obtained[30a] involving measured structure factors and pseudopotentials which, when evaluated for several amorphous alloys, shows the same Ω dependence found by Bergmann.

2. Transition Metals

Much less is known about transition metals because of the difficulty of obtaining good tunnel-junctions. Results of the systematic study of Collver and Hammond[31] of the occurrence of superconductivity in quench-condensed systems as the 4d band is filled are shown in Fig. IX.11.

The single maximum found for both Mo–Ru and Mo–Re quenched films is in contrast with the annealed films (dashed curve) and crystalline bulk alloys (Fig. VI.4). Lack of structure in the density-of-states curves of

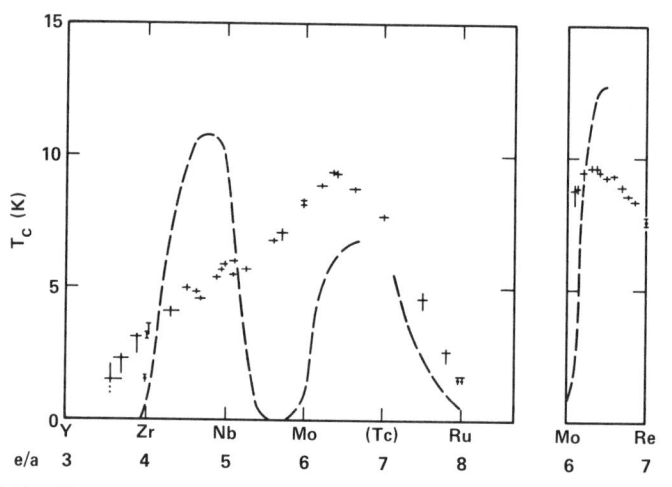

FIG. IX.11. The superconducting transition temperatures of vapor quenched 4d transition metal alloys (left) and a 4d-5d alloy (right) are shown as a function of the number of valence electrons per atom (e/a). The amorphous data on the left are for Y–Zr, Zr–Nb, Zr–Mo, and Mo–Ru alloys. The corresponding crystalline T_c data are indicated by the dashed curve.

[30a] S. J. Poon and T. H. Geballe, *Phys. Rev. B* **18**, 233 (1978).

[31] M. M. Collver and R. H. Hammond, *Phys. Rev. Lett.* **30**, 92 (1973); *Solid State Commun.* **22**, 55 (1977).

amorphous systems versus the two-peaked structure in crystalline systems seem to be the cause. The increase of T_c for Mo upon becoming amorphous has been studied by photoemission.[32] It was necessary to add 10^{-6} torr of N_2 to the high vacuum ($P < 10^{-10}$ torr) in which the evaporation of Mo onto a substrate at 77 K was taking place in order to obtain a Mo film whose diffraction pattern corresponded to a dense random packing of hard spheres. An increase in $N(E_F)$ of $\sim 40\%$ for the amorphous phase over the crystalline phase was observed. This increase together with a softening of the lattice $\langle \Omega^2 \rangle_a \sim 0.85 \langle \Omega^2 \rangle_{bcc}$ is sufficient to account for the increase in T_c from 0.9 to 8.5 K.

Iron, as an impurity in amorphous Mo, is not magnetic,[33] in contrast to the crystalline case (see Fig. VII.12). This is demonstrated by fact that $\sim 1\%$ Fe included in amorphous films of Mo has no effect upon T_c, but destroys T_c upon annealing and introduces a (Kondo) minimum in the resistance-temperature curve. The demagnetization of the Fe with increase of $N(E_F)$ is what is expected on the basis of the Anderson model for the magnetization of localized impurity levels (Chapter VII).

It is interesting to note that, unlike amorphous nontransition metal alloys, the values of $2\Delta/k_B T_c$ for amorphous transition metal-based superconductors[33a] all approach the weak coupling BCS value of 3.5. Thus the structural disorder alone does not lead to strong electron–phonon coupling.

In quenched alloys made from transition elements which are widely separated in the periodic system, and between which there is a sizeable charge transfer, T_c's are invariably lower than those shown in Fig. IX.11 for alloys having the same average electron-to-atom ratio. This behavior is like that observed for crystalline superconductors, and even for magnetic alloys [Slater–Pauling curve (Fig. VI.19)] when nonadjacent 3d elements in the periodic system are alloyed.

One of the features of an amorphous material that can affect its superconducting properties is its very short mean free path. This can be the order of an interatomic spacing placing the amorphous superconductor in the extreme dirty limit. One expects, and finds, where measured, large values of the Ginzburg–Landau parameter, κ, and large values of dH_{c2}/dT and H_{c2}. For amorphous Mo, $dH_{c2}/dT = 45$ kg/K,[34] which is

[32] B. Schröder, W. Grobman, W. L. Johnson, C. C. Tsuei, and P. Chaudhari, in "The Physics of Non-Crystalline Solids," Fourth Int. Conf., Clausthal-Zellerfeld 1976, (G. H. Frischat, ed.), p. 190. Trans Tech Pub., 1977.

[33] F. R. Gamble and T. H. Geballe, *Bull. Am. Phys. Soc.* [2] **15**, 343 (1970).

[33a] C. C. Tsuei, W. L. Johnson, R. B. Laibowitz, and J. M. Viggiano, *Solid State Commun.* **24**, 615 (1977).

[34] R. Keopke and G. Bergmann, *Solid State Commun.* **19**, 435 (1976).

50% greater than found for the high-T_c A-15 superconductors (Chapter VI). This gives a coherence length

$$\xi(0) = \left(\frac{\Phi_0}{2\pi T_{c_0}}\right)^{1/2} \left(-\frac{dH_{c2}}{dT}\right)^{-1} = 30 \text{ Å}, \qquad (9.14)$$

and, from the measured resistance, $\rho = 450 \ \mu \ \Omega \cdot \text{cm}^{-1}$, an upper limit of 1.7 Å for mean free path (assuming one conduction electron per Mo). The temperature coefficient of the resistivity is negative, which is not uncommon when the magnitude of the mean free path is unphysical (i.e., $\leqslant 1$ interatomic spacing).

E. AMORPHOUS MAGNETISM

In Chapter I we mentioned the spin glass. If the exchange coupling in such a dilute random alloy is increased to the extent that it overwhelms the local anisotropy fields conventional long range magnetic order occurs. This would be an amorphous magnet. When one speaks of amorphous magnets, however, one generally means concentrated systems. Amorphous antiferromagnets, such as phosphate glasses containing Fe or Co, do exist, but by far the most interesting from a technological point of view are the amorphous ferro- and ferrimagnets.

The first apparent experimental recognition of amorphous ferromagnetism was the work of Brenner et al.[9] in 1950 as we mentioned above. This work, however, did not receive wide circulation. In 1960 Gubanov[35] theoretically "predicted" the existence of ferromagnetism in an amorphous material. This prediction was subsequently "confirmed" by Tsuei and Duwez[36] who found amorphous palladium-base alloys, such as $Pd_{75}Fe_5Si_{20}$, prepared by splat cooling to be ferromagnetic.

A major task in understanding the properties of amorphous magnets is to understand the role of the structural disorder. For example, Gubanov[35] and Handrich[37] both considered, in the mean-field approximation, the effect of spatial fluctuations in the exchange on the magnetic properties. They both find a Curie temperature proportional to

$$T_c \sim \int r^2 dr g(r) J(r) \qquad (9.15)$$

where $r^2 g(r)$ is proportional to the radial distribution function. Harris et

[35] A. I. Gubanov, Fiz. Tverd. Tela 3, 502 (1960).
[36] C. C. Tsuei and P. Duwez, J. Appl. Phys. 37, 435 (1965).
[37] K. Handrich, Phys. Status Solid 32, K55 (1969).

al.,[38] on the other hand, assume fluctuations in the local anistropy field are the most important characteristic of the amorphous state. To resolve this issue it is helpful to systematically compare amorphous magnets with their crystalline counterparts. There are two classes of amorphous ferromagnets: (i) the quasibinary metal-metalloid alloys $(T_{1-x}M_x)_{80}B_{10}P_{10}$ where T represents Fe or Co and M represents V, Cr, Mn, Fe, Co or Ni; and (ii) the heavy rare earth-transition metal alloys. Unfortunately there are no known amorphous ferromagnets which are nonmetallic.

1. Moments

Studies of amorphous magnets show that the situation is not as simple as Eq. (9.15) would suggest. The moments of these amorphous magnets appear to be governed more by the presence of chemical disorder than structural disorder. In Fig. IX.12,[39] for example, we show the average moments of 3d atoms in the metal–metalloid amorphous ferromagnetic alloys. The dashed lines are the crystalline Slater–Pauling curve we presented in Fig. VI.19. Notice that the moments of the Fe and Co based amorphous alloys resemble those of Co and Ni based crystalline alloys. This is attributed to the transfer of electrons from the glass formers to the

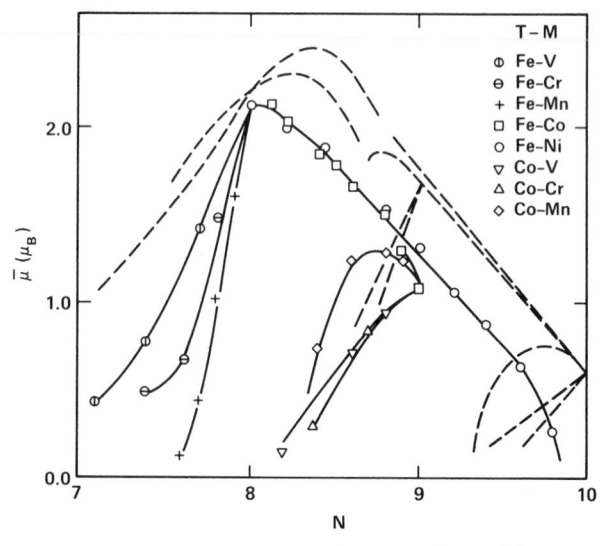

Fig. IX.12. The average magnetic moment of 3d atoms in quasibinary amorphous ferromagnetic alloys as a function of the average number of valence electrons. After Mizoguchi.[39]

[38] R. Harris, M. Plischke, and M. J. Zuckermann, *Phys. Rev. Lett.* **31,** 160 (1973).
[39] T. Mizoguchi, *AIP Conf. Proc.* **34,** 287 (1976).

transition metals. Kazama et al.[40] have prepared single-phase crystalline bcc Fe–P–C and hcp Co–Si–B by careful thermal treatment and find that the moments are essentially the same as in their amorphous state. This proves that the changes seen in Fig. IX.12 are associated with the glass formers themselves and not the structural disorder they stabilize.

Charge transfer also plays a role in determining the saturation moment of the rare earth-transition metal amorphous magnets. Consider $Gd_{0.33}Co_{0.67}$, for example.[41] If one assumes that the gadolinium moment remains 7 μ_B independent of its environment then one finds that the cobalt moment in crystalline $GdCo_2$ is 1.02 μ_B. Since this is less than that of crystalline cobalt it is argued from the Slater–Pauling curve that the Gd must have transferred charge to the Co. In the amorphous $Gd_{0.33}Co_{0.67}$, however, the Co moment is 1.4 μ_B. This suggests there is less charge transfer in the amorphous case. The reason for less charge transfer is presumably related to the lower density of the amorphous material (7.5g/cm³ vs. 9.7g/cm³ for the Laves phase $GdCo_2$).

If the rare earth has a nonzero angular momentum, then the variations in the local anisotropy have a strong effect on the spontaneous magnetization. This may be seen by comparing $GdFe_2(L = 0)$ and $TbFe_2(L = 3)$ (Fig. IX.13).[42] In both crystalline and amorphous materials the magnetiza-

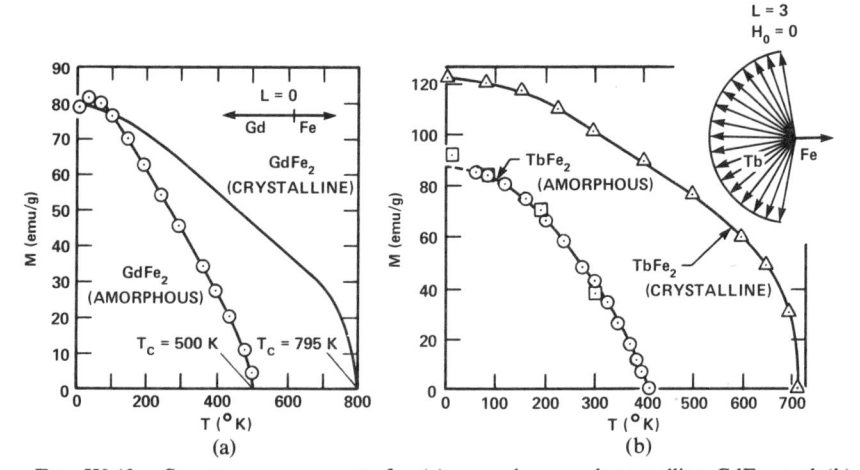

FIG. IX.13. Spontaneous moments for (a) amorphous and crystalline $GdFe_2$ and (b) $TbFe_2$ as functions of temperature. The inserts show probable spin alignments for the L = 0 (Gd) alloys and the L = 0 alloys having strong local anisotropy fields. After Rhyne.[42]

[40] N. Kazama, M. Kameda, and T. Masumoto, AIP Conf. Proc. 34, 307 (1976).

[41] L. J. Tao, S. Kirkpatrick, R. J. Gambino, and J. J. Cuomo, Solid State Commun. 13, 1491 (1973).

[42] J. Rhyne, in "Handbook on The Physics and Chemistry of Rare Earths" (K. A. Gschneider and L. Eyring, eds.), ch. 16. North-Holland Publ., Amsterdam, 1978.

tion is dominated by a strong Fe–Fe exchange which aligns the Fe moments. The RE–Fe exchange is antiferromagnetic but weaker. Thus in an amorphous material with $L \neq 0$, such as Tb, the rare earths will order in cones whose angle is determined by the relative directions and magnitudes of the anisotropy and exchange fields.

The temperature dependence of the magnetization in amorphous Fe and Co is shown in Fig. IX.14.[43,44] We see that the amorphous magnetization decreases more rapidly with temperature than the crystalline material. In both the crystalline and amorphous cases the decrease varies as $T^{3/2}$. This is the signature of spin waves and Birgeneau $et\ al.$[44a] have shown that in $Fe_{75}P_{16}B_6Al_3$, at least, the coefficient of the $T^{3/2}$ term is properly related to the spin wave dispersion. It is interesting to note that neutron scattering reveals the existence of "roton-like" magnetic excitations as shown in Fig. IX.15.[45,45a] These excitations are perhaps better described as heavily damped short wavelength excitations.

The temperature dependence of the magnetization of the rare earth–transition metal amorphous alloys is more complex owing to the presence of two magnetic components. Thus in some cases we find compensation points as illustrated by the Gd–Co system shown in Fig. IX.16.[46] It was

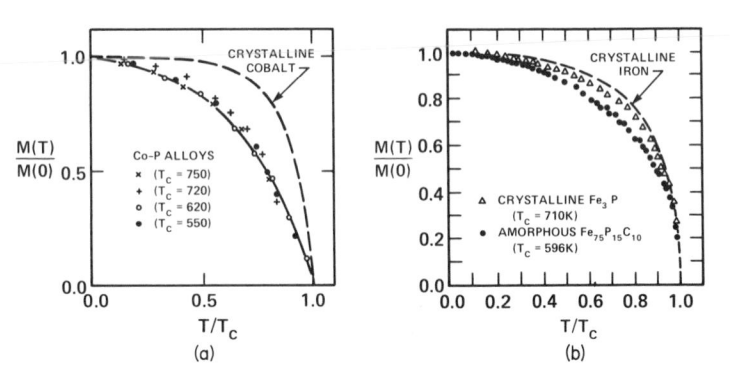

FIG. IX.14.   Temperature dependence of the magnetization in amorphous (a) Fe and (b) Co. Iron data are from Tsuei and Lilienthal[43]; that of cobalt from Cochrane and Cargill.[44]

[43] C. C. Tsuei and H. Lilienthal, $Phys.\ Rev.\ B$ **13**, 4899 (1976).

[44] R. W. Cochrane and G. S. Cargill, $Phys.\ Rev.\ Lett.$ **32**, 476 (1974).

[44a] R. J. Birgeneau, J. A. Tarvin, G. Shirane, E. M. Gyorgy, R. C. Sherwood, H. S. Chen, and C. Z. Chien, $Phys.\ Rev.\ B.$ **18**, 2192 (1978).

[45] H. Mook, N. Wakabayashi, and D. Pan, $Phys.\ Rev.\ Lett.$ **34**, 1029 (1975).

[45a] H. Mook and C. C. Tsuei, $Phys.\ Rev.\ B$ **16**, 2184 (1977).

[46] L. J. Tao, R. J. Gambino, S. Kirkpatrick, J. J. Cuomo, and H. Lilienthal, $AIP\ Conf.\ Proc.$ **18**, 641 (1973).

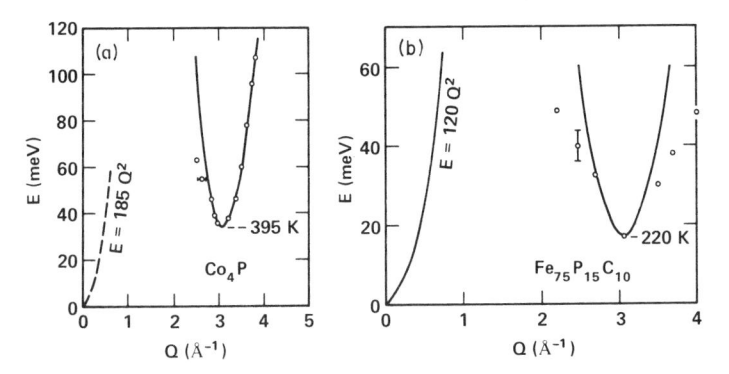

FIG. IX.15. Spin wave dispersion curves for amorphous (a) Co and (b) Fe showing roton-like excitations near the first peak in the static structure factor. (Cobalt data are from Mook et al.[45]; that of iron from Mook and Tsuei.[45a])

this possibility of designing a material to have a specified magnetization at room temperature that initially made these amorphous materials attractive as bubble materials. They can, in fact, support much smaller (submicron) bubbles than the garnets.

2. Transition Temperatures

There are two features associated with the transition temperatures of amorphous magnets. First they are generally lower than their crystalline counterparts, as illustrated by Fig. IX.17.[47] The only real exceptions are

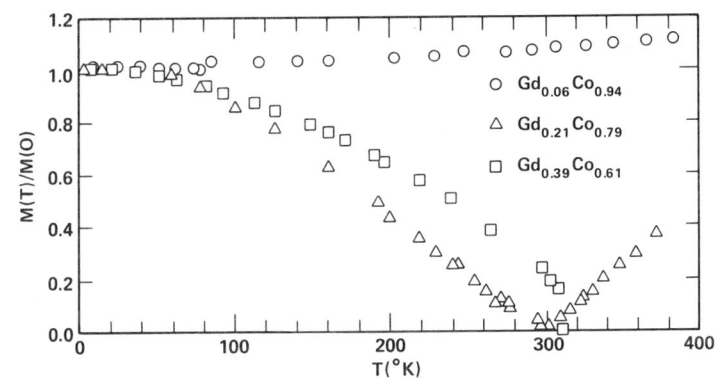

FIG. IX.16. Spontaneous magnetization as a function of temperature for three amorphous Gd–Co films. After Tao et al.[46]

[47] N. Heiman, K. Lee, and R. Potter, AIP Conf. Proc. 29, 130 (1975).

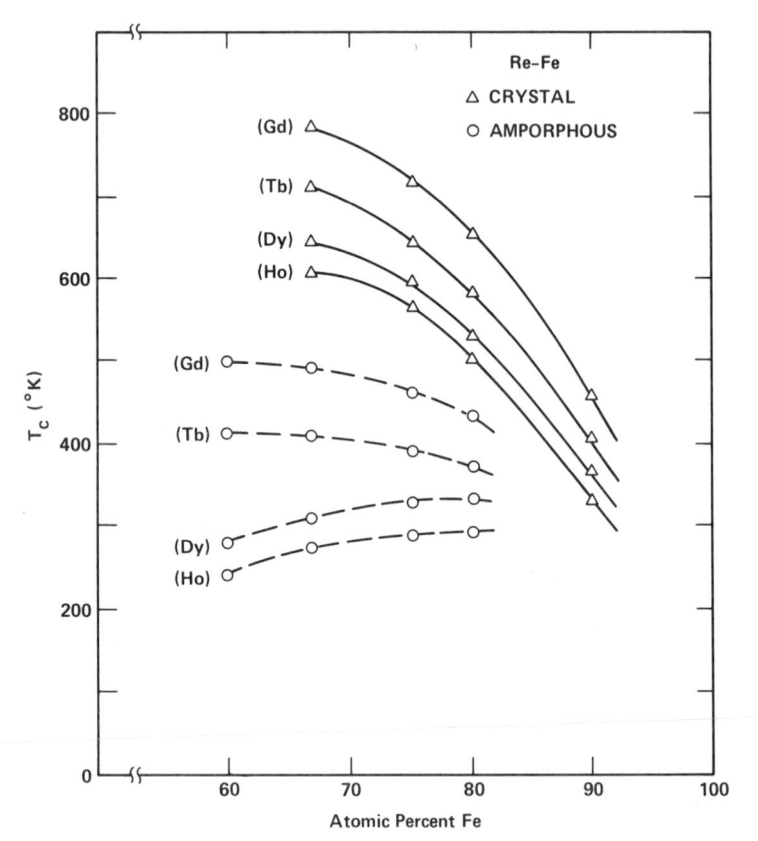

Fig. IX.17. Curie temperatures for crystalline and amorphous RE–Fe alloys. After Heiman *et al.*[47]

the cobalt rare earths in the composition range near $RECo_2$ as shown in Fig. IX.18. This is presumably associated with charge transfer as we indicated above. Heiman *et al.*[47] have shown that the dependence of T_c on composition shown in Fig. IX.18 can be entirely accounted for by the variation of the Fe spin as deduced independently from the hyperfine field at the ^{57}Fe nucleus. This suggests that the decrease of T_c relative to the crystalline materials is associated with a reduction in the Fe moment.

The second feature associated with the magnetic transition in an amorphous ferromagnet is that it is surprisingly sharp. Theoretical arguments based on the renormalization group which we mentioned in Chapter II indicate that the fixed point which, as we noted, characterizes the critical

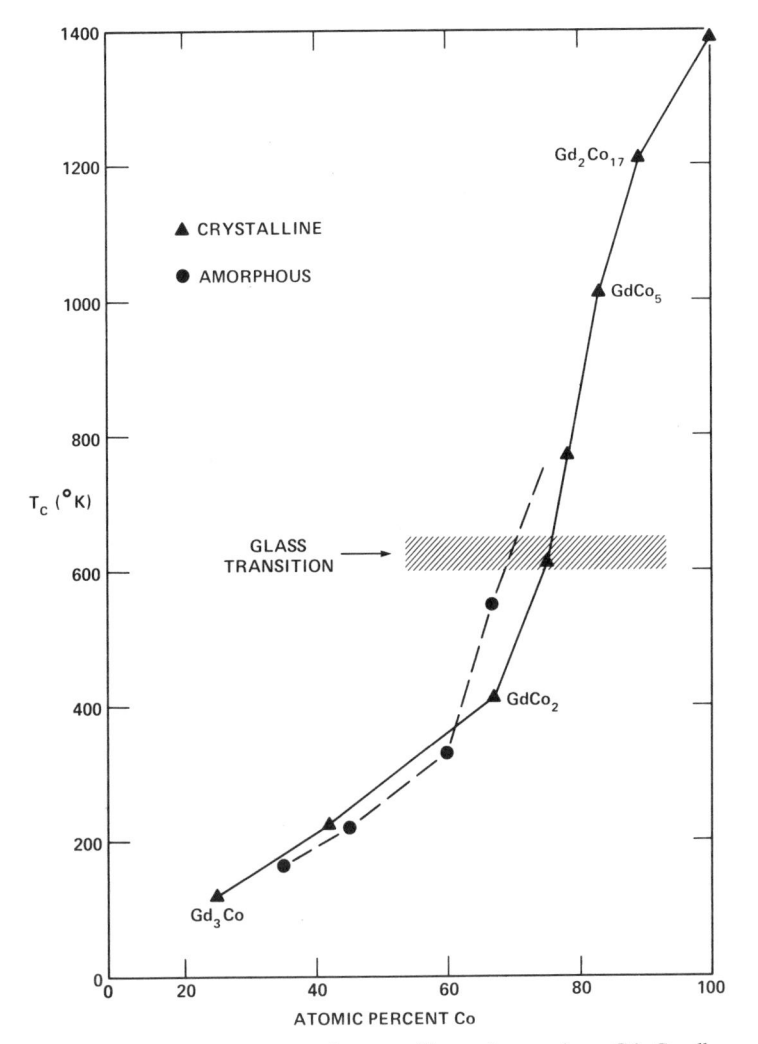

FIG. IX.18. Curie temperatures for crystalline and amorphous Gd–Co alloys.

behavior, is stable against disorder. This is supported by the specific heat data shown in Fig. IX.19.[48] The critical exponent was found to be $\alpha = -0.18$. The corresponding value for crystalline Fe is -0.12.

[48] L. J. Schowalter, M. B. Salamon, C. C. Tsuei, and R. A. Craven, *Solid State Commun.* **24,** 525 (1977).

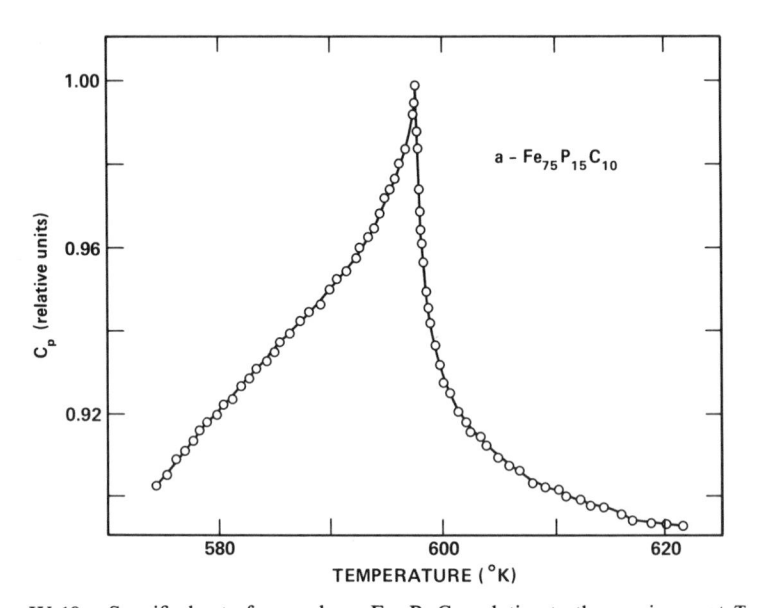

FIG. IX.19.   Specific heat of amorphous $Fe_{75}P_{15}C_{10}$ relative to the maximum at T_c as a function of temperature. After Schowalter *et al.*[48]

3. Anisotropy

Intuitively, we might expect an amorphous magnet to be magnetically isotropic. In fact, those materials that can be prepared in bulk do not exhibit magnetic anisotropy. Amorphous films, however, generally show a uniaxial anisotropy with the easy axis perpendicular to the film. This is believed to be associated with preferential pairing of certain atoms during the film growth. In Gd–Co films, for example, there appears to be approximately 1% more in-plane Co–Co pairs than out-of-plane pairs. It should be noted, however, that the net anisotropy is much smaller than in a similar crystalline material. This anisotropy is also sensitive to the presence of oxygen in the films.

4. Coercivity

If the anisotropy is very small in the amorphous material, then from our discussion in Chapter VIII we expect the coercivity also to be small. This is the case as illustrated by the hysteresis loops in Fig. IX.20. The coercivity is probably also reduced by the absence of grain boundaries and other structural defects. The prospect of producing "ribbons" of soft magnetic media for transformers, etc., is very promising.

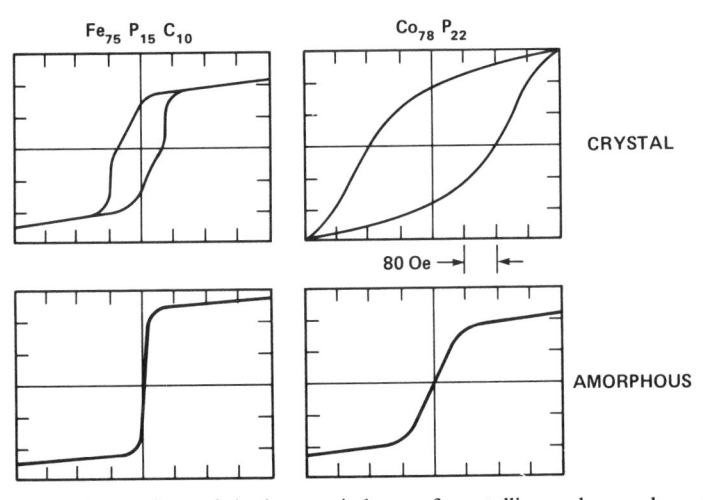

FIG. IX.20. Comparison of the hysteresis loops of crystalline and amorphous films of iron and cobalt.

Exceptions to the low coercivity of amorphous magnets are found in the rare earth-transition metal alloys in which the rare earth has a nonzero angular momentum, such as $TbFe_2$. These materials have high coercivities (100 Oe at room temperature) that increase dramatically at lower temperatures, reaching 30 K Oe at 4 K. This is a manifestation of the variation in direction of the local anistropy field.

5. Anomalous Hall Effect

The Hall effect is essentially the dc analog of the magneto-optical effects we discussed in Chapter VIII. In particular, assuming a conductivity tensor of the form given by Eq. (8.31), the electric field E_y appearing perpendicular to the current density j_x flowing in the presence of a magnetic field along the z-axis is

$$E_y = -\sigma_{xy} j_x / (\sigma_{xx}^2 + \sigma_{xy}^2).$$ (9.16)

The Hall resistivity ρ_H, defined by the ratio E_y/j_x, is therefore proportional to σ_{xy}. As we mentioned in Chapter VIII, σ_{xy} must be an odd function of the field or the magnetization. Thus, we write

$$\rho_H = R_0 B + R_s 4\pi M_s$$ (9.17)

where R_0 is the ordinary Hall coefficient and R_s is the anomalous or extraordinary coefficient.

Problem IX.1.

Using Eq. 8.35, show that the Faraday rotation, θ_F, is proportional to the ordinary Hall coefficient, R_0, and varies as $\omega^{1/2}$.

The physical origin of the anomalous Hall effect lies in the spin–orbit interaction. There has been a great deal of confusion in the literature, largely because the spin–orbit interaction enters the problem in different ways, all contributing to the effect. In the case of a semiconductor, where M_s is field-induced, the theory has been beautifully clarified by Nozières and Lewiner.[48a] They consider an s-like conduction band and a p-like valence band with a threefold orbital degeneracy. The electrons are assumed to scatter from spinless impurities. The scattering potential has the same form as that of the host, namely, a scalar part and a spin–orbit part. The spin–orbit part is only effective on the p-like valence electrons. Therefore, the impurity potential seen by the conduction electrons, V_1, will, in general, be different from that seen by the valence electrons, V_2. Impurity-induced interband scattering is neglected.

In dealing with transport in semiconductors it is convenient to transform the Hamiltonian into an effective Hamiltonian that does not involve interband coupling. For the conduction band this effective Hamiltonian becomes, after a somewhat lengthy procedure,[48a]

$$\mathcal{H}_{eff} = (\hbar^2 k^2/2m^*) + V_1 + \boldsymbol{\rho} \cdot \nabla V_2 - e\mathbf{E} \cdot (\mathbf{r} + \boldsymbol{\rho}). \qquad (9.18)$$

Here $\boldsymbol{\rho} = -\lambda \mathbf{k} \times \mathbf{s}$ where $\mathbf{s}$ is the electron spin and

$$\lambda = \left(1 - \frac{g^*}{2}\right) \frac{\hbar^2}{mE_g} \frac{2E_g + \Delta}{E_g + \Delta}. \qquad (9.19)$$

E_g is the energy gap, Δ the spin–orbit splitting of the valence band, and g^* the effective g-value. λ is proportional to the spin–orbit coupling through the factor $(2 - g^*)$. The first two terms correspond to the usual effective mass Hamiltonian. The two terms involving $\boldsymbol{\rho}$ are simply spin–orbit terms in which the electric field seen by the electron arises from host or impurity sites (the $\boldsymbol{\rho} \cdot \nabla V_2$ term) or externally (the $e\mathbf{E} \cdot \boldsymbol{\rho}$ term). Thus, $e\boldsymbol{\rho}$ may be thought of as an electric dipole moment arising from the spin–orbit interaction. The effective coordinate operator turns out[48a] to be $\mathbf{r} + \boldsymbol{\rho}$. Therefore, the velocity operator is (with $\hbar = 1$)

$$\mathbf{v} = \dot{\mathbf{r}} + \dot{\boldsymbol{\rho}} = -i[\mathbf{r}, \mathcal{H}_{eff}] + \dot{\boldsymbol{\rho}} = (\mathbf{k}/m^*)$$
$$+ \dot{\boldsymbol{\rho}} + \lambda[\nabla V_2 - e\mathbf{E}] \times \mathbf{s}. \qquad (9.20)$$

The current density is obtained by averaging the velocity operator with

[48a] P. Nozières and C. Lewiner, *J. Phys. (Paris)* **34**, 901 (1973).

respect to the electron distribution function, $f_{\mathbf{k}}$, which is the solution to the Boltzmann equation. The spin–orbit interaction leads to spin-dependent corrections to the current from both the driving terms and the scattering terms in this equation.

Consider, for example, the contribution to the current from $\dot{\boldsymbol{\rho}}$:

$$\mathbf{J}_1 = \sum_i \langle e\dot{\boldsymbol{\rho}}_i \rangle = -\lambda m^*[\mathbf{J}_0 \times \mathbf{s} + \mathbf{J}_0 \times \dot{\mathbf{s}}] \tag{9.21}$$

where $\mathbf{J}_0 = \Sigma_{\mathbf{k}}(e\mathbf{k}/m^*)f_{\mathbf{k}}$.

To lowest order, $\dot{\mathbf{s}}$ is proportional to $\mathbf{s} \times \mathbf{B}$. Therefore, for sufficiently small fields we may neglect the last term. $\mathbf{J}_0$ is obtained from the equation of motion, $\dot{\mathbf{k}} = -\dot{z}[\mathbf{k}, \mathscr{H}_{\text{eff}}]$, and the Boltzmann equation,

$$\partial f_{\mathbf{k}}/\partial t + eE \cdot (\partial f_{\mathbf{k}}/\partial \mathbf{k}) = -\sum_{\mathbf{k}'} W_{\mathbf{k}\mathbf{k}}(f_{\mathbf{k}} - f_{\mathbf{k}'}). \tag{9.22}$$

To lowest order, $\mathbf{k} = e\mathbf{E}$ which leads to the "driving" contribution to $\mathbf{J}_1$: $-ne^2\lambda\mathbf{E} \times \mathbf{s}$. Now consider the scattering contribution. The transition rate $W_{\mathbf{k}\mathbf{k}}$ is proportional to an energy-conserving delta function. The energies involved must include that associated with the dipole moment, $-e\mathbf{E} \cdot \boldsymbol{\rho}$. As a result there is now a contribution to the scattering term from the *equilibrium* distribution function $f_{\mathbf{k}}{}^0$. In particular, the expression on the right of Eq. (9.22) becomes

$$-\sum_{\mathbf{k}'} W_{\mathbf{k}\mathbf{k}'}(\partial f_{\mathbf{k}}{}^0/\partial \epsilon_{\mathbf{k}})\lambda e(\mathbf{E} \times \mathbf{s}) \cdot (\mathbf{k} - \mathbf{k}')$$

$$= -e(\Gamma m^*\lambda\mathbf{E} \times \mathbf{s}) \cdot (\partial f_{\mathbf{k}}{}^0/\partial \mathbf{k}) \tag{9.23}$$

where Γ is the scattering rate, i.e., $1/\tau$. This contributes to the Boltzmann equation like an effective electric field $\Gamma m^*\lambda\mathbf{E} \times \mathbf{s}$ which leads to the scattering contribution to $\mathbf{J}_1$:

$$\sigma_{xx}(\Gamma m^*\lambda\mathbf{E} \times \mathbf{s}). \tag{9.24}$$

Notice that when the electron scatters from $\mathbf{k}$ to $\mathbf{k} + \Delta\mathbf{k}$ the dipole moment changes by an amount $\Delta\boldsymbol{\rho} = -\lambda\Delta\mathbf{k} \times \mathbf{s}$. This may be interpreted as a *side jump* of the electron upon scattering. Since $\sigma_{xx} = ne^2\tau/m^*$, the two contributions to $\mathbf{J}_1$ exactly cancel each other. A *side jump* contribution, however, does lead to a nonzero transverse conductivity $\sigma_{xy} = -ne^2\lambda\langle 2s \rangle$ through the last term in $\boldsymbol{\nu}$. Notice that the scattering time τ does not appear in the final result for the side jump mechanism because it enters as the product $\sigma_{xx}\Gamma$. Therefore, from Eq. (9.16) we expect R_s to vary as $1/\sigma_{xx}^2$, i.e., as the square of the resistivity.

There is also a contribution which arises from the spin–orbit correction to the scattering matrix element itself. This is referred to as *skew scattering*. The anomalous current associated with this mechanism is proportional to $\sigma_{xx}^2\Gamma$, which leads to a linear dependence of R_s on resistivity.

The relative importance of these two mechanisms has been demonstrated for InSb by the very elegant experiments of Chazalviel and Solomon[48b] who employ electron spin resonance to "turn the magnetization on and off."

The real interest in the anomalous Hall effect is, of course, directed at magnetically ordered materials. The early theoretical work by Karplus and Luttinger[48c] assumed that the carriers were also responsible for the magnetization, i.e., an itinerant model. They obtain essentially the same result for σ_{xy} given above except that the bandgap is replaced by an average splitting of the d-bands. The side-jump mechanism appears to be responsible for the anomalous Hall effect in the 3d ferromagnetic metals as evidenced by a quadratic dependence of R_s on ρ. Even in *amorphous* iron where $\rho \simeq 200\ \mu\Omega \cdot$ cm as compared with a crystalline value of $\sim 10\ \mu\Omega \cdot$ cm the ratios of ρ_H/ρ^2 are nearly the same for the two materials.

In systems where the magnetization is associated with local moments a somewhat different treatment is necessary. Maranzana,[48d] for example, has considered nonmagnetic carriers interacting with local moments through an s–d exchange giving rise to spin–disorder resistivity. The skew scattering arises through a dipolar interaction between the carrier's orbital moment and the local spin. The resulting anomalous Hall resistivity is proportional to $\langle (M - \langle M \rangle)^3 \rangle$. This function has a maximum just below the Curie temperature and tends to zero with decreasing temperature. Such behavior is observed in high purity single crystals of Gd.

Because of the short mean free paths we expect the anomalous Hall coefficient to be unusually large in amorphous magnets. This is indeed the case, and has been exploited by Bergmann[48e] to measure the magnetic moments of very thin amorphous films of Fe, Co, and Ni. Figure IV.21 shows his results for Ni, deposited on a $Pb_{75}Bi_{25}$ substrate. The Ni atoms in films less then 2.5 atomic layers thick do not possess magnetic moments. Such films do have a Pauli susceptibility which may be obtained

[48b] J.-N. Chazalviel and I. Solomon, *Phys. Rev. Lett.* **29,** 1676 (1972).
[48c] R. Karplus and J. M. Luttinger, *Phys. Rev.* **95,** 1154 (1954).
[48d] F. E. Maranzana, *Phys. Rev.* **160,** 421 (1967).
[48e] G. Bergmann, *Phys. Rev. Lett.* **41,** 264 (1978).

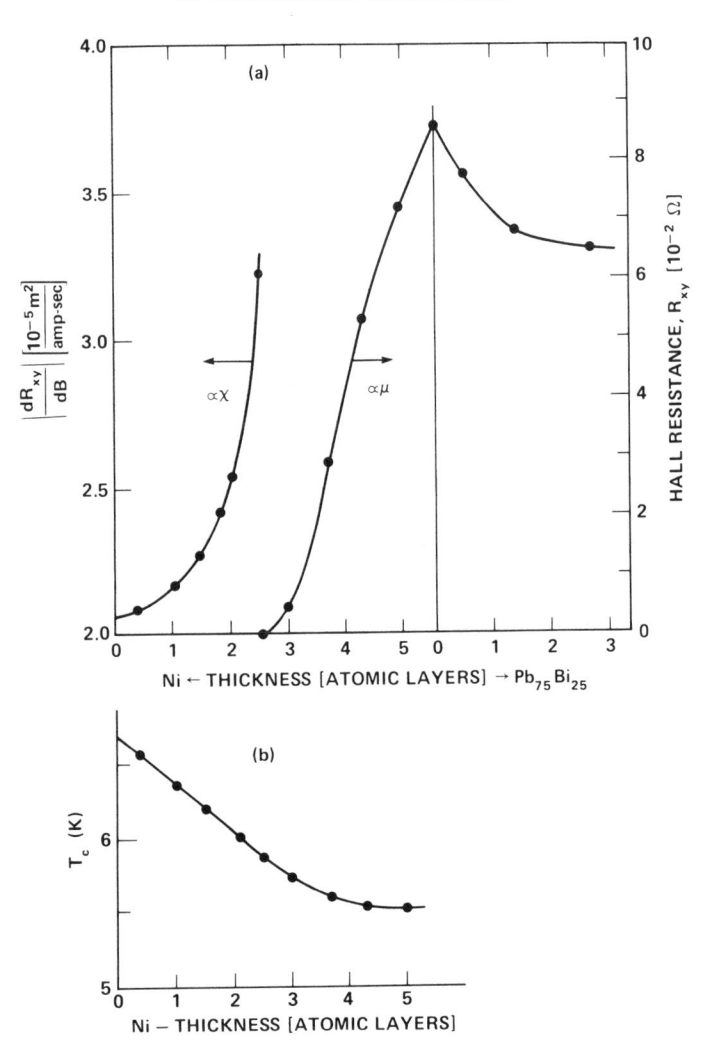

FIG. IX.21. (a) Hall resistance and its field derivative as functions of the Ni thickness. The Hall resistance is also shown for an additional condensation of $Pb_{75}Bi_{25}$ on top of the thickest Ni film. (b) The superconducting transition temperature of the sandwich as a function of Ni thickness. After Bergmann.[48e]

from the slope of the Hall resistance. From Fig. IX.21 we see that this slope increases rapidly with film thickness. This is interpreted as an increase in exchange enhancement. The lower part of the figure also shows the decrease in the superconducting transition temperature of the substrate. Fe and Co exhibit magnetic moments for average coverages less than one atomic layer. Notice that when $Pb_{75}Bi_{25}$ is deposited on top of 5.6 atomic layers of Ni the moment is suppressed. These results suggest that the hybridization which occurs at the interface reduces the exchange enhancement of the Ni to the extent that a moment cannot form.

Problem IX.2
    Discuss the implications of these results for Jaccarino and Walker's theory of local moment formation which we discussed in Chapter VII.

Large anomalous Hall effects have also been observed in the amorphous rare earth-transition metal magnets.[48f] The Gd-Co system is particularly interesting because the Gd and Co form anti-parallel magnetic "sublattices." Because the temperature dependence of each "sublattice" is different such a system shows a compensation point as we discussed in Chapter VI with respect to the garnets. Figure IX.22 shows that the anomalous Hall coefficient reverses sign at the compensation point. Since the two sublattices reverse direction relative to the applied field at this point,

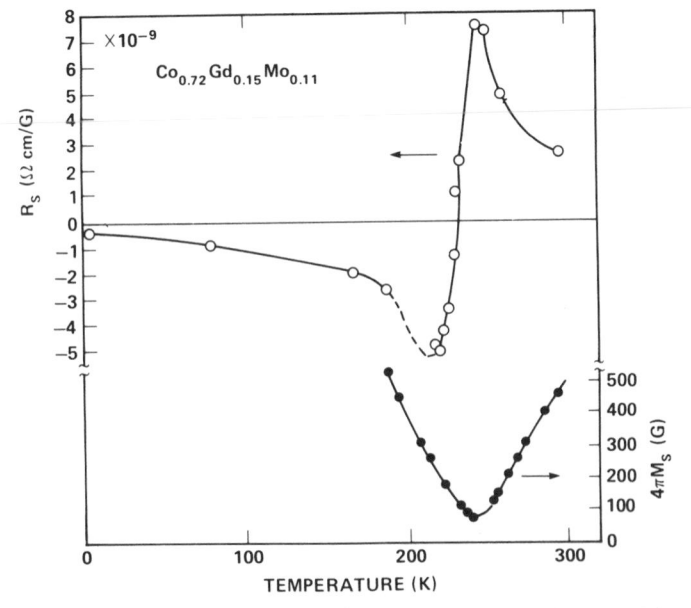

FIG. IX.22.   Temperature dependence of the anomalous Hall coefficient and the magnetization in the vicinity of the compensation point. After McGuire *et al.*[48f]

[48f] See, for example, T. R. McGuire, R. J. Gambino, and R. C. Taylor, *J. Appl. Phys.* **48**, 2945 (1977).

this reversal in R_s indicates that the effect is being dominated by one of the sublattices.

F. GRANULAR METALS

Granular metals, sometimes called ceramic metals, or cermets, present an interesting state of matter intermediate between the bulk crystalline state and the "microcrystalline" amorphous state. They consist of small metal grains mixed with a second immiscible insulating phase. We refer to them as a "state of matter" because they can be reproducibly and uniformly prepared. Depending upon the volume fraction x of metal, the composite can either be a dirty metal, with a conductivity orders of magnitude less than the pure crystalline state, or an insulator with tunneling between isolated metal particles. Very thin films which consist of islands of disconnected metal particles on a substrate form two-dimensional granular systems. We shall mention methods of the preparation and then discuss the resulting microstructure and how it relates to superconducting and magnetic long range order. The reader is referred to review articles by Abeles and coworkers[49,49a] for further details.

A large number of metals have been made into granular films, including the 3d, 4d, and 5d transition metals, as well as Cu, Ag, Au, Be, Al, In, Sn, and La. The insulators used include Al_2O_3, SiO_2, MgF_2, and MgO. Many more composites can obviously be fabricated. Depending upon the processing, the metal grains can be made extremely small, down to the limit of a single atom in the special case of rare gas–alkali metal composites such as Cs–Xe.

The most common method of preparing granular metals is by physical vapor deposition, either by co-evaporation or co-sputtering of metals and insulators, or by reactive evaporation or sputtering of oxidizeable metals. A number of other methods can also be used including evaporation of particles (slightly oxidized metals) in an inert gas or the introduction of metal particles into an insulator by physical or chemical means (e.g., ion implantation). Very thin films of disconnected metal islands can be prepared simply by quench-condensing the metal onto a liquid helium cooled substrate.

1. Microstructure

As the volume fraction of metal, x, goes from 1 to 0, the composite changes from having a continuum of connected metal grains with im-

[49] B. Abeles, P. Sheng, M. D. Coutts, and Y. Aire, *Adv. Phys.* **24**, 407 (1975).
[49a] B. Abeles, *Appl. Solid State Sci.* **6**, 1 (1976).

bedded dielectric to the inverse, namely a dielectric continuum with imbedded metal grains. In the transition region the metal and insulator form, as might be expected, labyrinth structures with both the metal and insulating grains partly connected. The size and shape of the grains can be controlled by the parameters of the deposition, i.e., rate of deposition and temperature of substrate, as well as the volume fraction of metal, x, and subsequent annealing procedures. The grain size, d, decreases with x, and with the ratio of the substrate temperature to the melting point of the metal. For instance, the average size, d, of sputtered Ni and Au grains decreases from ~ 40 Å to ~ 20 Å when x is reduced from 0.5 to 0.2 in films prepared by sputtering. The size distribution is broadened but not excessively so as d becomes small, as illustrated for Pt–SiO$_2$ in Fig. IX.23.[50] Deposition onto high-temperature substrates, or annealing subsequent to deposition tends to increase the grain size and produce rounding of the grains. As a result the metal–insulator transition as a function of x becomes sharper. Annealing W–Al$_2$O$_3$ composites at 1100 K for eight hours, for example, increases the grain size at the composition $x = 0.5$ from ~ 30 Å to 120 Å. On the other hand, d remains unchanged at ~ 20 Å

SIZE DISTRIBUTIONS FOR Pt-SiO$_2$ SAMPLES

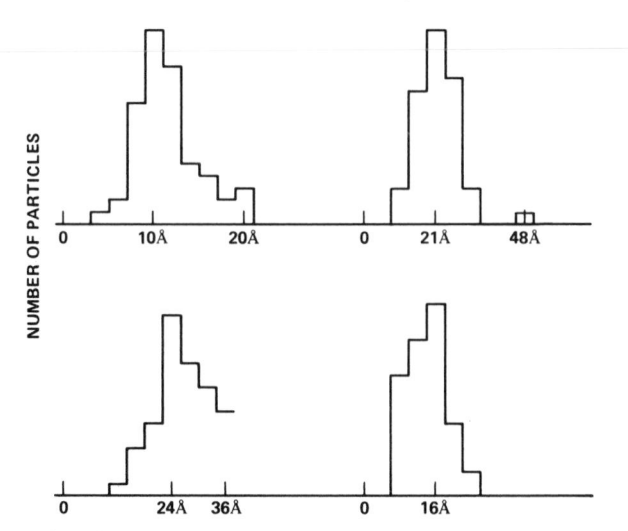

FIG. IX.23.   Size distribution histograms constructed from micrographs of four Pt–SiO$_2$ films. Particle size was varied by varying the amount of Pt on the sputtering target. After Stewart.[50]

[50] G. R. Stewart, *Phys. Rev.* **15**, 1143 (1977).

for $x = 0.2$ under the same treatment. X-Ray and electron diffraction studies in most instances show that the metal is crystalline and the insulators are amorphous for room temperature depositions. MgO, however, is crystalline, and for higher temperature depositions other insulators should be also.

2. Resistivity

Most granular metals have metallic conductivities until the volume fraction of metal decreases to a composition, x_0, where the temperature coefficient of resistance becomes zero. At $x = x_0$, which is generally of the order of 0.5, the contribution to the conductivity by tunneling from one grain to another becomes comparable to that due to percolation along a metallic maze. It turns out that there is a critical concentration of metallic phase above which conduction occurs. This threshold has been calculated for various models.[51] The "bond percolation" model, for example, consists of a regular lattice of sites connected by bonds, some known fraction of which are randomly missing. The mean number of allowed bonds per site at threshold is determined almost entirely by the dimensionality, d, according to $n_c \sim d/(d - 1)$. Thus, for a three-dimensional simple cubic lattice the *fraction* of bonds required for connectivity is $1.5/6 = 0.25$, while, for a two-dimensional square lattice, it would be 0.5. Experimentally, it is found that the percolation thresholds for cermets are considerably larger. This is thought to be due to the irregular shapes or to the coating of the metal grains. When symmetrical grains are obtained, as in the rare gas-alkali metal composites, the percolation thresholds are closer to those predicted.

In the transition region x just $>x_0$, the temperature coefficient of resistance is positive, although the path length is longer than that calculated for sample dimensions. The increased resistivity over the bulk metal is due partly to the increased path length and partly due to increased scattering processes.

A typical resistivity behavior is illustrated in Fig. IX.24. Below $x = x_0 = 0.55$ (at which composition the temperature coefficient of resistivity is assumed to be zero on the basis of the two curves shown), the dc resistivity increases rapidly with both composition and temperature. In this region, the transport is by tunneling from one grain to another. If the two grains are initially neutral, as they would be at very low temperatures, such tunneling involves a "charging" energy of the order of $2e^2/C$ where C is the capacitance of a grain. C is proportional to the average size

[51] See, for example, S. Kirkpatrick, *Rev. Mod. Phys.* **45**, 574 (1973).

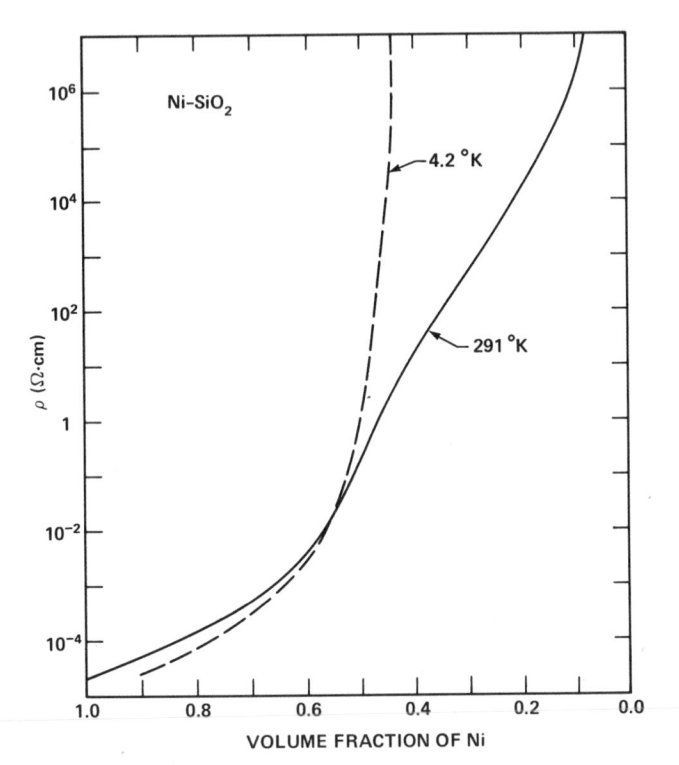

Fig. IX.24. Low-field resistivity at 4.2 K and 291 K as a function of the volume fraction of Ni in sputtered Ni–SiO$_2$ films. After Abeles *et al.*[49]

of the grain, d, i.e., $C \sim \epsilon d$ where ϵ is the effective dielectric constant of the metal. When $k_B T > 2e^2/C$, the charges are thermally activated and the resistivity is lower as Fig. IX.24 shows. At very low temperatures in very small grains the quantization of the electronic states also contributes to the activation energy. At low temperatures an electrical current can only be generated by electric fields high enough to cause field emission.

The ac conductivity of granular films of Al–Al$_2$O$_3$ has been found to increase as $\omega^{0.7}$ and for Sn–SnO$_2$ to increase as $\omega^{0.8}$, both in the range up to 10^6 Hz. Discontinuous very thin films approach $\omega^{1.0}$. This nearly linear dependence on frequency is what is to be expected for any ensemble with a distribution of relaxation times, which, in the present case, is due to a distribution of tunneling times. Assuming the polarizability α of a grain may be characterized by a Debye relaxation, i.e., $\alpha = \alpha_0/(1 - i\omega\tau)$, the conductivity is proportional to $\omega^2\tau/(1 + \omega^2\tau^2)$. If this is averaged over a broad distribution of relaxation times one finds that $\sigma(\omega) \sim |\omega|$.

3. Superconductivity

The superconducting behavior of granular superconductors, as typified in Fig. IX.25,[52] can also be understood in terms of microstructure. In the metallic region, $(x > 0.7)$ T_c is enhanced by almost a factor of 3 over bulk Al. According to Eq. (3.26), $\lambda = N(0)\langle I^2 \rangle / M\langle \Omega^2 \rangle$, the increase in λ must be explainable by some combination of phonon softening and increase in $N(0)\langle I^2 \rangle$. Heat capacity results of King et al.[53] suggest that $N(0)$ does not

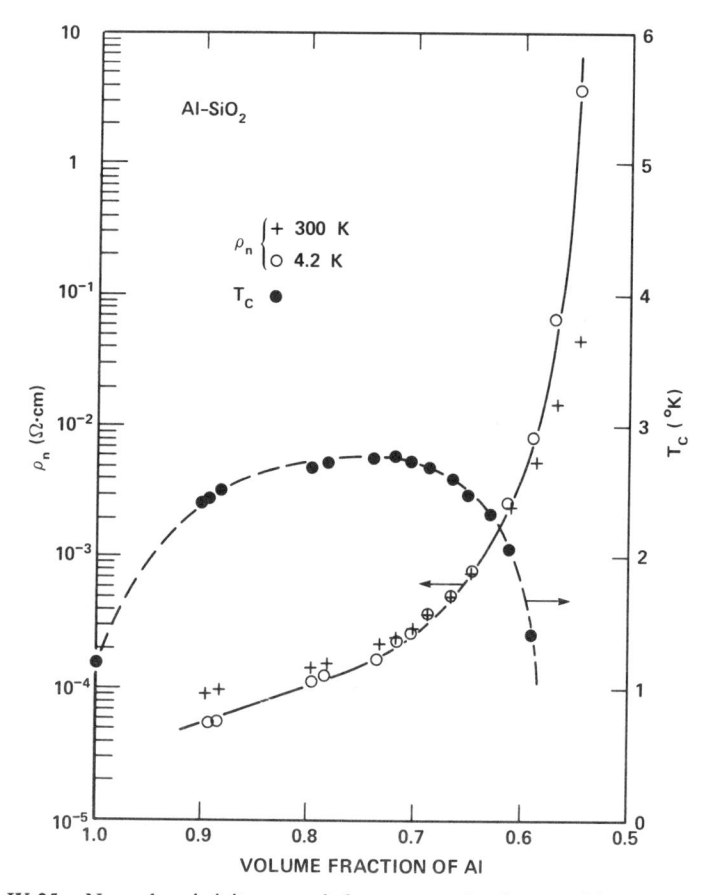

FIG. IX.25. Normal resistivity ρ_n and the superconducting transition temperature of sputtered Al–SiO$_2$ films as a function of the volume fraction of Al. Notice that the ρ_n curve for 300 K crosses that for 4.2 K near $X_0 = 0.6$. After Abeles and Hanak.[52]

[52] B. Abeles and J. J. Hanak, Phys. Lett. A **34**, 165 (1971).
[53] C. N. King, R. B. Zubeck, and R. L. Greene, Proc. Int. Conf. Low Temp. Phys., 13th, 1972 Vol. IV, p. 626 (1974).

change but that θ_D is reduced by 28%. However, Filler et al.[54] suggest that part, if not all of this reduction may be due to the amorphous oxide. Furthermore, tunneling data[55] indicate that $\alpha^2 F(\Omega)$ does not appear to change. They suggest that the enhancement of T_c is due to an increase in the electron–phonon interaction, i.e., $\langle I^2 \rangle$.

It can be seen for $x < 0.6$, in Fig. IX.25, that the films no longer exhibit a superconducting transition. The isolated grains of Al may still be superconducting providing the grains are large enough. Fluctuations in Al particles less than approximately 30 Å suppress the superconductivity. Individual grains of Al greater than 30 Å will develop order parameters but with no coherence between the grains. Such films are called "zero-dimensional." The cross over between three-dimensional to zero-dimensional behavior can result from thermal quenching of Josephson coupling between the grains[56] as the films are heated towards T_c.

Dynes, Garno, and Rowell[57] have studied very thin "island" films of Pb, Sn, Au, Al, and Cu quench-condensed onto substrates. Universally, films with resistances $\geq 30{,}000$ ohms per unit area ($\square$) have a negative coefficient of resistance and are not metallic. This can be seen in Fig. IX.26 for Pb. The thicker film shows a relatively sharp transition into the superconducting state near 7 K. The thinner films show a marked increase in their negative coefficient of resistance below 7 K. Also, surprisingly, the films become more resistive when superconducting. At T_c the individual islands develop superconducting energy gaps which evidently act in consort with the already present charging energy and possibly size quantization mismatch to make the intergrain tunneling more difficult. On the other hand, in the thick films the Josephson coupling between the grains is sufficient to couple all the grains with a single order parameter and make supercurrent flow possible.

Dynes et al.[57] find evidence in all the films that the transition from metallic to insulating behavior occurs in the vicinity of 30,000 ohms per square. This suggests that when the electronic mean free path approaches the Fermi wave length, $\rho \sim 0.12\, e^2/\hbar$ referred to as (the Ioffe–Regel condition[58]), the condition for minimum metallic conductivity has been reached.

Granular films provide likely candidates for excitonic-induced supercon-

[54] R. L. Filler, P. Lindenfield, and G. Deutscher, Proc. Int. Conf. Low Temp. Phys., 14th, 1975 Vol. 2, p. 105 (1975).

[55] M. Dayan and G. Deuscher, Proc. Int. Conf. Low Temp. Phys., 14th, 1975 Vol. 2, p. 421 (1975).

[56] G. Deutscher, Y. Imry, and L. Gunther, Phys. Rev. B 10, 4598 (1974).

[57] R. C. Dynes, J. P. Garno, and J. M. Rowell, Phys. Rev. Lett. 40, 479 (1978).

[58] A. F. Ioffe and A. R. Regel, Prog. Semicond. 4, 237 (1960).

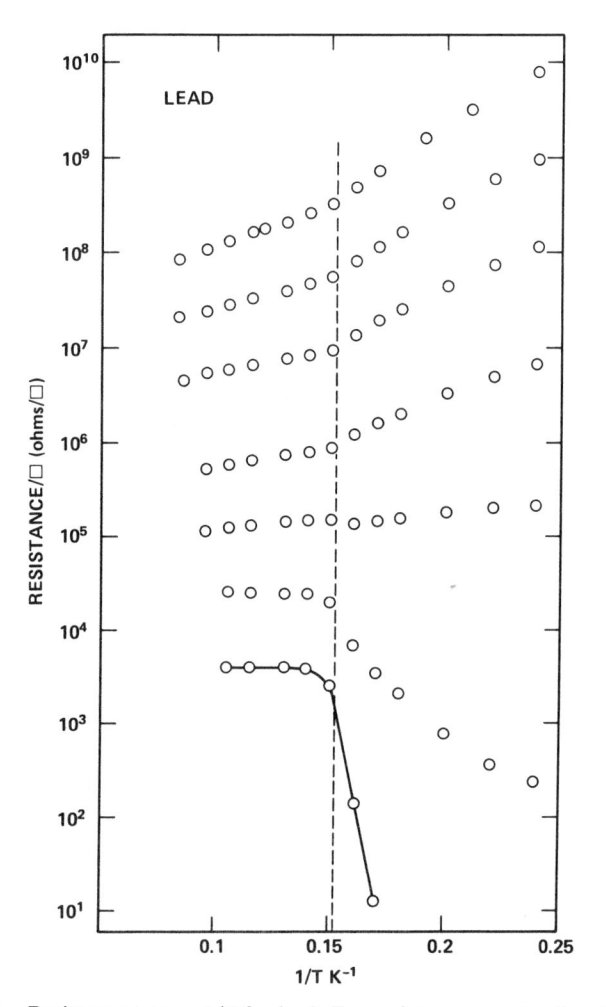

FIG. IX.26.   Resistance versus $1/T$ for lead. For resistances greater than 30,000 $\Omega/\square$, the temperature dependence is activated with an increase at T_c. Below 30,000 $\Omega/\square$, "metallic" behavior is observed with superconductivity below T_c. After Dynes *et al.*[57]

ductivity of the type postulated by Ginzburg that might occur at a metal-dielectric interface (discussed in Chapter III). No evidence exists for such a mechanism at present, although a wide variety of interfaces have been examined.

4. Ferromagnetism

Granular ferromagnets show ordering sequences similar to those seen in superconductors. In particular, as Fig. IX.27 shows, near a volume

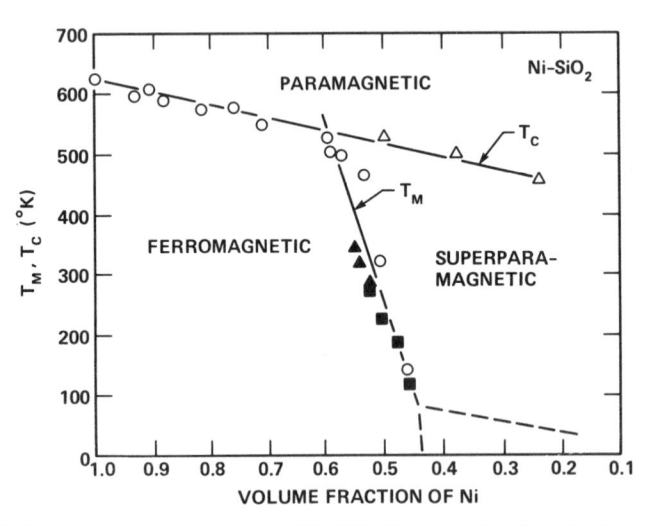

Fig. IX.27.   Ordering temperatures of Ni–SiO₂ films as a function of volume fraction of Ni. Open circles and triangles were obtained from susceptibility measurements, full squares from magnetoresistance, and full triangles from magneto-optic Kerr effect. The three magnetic states (paramagnetic, ferromagnetic, and superparamagnetic) are indicated; T_m and T_c are the inter- and intragrain ordering temperatures, respectively. The dashed lines indicate that below some "blocking" temperature the system cannot reach equilibrium. After Abeles.[49a]

fraction of 0.5 the Ni–SiO₂ system exhibits two ordering temperatures: one at which the grains order individually, and a lower one at which the grains themselves order. The latter ordering is due to the exchange interaction between those electrons whose wave functions penetrate into the tunneling barrier.

Consider the region where the individual grains are ferromagnetic, but are not ordered throughout the sample. When a magnetic field is applied the magnetization M_s within each grain will attempt to align itself with this field. In general, there will be a crystalline or shape anisotropy that will tend to oppose such alignment. This represents a barrier that must be overcome in order to reach equilibrium in the presence of the field. When $k_B T$ exceeds this barrier the grains will be able to equilibrate and give a Curie susceptibility of the form

$$\chi_{\text{super}} = M_s^2 vx/k_B T \tag{9.25}$$

where v is the average grain volume and x the volume fraction of ferromagnetic metal. This situation is referred to as *superparamagnetism*. Large grains or low temperatures will lead to a nonequilibrium state which is the granular analog of the spin glass.

Epilogue

The present state of science is evermore able to deal with real systems as evidenced by our concluding discussion of granular metals. This is a far cry from less than a generation ago when pure single crystals were the focus, and initial studies of superconductivity in the A-15 compounds were characterized as "smutch physics." We believe that in meeting the challenge of real systems—those that occur naturally as well as those synthesized as metastable structures in the laboratory—new stepping stones for future advances will be uncovered.

Already with granular metals, for example, we are dealing with cooperative effects involving numbers of particles that are very small with respect to Avogadro's number yet still large with respect to unity. This represents a scale that encompasses the number of atoms comprising future ultrasmall circuit elements (be they semiconducting, superconducting, or magnetic) on the one hand, and the macromolecules of biochemistry on the other. As scientists take up one of our most challenging problems, that of understanding the operation of the human brain—a bootstrap operation in the truest sense—they may find guidance in the combination of empiricism, formalism, and intuition arising from the study of long range order.

Author Index

Numbers in parentheses are reference numbers and indicate that an author's work is referred to although his name is not cited in the text.

A

Abbar, D. C., 286
Abeles, B., 383, 386, 387, 390
Abrikosov, A. A., 264, 286, 287, 290, 291, 295
Aire, Y., 383, 386(49)
Akoh, H., 340, 342
Aldred, A. T., 300
Allen, J., 144, 145
Allen, J. W., 134, 165, 363, 364(23)
Allen, P. B., 114, 119, 120, 222, 238
Allender, D., 121
Almquist, L., 110, 111(18)
Als-Nielsen, J., 77, 78, 141, 190
Alvarado, S. F., 67, 68
Ambegaokar, V., 64, 65, 182
Anderson, P. W., 18, 21, 52, 53, 63, 102, 110, 118, 135, 138, 140, 264, 268, 269, 270, 271, 272, 274, 276, 277, 285, 334
Andres, K., 168, 169(4), 243
Argyres, P. N., 323
Argue, C. R., 20, 21
Arrott, A., 156, 173, 174
Asano, S., 156
Ashley, C. A., 360
Aslamosov, G., 183
Aspens, D., 79
Astrov, D. N., 263
Aton, T. J., 286
Axe, J. D., 41, 43(38), 240, 241

B

Bader, S. D., 238
Bagley, B. C., 366
Bak, P., 83

Baker, A. G., 53
Baker, J. M., 144, 145(37)
Balberg, I., 257
Bardeen, J., 92, 93, 101, 121, 196, 202
Barrett, C. S., 238
Barth, N., 365, 366
Bateman, T. B., 239
Batterman, B. W., 238, 239
Béal-Mond, M. T., 274
Bean, C. P., 333, 334
Beasley, M. R., 177, 179, 244, 331
Belov, K. P., 173
Benda, J. A., 234
Benedek, G., 79
Beni, G., 361
Bennett, C. H., 358
Bennett, H. S., 323
Bergmann, G., 120, 365, 367, 368, 380, 381
Berlinger, W., 29, 30(23)
Bernal, J. D., 352
Bertaut, F., 258
Berthier, C., 197, 198, 301
Betbeder-Matibet, O., 288, 289(22), 296(22)
Bethe, H. A., 134
Bhatt, R. N., 242
Binder, K., 339, 340, 349
Birgeneau, R. J., 26, 27(21), 41, 43, 77, 78, 144, 145
Bishop, A. R., 347
Blaugher, R. D., 224, 230, 231
Bloch, D., 142
Bloch, F., 132, 133, 138, 146, 148, 159
Bloomfield, P. E., 281
Blount, E. I., 52, 53, 116
Blume, M., 188
Bobeck, A. H., 302, 303, 315
Bobell, F., 366

Subject Index

A

B

C

Substance Index